WIRELESS

WIRELESS

A TREATISE
ON THE THEORY AND PRACTICE OF
HIGH-FREQUENCY ELECTRIC
SIGNALLING

BY

L. B. TURNER,
M.A., M.I.E.E.

*Fellow of King's College, and University Lecturer
in Engineering, Cambridge*

CAMBRIDGE
AT THE UNIVERSITY PRESS
1931

CAMBRIDGE UNIVERSITY PRESS
Cambridge, New York, Melbourne, Madrid, Cape Town,
Singapore, São Paulo, Delhi, Mexico City

Cambridge University Press
The Edinburgh Building, Cambridge CB2 8RU, UK

Published in the United States of America by Cambridge University Press, New York

www.cambridge.org
Information on this title: www.cambridge.org/9781107636187

© Cambridge University Press 1931

This publication is in copyright. Subject to statutory exception
and to the provisions of relevant collective licensing agreements,
no reproduction of any part may take place without the written
permission of Cambridge University Press.

First published 1931
First paperback edition 2013

A catalogue record for this publication is available from the British Library

ISBN 978-1-107-63618-7 Paperback

Cambridge University Press has no responsibility for the persistence or
accuracy of URLs for external or third-party internet websites referred to in
this publication, and does not guarantee that any content on such websites is,
or will remain, accurate or appropriate.

PREFACE

THIS BOOK is a development of the author's *Outline of Wireless*, published in 1921 and now for some years out of print. The scope and intention of that small volume were described in the Preface in the following words. "The book is an outline only of the framework of a great and growing subject; but it is hoped that a careful perusal will enable the intelligent reader to appreciate the problems which are presenting themselves, and to read, if he so desire, with understanding any of the discussions of these problems appearing in the technical press." The aim was to present a brief account of wireless theory and practice which should be readable by a competent electrical engineer who had not studied high-frequency phenomena. This aim has continued in the present work to influence the choice of material and the manner of treatment. But in the intervening years engineering achievement and exact scientific research have far advanced, and if the book was to retain its former relation to the subject a corresponding expansion was necessary. Although new matter, rather than alteration of the old, seemed called for, the unity of the book would have suffered if only a partial revision had been made. In the event, the present volume is nearly three times as large as the previous; and apart from a few diagrams but little of the latter remains.

The field over which the book ranges is intended to be that of wireless telegraphy and telephony—the principles and the more prominent present-day practice. I have not included picture telegraphy, tele-kinematography and television, although much of the book is apposite as an introduction to these and other outgrowths of the thermionic technique. Further, I have gladly constricted my task by omitting all discussion of the methods of high-frequency measurement. This subject has been ably dealt with by Mr E. B. Moullin in his *Radio Frequency Measurements* (1926), of which a new edition has just appeared as I write this preface.

In the author's opinion, those readers who seek a general knowledge of modern wireless should study the whole of the book. But if severe restriction to matters of direct engineering application is desired,

Sections 4 and 5 of Chapter III, Sections 2 to 5 of Chapter VII, and Chapter XVII may be omitted. These contain respectively sketches of the theory of propagation in the upper atmosphere, of the internal physics of the thermionic tube, and of the nature and effect of atmospherics. The generalised treatment of triode circuits in Chapter X may be referred to on occasion rather than read. The rigorous division of the Chapters into titled Sections and Subsections, as well as the full index, will facilitate the use of the book for reference by those who have passed the stage of consecutive text-book reading. Students who, unassisted, are making a first study of the subject are advised to leave certain portions of the book until the remainder has been read. These portions are marked with an asterisk in the Table of Contents.

The effects dealt with in electrical engineering—unlike mechanical—are remote from the sense-perceptions and concepts of ordinary daily life. The student can acquire the power of visualisation, and that sense of the naturalness of the phenomena he meets which is a pre-requisite of engineering efficiency, only by supplementing his theoretical studies with much experimental exercise in the laboratory or elsewhere. Particularly in high-frequency work does a lack of prolonged practical experience make itself apparent. Proficiency with mathematical equations and circuit diagrams is indispensible, but is not sufficient to make the student effective as an engineer. He must manipulate the things whose idealised qualities are portrayed in the circuit diagrams; and often a rather long apprenticeship proves necessary. Fortunately, facilities for wireless experimental work of the right character—which is, in a word, metrical—are exceptionally easily provided.

Many years of learning and of teaching wireless have shown me the importance of maintaining always a close touch with numerical values. A circuit disposition of even moderate complexity, supposed to comprise parts A, B, C, D, may behave in diverse ways. The conceivable actions and interactions of the components are so many that one can offer *prima facie* explanations of almost any phenomenon reported. Diagnosis of faults and removal of seeming anomalies must depend on the *sizes* of A, B, C, D; and upon whether there may not, in fact, be other components a, b, c, d, intruding in appreciable amount. Not the least boon conferred by the thermionic tube is the means it has brought for exact measurement of all sorts, with the consequent opening for fruitful analysis. But while mathematical exploration is of the greatest service, real wireless dispositions are peculiarly liable to contain factors left out of account: sometimes because they were unsuspected; sometimes as an approximation to reduce the complexity of the analysis. Unlike, therefore, the mathematician of Mr Bertrand Russell*, who is not concerned to know what he is talking about or whether his statements are true, the wireless engineer must not only know precisely what physical quantities his symbols stand for; he must have a good notion how large those quantities are. For pointing a moral arithmetic will often be more effective than algebra. Accordingly throughout this book I have tried to convey some idea of size by taking numerical examples: sometimes in the form of an actual instrument or installation; sometimes the results of my own measurements; sometimes as a more or less arbitrary choice intended to typify practical conditions.

In preparing a book of the present character, the author's work is the selection and coordinated presentation of material which is, for the most part, already well known to his readers collectively. He mingles with what others have taught him notions he feels are his own, in the hope that a useful coherent message may emerge. In this process I have doubtless been influenced by the publications and the talk of many workers, not all of whom are referred to explicitly in the text. But in concluding this arduous, if often absorbing, task I wish to acknowledge generally the inspiration I have received on many occasions from contact with two scholarly and original minds. For their writings and their conversation I owe much to Dr W. H. Eccles, F.R.S., with whom I have been associated on the Wireless Telegraphy Commission of 1920–26 and other bodies; and to Mr E. B. Moullin, M.A., until lately my colleague at Cambridge.

I have received technical information, photographs, and permission to reproduce diagrams, from numerous sources, all of which, I trust, are specifically acknowledged in the text. I am grateful for opportunities to examine examples of the most recent wireless practice accorded by the German Post Office and the Telefunken Company, by the British Broadcasting Corporation and the Marconi Company,

* *Mysticism and logic* (1918), p. 75.

and especially by my former colleagues of the Engineering Department of the British Post Office.

My thanks are due to Mr E. St J. Thackeray, B.A., for valuable help in reading the proofs. I hope that few mistakes remain; but the book contains many fresh calculations, and amongst them some errors must be expected. I shall be grateful to readers who point them out.

L. B. T.

ENGINEERING LABORATORY
    CAMBRIDGE
      April, 1931

TABLE OF CONTENTS

Students making a first study of wireless are recommended to omit, until after the rest of the book has been read, those portions marked in this Table with an asterisk.

Chapter I. Introduction

SECTION		PAGE
1	Signalling systems	1
2	Waves	3
3	Special features of wireless telegraphy as a signalling system	6

Chapter II. Electromagnetic radiation

1	Radiation occurs whenever current changes	11
2*	Maxwell's displacement current	14
3*	Propagation of electric disturbance in a homogeneous isotropic medium (including free space as a particular case)	16
4*	Radiation from a vertical antenna	23

    (a) The field of an accelerated electric charge at a large distance
    (b) Application to an antenna
    (c) Radiation resistance of antenna
    (d) Numerical example of antenna radiation

Chapter III. Propagation of wireless waves round the Earth

1	Curvature of the ground; the Heaviside layer	35
2*	Degradation of radiated energy	41
3*	Complications in propagation phenomena	43

SECTION	PAGE
4* Theory of the Heaviside layer	46
(a) Free electrons in the upper atmosphere	
(b) Absorption in the Heaviside layer	
(c) The Earth's magnetic field	
(d) Polarisation effects	
5* Experimental knowledge of the Heaviside layer . .	56
(a) Phase and group velocities	
(b) Methods of estimating the height of the Heaviside layer	
(c) Beyond the Heaviside layer	

Chapter IV. Oscillatory circuits

1	Reactances and resistances	63
2	Undamped impressed E.M.F.	65
	(a) General case	
	(b) The decrement	
	(c) Certain initial conditions	
	(d) Resonance, with the same initial conditions	
	(e) The forced oscillation alone	
	(f) Decrement and power factor	
3	The wavemeter	74
4	The antenna as oscillatory circuit	76
5	Damped impressed E.M.F.	83
	(a) Free oscillation	
	(b) Loosely coupled circuit	
	(c) Resonance and damped E.M.F.	
	(d) Practical equivalence of methods of distuning	
6	Tight coupling	89
7	Methods of coupling circuits	93
8	The steady state in compound circuits	99
	(a) Method of solution	
	(b) Mutual coupling for maximum secondary current	
	(c) Impedance of a rejector circuit	
9	Mechanical model of coupled oscillatory circuits . .	105

TABLE OF CONTENTS

Chapter V. The production of high-frequency currents

SECTION   PAGE
1. Spark transmitters 107
   - (a) General
   - (b) Synchronous sparking
   - (c) Quenched spark
2. Excitation by buzzer 117
3. Continuous wave methods 118
4. Arc generators 120
   - (a) The arc as a conductor
   - (b) Duddell's singing arc
   - (c) The Poulsen arc
5. High-frequency alternators 132
   - (a) General
   - (b) Inductor alternators
   - (c) The Alexanderson alternator
   - (d) Bethenod-Latour alternators
   - (e) Frequency-multiplying alternators
   - (f) Saturated-iron doublers and triplers
   - (g) Multipliers with very high saturation

Chapter VI. The detection of high-frequency currents

1. Methods of registering low-frequency currents . . 144
2. Rectification by non-ohmic conductors 145
   - (a) Without external resistance
   - (b) With external resistance
3. The detector in relation to the oscillatory circuit . . 150
   - (a) Coefficients of performance of the detector
   - (b) Connection to the oscillatory circuit
4. Practical detectors 153
5. Rectification of strong signals 158
   - (a) The output of the detector
   - (b) The effective resistance of the detector
   - (c) The requisite shunting capacitance
6. Tests of actual rectifiers 162
7. Continuous wave reception 167

Chapter VII. Thermionic tubes: general properties

SECTION		PAGE
1	Introduction	172
2*	Thermionic emission	174
	(a) Rates of emission	
	(b) Low-temperature emitters	
	(c) Speeds of emitted electrons	
3	Thermionic tubes: diodes	185
4	The space-charge	188
5	Introduction of a third electrode	192
6	Triode metrical relations	195
	(a) Anode current a function of anode and grid potentials	
	(b) Characteristic curves	
	(c) Simple equivalent circuit	
7	Modern patterns of triodes	204
	(a) Low power	
	(b) High power	
8	Grid current	212
9	Multi-grid tubes	214
	(a) Two grids as control electrodes	
	(b) Fixed-potential space-charge grid	
	(c) Fixed-potential screen grid	
	(d) Screen-grid tetrode for preventing retroaction	
	(e) Negative conductance with screen-grid tetrode	
	(f) Pentodes	

Chapter VIII. The triode as amplifier

1	Output of the triode	225
2	Analysis of amplifier with mixed impedance	231
3	Construction of the impedance	233
4	Cascade amplifiers	237
5	Transformer connection between triodes	240
	(a) General	
	(b) Dependence of amplification upon frequency (acoustic)	
	(c) Measurements of a transformer, and of a three-stage amplifier, for acoustic frequencies	

SECTION		PAGE
6	High-frequency amplifiers	250
	(a) Stray capacitance in resistance amplifiers	
	(b) Rejector-circuit impedance	
7	Capacitance between anode and grid	255
	(a) Effects of anode-grid capacitance	
	(b) Numerical example	
	(c) Methods of preventing anode-grid capacitance current	
8	Strictly aperiodic amplifiers giving high amplification	260

Chapter IX. The triode as oscillator

1	General principle of retroaction in triodes	264
2	Analysis of a usual circuit from first principles	267
	(a) Inception of oscillation	
	(b) Amplitude of steady oscillation	
	(c) Output and efficiency in the linear régime	
	(d) Numerical examples of linear régime	
3	Other retroactive oscillator circuits	278
4	Oscillators of high efficiency	281
	(a) The high-efficiency cycle	
	(b) Harmonics produced	
5	The triode telegraph transmitter	289
	(a) Connection of oscillator to antenna	
	(b) Master-oscillator system	
	(c) Keying	
	(d) Example of short-wave telegraph transmitter	
6	Other types of triode oscillator	295
	(a) Negative conductance without retroaction	
	(b) Tuning-fork master oscillator	
	(c) Quartz-crystal master oscillator	
7	Constancy of frequency	306

Chapter X*. Generalised treatment of complex triode circuit in steady-state rectilinear operation

SECTION | PAGE
1 Most general circuit considered 310
2 Nature of impedances specified; anode and grid circuits both single 312
3 Case of $\gamma = \infty$. (No anode-grid path) 313
   (a) Nature of impedances not specified
   (b) Nature of impedances specified; anode and grid circuits both double
   (c) Nature of impedances specified; anode circuit double; grid circuit single
   (d) Nature of impedances specified; anode circuit single; grid circuit double
4 Case of $\delta = 0$. (No anode-grid mutual inductance) 316
   (a) Nature of impedances not specified
   (b) Nature of γ path specified
   (c) Nature of impedances specified; anode and grid circuits both single

Chaper XI. The triode as rectifier

1 The triode in relation to other rectifiers . . . 319
2 The anode rectifier 321
3 The grid rectifier 326
4* Closer analysis of the grid rectifier 332
   (a) The higher differential coefficients
   (b) Exponential characteristics
   (c) Quartic characteristics
   (d) Subsidiary effects
5 Load on the input circuit 341
   (a) Load due to grid current
   (b) Load due to anode-grid capacitance
6 Approach to linear performance 347
7 Summary comparison of triode rectifiers . . . 351

Chapter XII. Retroactive amplifiers, and self-oscillating receivers

SECTION		PAGE
1	Reduction of damping by retroaction	353
2	Retroactive amplifiers	358
3	The autoheterodyne	359
4	The superheterodyne	362
5*	Super-regenerative reception	365
6	Example of triode receiver	370

Chapter XIII. Telephony

1	The transmission unit	372
2	Telegraphy and telephony compared	373
3	Microphonic modulation at the transmitter . . .	376
	(a) Methods not dependent on the use of triodes	
	(b) Methods dependent on the use of triodes	
	(c) Analysis of anode-choke control	
4	The modulated oscillation	389
	(a) As first produced	
	(b)* Suppression of the carrier-wave	
	(c)* Propagation between transmitter and receiver	
	(d)* Change of phase of one side-wave with respect to the other	
	(e)* Change of phase of carrier-wave with respect to side-waves	
	(f)* Suppression of one side-wave	
	(g)* Single-side-band transmitter	
5	Distortions in the receiver	405
	(a) General	
	(b) In the high-frequency circuits	
	(c) Non-linear rectification	
	(d) Rectifier effects of finite modulation frequency	
	(e) Push-pull grid rectifier	

xvi TABLE OF CONTENTS

SECTION PAGE
6* Electromechanical relations 415
   (a) General
   (b) Moving-coil loud-speaker
   (c) Impedance of coil movable with elastic restraint in magnetic field
   (d) Moving-iron speakers
7 Examples of telephone installations 423
   (a) A broadcasting transmitter
   (b) An inter-continental short-wave receiver

Chapter XIV. Antennas and antenna combinations

1 Transmitting antennas 432
   (a) Long waves
   (b) Short waves
   (c)* Harmonic excitation
2 Receiving antennas 439
3* Directive antennas 442
4* The Beverage antenna 444
5* Loop antennas 448
6* Direction finding 453
   (a) Rotating beacons
   (b) D.F. receivers
   (c) Night effect
7* Two spaced antennas 462
8* Antenna arrays 468
   (a) Nature of an array
   (b) Closely packed array
   (c) Any spacing
   (d) Reflecting array
   (e) Height exceeding the half-wavelength
   (f) Example of antenna array

Chapter XV*. Distribution of high-frequency current in conductors

SECTION		PAGE
1	Slabs	477
2	Screening envelopes	479
3	Skin effect in a wire	480
4	Proximity effect in wires	486
5	Examples of inductance coils	488

Chapter XVI*. Filters

1	General	492
2	Pure-reactance elements	495
3	Low-pass filter	497
4	High-pass filter	499
5	The terminal impedance	500
6	Band-pass filter	502
7	Mixed filters	506

Chapter XVII. Atmospherics

1	Nature of atmospherics	507
2*	The atmospheric as a pulse	508
3*	The atmospheric as a continuous spectrum	513

INDEX	.	519
PLATES I—XXXI		*between* pp. 256 *and* 257

NOTES

Throughout the book letters standing for quantities are in italic type. Also, in the case of an article such as a condenser or an inductance coil, an italic letter is used to designate both the article and its size. Thus:

C_1 is a condenser of capacitance C_1
L_2 is a coil of inductance L_2

Where equations or formulas are numbered for identification, the numbers run consecutively throughout the Chapter.

All the Plates are placed together at the middle of the book.

In accordance with the recommendations of the International Electrotechnical Commission (Publication No. 27), the following signs for names of units are used:

A for ampere	F for farad
V „ volt	H „ henry
Ω „ ohm	m „ metre
W „ watt	

Prefix $\mu \equiv$ micro-, e.g. 1 $\mu\mu$F $\equiv$ 1 micromicrofarad $\equiv 10^{-12}$ farad
„ m $\equiv$ milli-, „ 1 mA $\equiv$ 1 milliampere $\equiv 10^{-3}$ ampere
„ k $\equiv$ kilo-, „ 1 kW $\equiv$ 1 kilowatt $\equiv 10^{3}$ watts
„ M $\equiv$ mega-, „ 1 MΩ $\equiv$ 1 megohm $\equiv 10^{6}$ ohms

In addition, db is used for decibel.

In accordance with the recommendation of the International Radio Technical Committee (Hague, Sept. 1929) the following designations of wavelengths have been observed:

Long	= above 3000 m
Medium	= 3000—200 m
Intermediate	= 200—50 m
Short	= 50—10 m
Very short	= below 10 m

CHAPTER I

INTRODUCTION

1. SIGNALLING SYSTEMS

FROM the engineering aspect any signalling system is a power transmission system; but whereas in most power transmission systems ordinarily so called it is essential that the power received shall be a large fraction (approaching unity) of the power transmitted, and the switching or control of power is of secondary importance; in a signalling system the switching or control processes are of primary importance, and the received power may be ever so small a fraction provided only it be perceptible.

In every power transmission system energy is conveyed from a *transmitter* T to a more or less distant *receiver* R through a connecting *medium* M. We may picture an identifiable packet of energy entering M at T, and part of it being subsequently handed over at R. We say *subsequently* because, however little may be the practical importance of the time of transit from T to R, we believe that *some* time must elapse before energy located at one place reaches another place*.

An extreme example of a power transmission system is provided by the familiar agricultural wind-driven pump. Here the wind motor T at the top of a tower transmits energy to the vertical rod M connecting it to the pump R in a well underneath. It is true that we are apt to think of the energy which leaves T as passing directly to R; but we do so only because the rod is short—or, more precisely, because the time taken for a stress to be propagated from end to end of the rod is short compared with the period of rotation of the crank. The action of the machine would remain unchanged in principle if the wood rod M joined to cranks T and R were replaced by a column of water filling a pipe joined to cylinders with

---

* Whether in the new physics it is necessary or nonsensical to think of energy as possessing the attribute of position, and although the aether of Faraday and Maxwell threatens to introduce intellectual difficulties of a kind it was expressly invented to dispel, it seems certain that for many years to come engineers—even those concerned with free electrons and with radiation—must continue to employ the language and conceptions of the classical mechanics, of Newton and of Maxwell.

pistons. But if now the pipe were much lengthened, or if the piston T were made to vibrate much more rapidly, the machine would operate on the Constantinesco water-wave power-transmission principle, and we should have no hesitation in thinking of the transmission of energy from T to M as an event separate from the subsequent reception of energy by R from M. The significant change is merely the increase in the ratio between the time of propagation of a disturbance along M and the period of T, or (as we shall see later) in the ratio between the distance TR and the wavelength of the disturbance propagated in M.

If the distance TR were lengthened sufficiently, the reaction of R on T would become indefinitely reduced; and with a very long water column the transmitter T would deliver its energy to the medium M without being affected by the conditions at R—indeed, whether or not there were a receiver at R at all.

An ordinary short-distance telegraph or telephone installation is an electrical power transmission system analogous to our mechanical example with a short wood rod or water column. The P.D. at the receiving end of the line is approximately a copy of the P.D. applied at the transmitting end; and if the received power were cut off by opening the circuit at R, the transmitted power would also fall approximately to zero. But as the connecting line is increased in length, an electrical system is approached which is analogous to the Constantinesco system arrived at on sufficiently lengthening the water column in the mechanical example. Thus in a long submarine telegraph cable, the transmitter T may deliver its spurts of energy to the cable M almost unaffected by the conditions at the receiver R. Telephone lines are of all lengths, so that telephone installations may be encountered occupying every position from the indefinitely short line, analogous to the wind pump with the usual short wood rod, to the "infinite" line in which the conditions at the receiving end have no perceptible influence at the transmitting end.

On passing from electrical power transmission with conducting lines as the connecting medium to wireless telegraphy in which the conductors are dispensed with, the same gradual transition from the seemingly direct to the obviously indirect may be discerned. An early demonstration of telegraphy in which conduction currents between transmitter and receiver were clearly not instrumental in effecting communication was made by W. H. Preece across the mouth of the Severn in 1886*. The short stretch (four miles or so) of aether as the medium M between his transmitting and receiving loops T and R corresponds with the wood rod of the wind pump or the short line of a local telephone connection. Nevertheless, energy passed from T to R only by way of a stress propagated across M. To-day a broadcast transmitter T situated in the heart of a city transmits to receivers R situated at all distances ranging from hundreds of feet to hundreds of miles. Whether the receiver be near or far, the energy is communicated from T to R only by propagation across the connecting medium M.

Thus no sharp physical distinction can be drawn between direct and indirect, between short range and long range; in every case a wave of disturbance (mechanical or electrical) is propagated across the connecting medium, be it wooden rod, the atmosphere, or pure aether. Differentiation is a question of practical emphasis only, and appears to depend on the relative lengths of the time of propagation across the connecting medium and the period of the disturbances propagated. The agricultural mechanic is not forced to dwell upon the propagation of mechanical stress along his wood rod, nor the electrician upon the propagation of electrical stress along his bell wires; but the telephone engineer *is* forced to consider how electric disturbances pass along his cables, and still more the wireless engineer how they pass across the aether connecting his transmitter and receiver.

The mechanism of the propagation of the disturbance within the medium is designated wave motion, and the disturbance is called a wave.

2. Waves

It is not easy to give a succinct and general verbal definition of the phenomenon to which the term wave or wave motion is applied. "Speaking generally, we may say that it denotes a process in which a particular *state* is continually handed on without change, or with only gradual change, from one part of a medium to another†." Let

$t =$ time from an arbitrary instant;

$r =$ distance from some fixed point;

$\chi =$ displacement (or stress or strain or other disturbance) in the medium at time t and position r.

* See J. J. Fahie, *History of Wireless Telegraphy* (1901), p. 145.
† *Encyclopaedia Britannica*.

I. INTRODUCTION

Then the medium is the seat of a wave if

$$\chi = f(r + vt),$$

where f stands for any function, and v is a constant. The physical meaning of the equation is this: if an explorer, being situated at a definite point in the space-time continuum, sets out to examine neighbouring regions, he finds that the disturbance χ varies by the same amount if he advances spatially to a position r' (i.e. moves a distance $(r' - r)$) as if he stays where he is and lets time flow past him by an amount $(t' - t)$, provided that

$$r' - r = v(t' - t).$$

At any spot, χ varies with time t; and at any instant, χ varies with distance r; but if, as time passes from t to t', the explorer also retires through the appropriate distance $(r' - r) = v(t' - t)$, the disturbance χ under his observation remains unchanged.

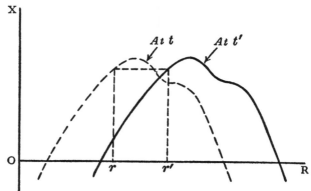

Fig. 1. A wave.

In Fig. 1 the full-line curve is supposed to be the plot of χ as a function of r at a particular instant t. This plot must be shifted along RO at the speed v if it is to show the relation between χ and r at subsequent instants. A wave of χ-ishness (whatever χ may be), whose shape depends on the form of the function f, moves through the medium (in the negative direction of r) with speed v. In the same way, $\chi = f(r - vt)$ is a like disturbance propagated in the positive direction with the same speed v*.

---

* The two waves $\chi = f(r + vt)$ and $\chi = f(r - vt)$ are the solution of the characteristic wave equation $\frac{\partial^2 \chi}{\partial t^2} = v^2 \frac{\partial^2 \chi}{\partial r^2}$, as may be verified by substitution.

2. WAVES

In wireless telegraphy (and in telephony) the chief interest is confined to *persistent waves*, that is, those in which the function f is a periodic function. As the simplest persistent wave, we take

$$\chi = A \sin 2\pi \left(\frac{r}{\lambda} - \frac{t}{T}\right).$$

Here χ repeats itself at every increment λ in distance and at every increment T in time. λ is called the *wavelength*; T is called the *period*, and the reciprocal $\frac{1}{T} \equiv n$ is called the *frequency*; A is called the *amplitude* of the wave. As before the *velocity of propagation** is $v = \frac{\lambda}{T} = \lambda n$.

The wave $\chi = A \sin 2\pi \left(\frac{r}{\lambda} - \frac{t}{T}\right)$ is an ideal extreme case which in the real world can only be approached, for it continues wholly unchanged over an infinite extent of space r and time t. If a steady tone is sounded continuously at one end of an ideal speaking tube whose other end is infinitely remote and whose walls neither transmit nor absorb energy from the air within, then whenever and wherever along the tube an observation is made it is found that the air pressure within the tube goes through the same cycle. But in any real tube, the amplitude must get smaller the further is the point from the source of sound; for the tube does not in fact perfectly confine and preserve the sound energy, but allows some to penetrate to the outer air, and converts some into thermal energy in the metal. The amplitude A is then not a constant, but diminishes as r increases (although the change may be slight for increments of r of the order of a wavelength)†. Such a wave is said to have *attenuation*, and does not conform strictly to our definition $\chi = f(r + vt)$; but the likeness of a moderately attenuated wave to a true wave is as obvious as the likeness of a moderately "damped harmonic motion" to a true (undamped) harmonic motion; and in each case much understanding of the properties of the former is conferred by a study of the latter.

In addition to spatial attenuation—i.e. A not quite independent of r—it is obvious that in real waves A can be taken as independent

---

* More precisely, the *phase velocity*. See III–5 (a).
† If the tube had a bore which increased with distance from the source— and especially if the directing tube were absent altogether—the amplitude would diminish still more rapidly with increase of distance from the source.

of t only over a limited range of time; and indeed in wireless telegraphy and telephony it is the rise and fall of A with time which constitutes the signal. Thus the electrical disturbance set up around (say) the London (Oxford Street) broadcasting station is a modulated attenuated wave which—in the absence of various complicating factors here ignored—may be written

$$\chi = \frac{a + \alpha \sin 2\pi \frac{t}{\tau}}{r} \sin 2\pi \left(\frac{r}{\lambda} - \frac{t}{T}\right),$$

where the former constant A is replaced by a quantity

$$\frac{1}{r}\left(a + \alpha \sin \frac{2\pi t}{\tau}\right).$$

At any spot—e.g. Hitchin or Huntingdon—this quantity fluctuates with an acoustic period τ (e.g. $\tau = \frac{1}{1000}$ sec.); and its maxima and minima are roughly twice as great at Hitchin as at Huntingdon. Nevertheless, since at Hitchin and Huntingdon $r \gg \lambda$, and since $\tau \gg T$, in considering the nature of the propagation of energy it is useful to treat the disturbance as a true persistent unattenuated wave

$$\chi = A \sin 2\pi \left(\frac{r}{\lambda} - \frac{t}{T}\right).$$

The inconstancy of A can subsequently be examined, and the necessary corrections be applied.

3. Special Features of Wireless Telegraphy as a Signalling System

In wireless the medium M is the aether, and the wave of χ-ishness is an electric wave; so that χ might stand for the strength of an electric or magnetic field at the instant t and at the spot r in the medium. The essentials of a complete wireless installation are (i) a collection or box of instruments (often of very simple construction) in which alternating currents of very high frequency are produced from some local source of electric energy; (ii) an antenna, or electric circuit of such a geometrical form that high-frequency electric currents in it are accompanied by a marked radiation of energy into surrounding space; and elsewhere on or near the surface

3. SPECIAL FEATURES OF WIRELESS TELEGRAPHY

of the earth, (iii) another antenna, with (iv) another box of instruments in which alternating currents are produced by the tiny fraction of the radiated power which reaches them from the sending station.

Electromagnetic radiation, by which energy is transferred from the sending to the receiving antenna, is a phenomenon familiar to all who have eyes to see, or a skin to feel, the Sun's rays. It is, we believe, by electromagnetic radiation that light reaches us; and the same mechanism of the transference of energy with the same mathematical analysis suffices to describe radiation in the case of visible light, where the frequency is about 5×10^{14} cycles per second, as in the case of wireless telegraphy, where the frequency commonly lies between 5×10^6 cycles per second and one-hundredth of that number.

There are these contrasts to be noted, however. Firstly, in the case of light the frequency is so large, and therefore the wavelength and the size of the radiating oscillator so small, that physicists have as yet been able to do little in the way of arbitrarily constructing and disposing their luminous oscillators, but must take them as they find them in the atom; whereas the wireless engineer builds his own radiator, his antenna, long or short, high or low, of this shape or that; for it is, as it were, large enough to give room for his fingers. Secondly, in wireless we are not so much concerned with radiation through free space as along Earth's surface. Even the aeroplane cannot get far enough from the ground to be regarded as unaffected by it. There is, moreover, the further complication of another conducting surface, the ionised upper atmosphere. So radiation in wireless telegraphy is not through free space; or even over a plane conducting surface bounding free space, though this is sometimes a convenient approximation to the actual conditions. It occurs between an uneven heterogeneous spheroidal solid and liquid body, and the even less well-defined gaseous conducting layer. Consequently the exact mechanism of this radiation, whether at the antenna or far away between the antennas, is imperfectly understood and hard to ascertain.

The processes occurring within the boxes of instruments associated with the antennas can be analysed with greater precision and detail. They are those encountered in ordinary alternating current theory, with only such quantitative differences as follow from the much higher frequencies of the currents to be handled.

I. INTRODUCTION

The power engineer who has studied alternating currents, including transient phenomena, already knows much of the theory of wireless circuits. But as the admittance of a condenser and the impedance of a coil are proportional to the frequency, with the vastly greater frequencies of wireless tiny capacitances and tiny inductances become important which would be utterly ignored in alternating current power circuits, or even in telephone circuits.

It is well to get an idea of actual values. In wireless, particularly of course in short-wave work, a capacitance of (say) 1 micromicrofarad may be very perceptible. This is the capacitance in air between two parallel sixpenny pieces spaced about twice their total thickness apart ($\frac{3}{32}$ inch), or between the earth and a distant sphere of about 1 cm radius. At 3×10^6 cycles per second (which corresponds with a wavelength of 100 metres), an inductance of 1 microhenry might be of equal importance; and this would be provided by two or three close turns of wire round an ordinary glass tumbler. Now the 50-cycle engineer does not worry about a micromicrofarad, for three thousand million of his volts would be needed to drive an ampere through it; nor does he much appreciate a microhenry, for three thousand of his amperes would produce a P.D. of only 1 volt across it. The efficient wireless experimenter, however, must be constantly alive to the effects of such small capacitances and inductances. He develops a habit of mind which classifies the points of a circuit as sacred, and profane or earthy, from the high-frequency aspect. The sacred point is one at which high-frequency potentials are developed, and no liberties must be taken there; the profane or earthy point is one at which no high-frequency potentials should be developed, and if any necessarily earthy instrument, such as a pair of headgear telephone receivers or a bulky battery, is to be inserted in the circuit it should be at this point. Fig. 2 illustrates this in the very simple case of a receiving circuit comprising the antenna A, the rectifier R and the telephone T. Capacitance between telephone and earth would be without effect in the Right arrangement, but in the Wrong would shunt the rectifier and distune the antenna.

Regarded as a system of power transmission, wireless telegraphy occupies a peculiar place in the extreme smallness of the fraction of the transmitted power which reaches the receiver. In the ordinary electric power line, energy may be poured into one end at an enormous rate, but at the other end the power is on a corre-

3. SPECIAL FEATURES OF WIRELESS TELEGRAPHY

spondingly large scale. In submarine telegraphy moderate power is transmitted and very small power received; in ordinary telephony, the power transmitted is only a small fraction of a watt, and, over a line with the conventional commercial limit of attenuation ($\alpha l \fallingdotseq 5$), only about one ten-thousandth of this is received at

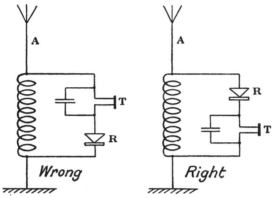

Fig. 2. The importance of small stray capacitances.

the other end; but in both these cases the power received much exceeds that in a wireless receiver working at full commercial range. The telephone receiver, although itself a machine of very poor efficiency, in conjunction with the human ear is a marvellously sensitive indicator, and is used as such in wireless telegraphy. In telephony it must reproduce intelligible speech, and this demands much greater power than is necessary for the merely audible buzz

	Watts transmitted	Watts received	Ratio
Power line	say 10^6	10^6	1
Submarine telegraph	5	5×10^{-7}	10^{-7}
Telephone	10^{-2}	10^{-6}	10^{-4}
Wireless telegraph...	say 10^5	10^{-12}	10^{-17}

required in the wireless telegraph receiver. Consequently in wireless telegraphy the ratio between power received and power transmitted may reach a degree of smallness quite unapproached in telephony. The Table shows in a rough illustrative way the orders of magnitude of the powers transmitted and received in the several signalling systems named, " watts received " signifying the power as taken from the connecting medium and before conversion to any other form.

In the wireless telegraph receiver of Fig. 2, the overall efficiency of antenna, rectifier and telephone, as a converter from electrical high-frequency input to acoustic output, is itself excessively small. It has been estimated as 0·015 per cent. for just perceptible signals*.

* W. H. Eccles, *Continuous Wave Wireless Telegraphy*, Part I (1921), p. 18.

CHAPTER II
ELECTROMAGNETIC RADIATION*

1. Radiation occurs whenever current changes

At a wireless station energy in the form of electromagnetic waves is radiated into space by the sending antenna, and conversely is absorbed from space by the receiving antenna. On first contemplating the fact of radiation electrical engineers are sometimes puzzled to see how this new thing crops up. Have they not dealt, in practice and in theory, with multifarious alternating current circuits, with never a thought of radiation nor loophole for its inclusion in the calculations made! In this Section an attempt is therefore made to convince such a reader—to persuade him rather than to prove to him—by reference to none but familiar conceptions that radiation of energy must and does occur whenever the current in a circuit changes; and to show why this radiation effect is large in wireless antennas although negligibly small in our ordinary circuits. When once he apprehends that no new phenomena are involved, and that the unfamiliar differ only quantitatively from the familiar, that they are both described by the same natural laws, any irritating sense of mystery should be allayed.

Consider a circuit (Fig. 3) of resistance R carrying an alternating current $i = I \sin pt$, and producing at any instant a magnetic field such that ϕ lines of flux thread the circuit. We are accustomed to write

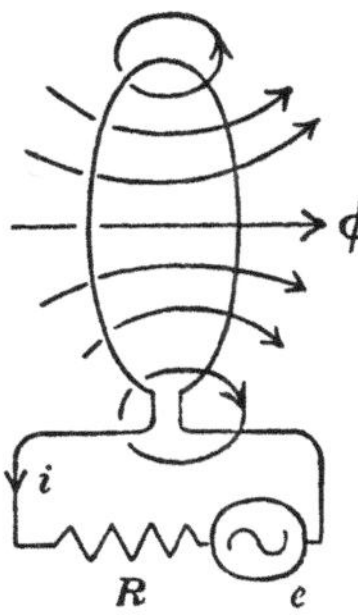

$$\phi = Li = LI \sin pt,$$

Fig. 3.

where L is a constant (in the absence of iron) known as the self-inductance of the circuit. This is strictly true if i changes indefinitely slowly, and it could therefore be true for rapid changes of i only if

---

* The full analysis of electromagneic radiation makes heavy calls upon advanced mathematics, and must be studied in standard works on the mathematical theory of electricity, such as J. H. Jeans' *Electricity and Magnetism* (1925) or G. W. Pierce's *Electric Oscillations and Electric Waves* (1920). E. B. Moullin's important paper, "The fields close to a radiating aerial," *Camb. Phil. Soc. Proc.* Vol. xxv (1929), p. 491, should be consulted. It is possible, however, to acquire a useful, if incompletely detailed, understanding by less sublimated methods. The present Chapter follows such a course.

a change of current produced immediately the corresponding change of magnetic field. This would involve either action at a distance or infinite speed of motion of the magnetic lines—conceptions sufficiently repulsive to most persons, with or without a mathematical training, for us to reject them. It follows that as i changes, the value of ϕ at any instant must diverge from the value of Li at that instant towards the value of Li during the preceding instants. As we have taken the case of a sinoidal current i, we may make some approach to this condition by supposing that ϕ is also sinoidal and lagging by some phase angle θ behind the current. Thus let us take

$$\phi = LI \sin(pt - \theta)$$

as indicated in Fig. 4.

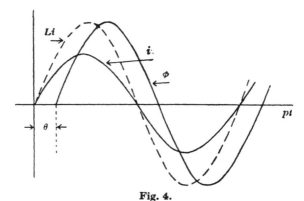

Fig. 4.

By the laws of Ohm and Faraday we then have for the impressed E.M.F.

$$e = Ri + \frac{d\phi}{dt},$$
$$= RI \sin pt + pLI \cos(pt - \theta),$$
$$= \cos pt \, (pL \cos \theta) I + \sin pt \, (R + pL \sin \theta) I.$$

The first of these terms is the quadrature or reactance component of the E.M.F.; the second is the in-phase or resistance component. The total power delivered to the circuit by the alternator is obviously

$$\frac{I^2}{2}(R + pL \sin \theta).$$

Of this, $\frac{I^2}{2} R$ is converted into heat within the circuit, and the

1. RADIATION OCCURS WHENEVER CURRENT CHANGES

remainder, $\frac{I^2}{2}(pL \sin \theta)$, must therefore travel away from the circuit into space. $pL \sin \theta$ is thus the "radiation resistance" of the circuit; i.e. that quantity which, multiplied by the square of current, gives power radiated. We have now to see how the radiation resistance may become large.

With given p, the expression $pL \sin \theta$ for the radiation resistance is increased by increasing L or θ. L is increased by enlarging the loop or by winding on more turns of wire; and with given L, θ is increased by spreading out the field, i.e. by providing the given inductance L in a large coil of few turns rather than in a compact coil of many turns. An open antenna, consisting of a vertical straight wire one of whose ends is earthed and the other insulated and elevated, may be regarded as an extreme case giving maximum spread of field for given inductance. Again, remembering that θ is the phase lag of the field behind the current producing it, we see that increase of frequency will rapidly increase the radiation resistance, since both p and θ are then increased.

Similar considerations with regard to the electrostatic field instead of the magnetic field would lead to similar conclusions. We can see, therefore, that the powerful radiation from an antenna is due to the exceptionally spread-out configuration of its magnetic and electric fields, and especially to the very high frequency of the current in it.

It will be interesting at this stage to compare these rough conclusions as to conditions for marked radiation with formulas arrived at by mathematical analysis. At a frequency of n cycles per second the radiation resistance of an isolated loop consisting of T close turns of area S sq. cm, if the current were uniform along the wire, is

$$3 \cdot 9 \times 10^{-38} n^4 S^2 T^2 \text{ ohms}*.$$

The corresponding expression for the radiation resistance of an open antenna consisting of an earthed vertical wire of length l cm is found in Section 4 of this Chapter; it is

$$7 \cdot 1 \times 10^{-19} n^2 l^2 \text{ ohms}.$$

An examination of each of these expressions confirms our conclusions that the radiation resistance of a circuit increases as the magnetic or electric field is more widely spread out, and as the frequency is raised.

* See E. B. Moullin, *loc. cit.* p. 11.

2. Maxwell's displacement current

In framing a theory of electromagnetism it is necessary to lay down some hypothesis serving to connect magnetic charge (pole) with the motion of electric charge (current). The form of the hypothesis probably most familiar to electrical engineers, sometimes known as Ampère's Law, is the following.
If ABC (Fig. 5) is any closed line linked by an electric current i, the work done on unit pole in taking it once round the line ABC is $4\pi i$; or

M.M.F. round ABC = $4\pi i$.

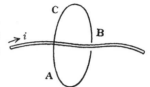

Fig. 5. Ampère's law.

An equivalent hypothesis is that the magnetic field set up by any circuital current i is the same as that of a magnetic shell of uniform strength i bounded by the circuit.

For specifying the magnetic field set up by a steady current Ampère's law is entirely adequate, and its truth has been well established; but when unsteady currents are dealt with, it is clear on any of the following three grounds that the law requires alteration or extension.

(i) If the propagation of light and other disturbance across space devoid of matter is to receive an electromagnetic explanation—and no other is known—a magnetic field must be possible which is not due to the proximity of either currents or magnetic poles.

(ii) When the current is varying, Ampère's law would presumably express an instantaneous relation between current and magnetic field. This would imply instantaneous action at a distance.

(iii) Steady currents are necessarily circuital, so that there is no obscurity in the current *linking* the line ABC (Fig. 5), or in *bounding* the equivalent magnetic shell. Consider, however, the Hertz oscillator PQ in Fig. 6. When the spark occurs at R, electrons stream across from P to Q: there is a current along PQ. But there is no path for their circulation back from Q to P, and they accumulate at Q. It is therefore impossible to state what current is threading the line ABC at any instant.

The requisite amendment or generalisation of Ampère's law was provided by Clerk Maxwell's conception (1865) that the imaginary current consisting of changing electric strain in a dielectric is

2. MAXWELL'S DISPLACEMENT CURRENT

precisely equivalent in its magnetic effect to a real current. The real current along PQ in Fig. 6 is thus rendered circuital by the addition of the return "dielectric current" from Q to P imagined to flow along the dotted lines in Fig. 7. With this extension of the connotation of the word *current*—that is, to include both conduction current and dielectric current—Ampère's law is freed from all restriction and ambiguity. Its usual mathematical

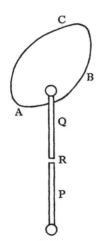

Fig. 6. Conduction current not circuital.

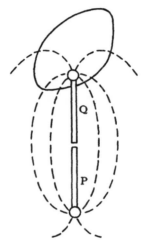

Fig. 7. Displacement current.

expression in terms of components along three rectangular axes is easily found to be

$$4\pi \left(\sigma_x + \frac{\kappa}{c} \frac{\partial F_x}{\partial t} \right) = \frac{\partial H_y}{\partial z} - \frac{\partial H_z}{\partial y},$$

and the two similar equations,

where σ = density of conduction current (in E.M. units),
 F = electric field strength,
 H = magnetic field strength,
 κ = dielectric constant, or permittivity, of the medium,
 c = number of E.S. units of current in 1 E.M. unit.

These are the famous Maxwell equations[*]. By their aid Maxwell formulated an electrical theory of the nature of light or any other

---

[*] See Jeans or Pierce, *loc. cit.* p. 11; or, briefer, F. B. Pidduck's *Treatise on Electricity* (1925).

16   II. ELECTROMAGNETIC RADIATION

disturbance in the aether—a theory which received dramatic confirmation when H. R. Hertz a quarter of a century later (1888)* produced aether waves by electrical means, and found experimentally that they possessed the properties predicted.

Limiting the investigation to propagation through a homogeneous isotropic medium of zero electric conductivity—of which free space is a particular example—it is possible to establish the implications of the Maxwell equations without making any tax on the mathematical agility of the reader, at the same time holding close to those physical concepts which the engineer is always loth to abandon in favour of abstract functions. This is done in the next Section.

3. Propagation of electric disturbance in a homogeneous isotropic medium (including free space as a particular case)†

The magnetic pole as a physical entity will doubtless be abandoned in electrical theory; and what is now called a magnetic field, recognised and measured by the force exerted on a stationary magnetic pole, will be recognised and measured by the force exerted on a moving electron. At present, in formulating a simple electromagnetic scheme, it is convenient—probably chiefly because it is customary—to retain the conception of a magnetic pole, despite its artificiality and that most unhappily inconstant "physical constant" permeability which follows in its train. In the ensuing *résumé* the two conventional fundamentals, electric charge and magnetic pole, and some of their derivatives, are set out in parallel columns to lead to a symmetrical statement of the two laws of electrodynamics. From these we pass very briefly and simply to the nature of radiation, and the relations between the electrostatic (E.S.) and electromagnetic (E.M.) systems of units.

* But D. E. Hughes, the inventor of the microphone, in London in 1879 had undoubtedly already achieved wireless (radiation) telegraphy. His many successful experiments with spark radiators and coherer receivers were recorded in his private notes and were demonstrated to Sir G. Stokes and others; but they were not published until 1899, long after Hertz and Branly, in ignorance of Hughes' work, had repeated his discoveries. See Appendix D of Fahie's book, *loc. cit.* p. 3.

† This Section will be useful as a sort of reminder course to readers who have not a coherent system of electromagnetic fundamentals clearly in mind. It may well be distasteful to others more fortunate, and they should pass it by. Its preparation was much aided by the chapter "Electrostatics and electrodynamics" in W. H. Eccles' book cited on p. 10.

3. PROPAGATION OF ELECTRIC DISTURBANCE

RÉSUMÉ OF ELECTROSTATICS AND MAGNETOSTATICS

Electrostatics	Magnetostatics
\(a\) *Amount of substance.*	
e units of electric charge.	m units of magnetic charge (pole strength).
The electron is well known to us as the atom of electricity and is negative in sign.	The northern region of Earth exhibits charge of negative sign.

These substances are to be regarded as indestructible. Each is measured by the radial repulsive force

$\dfrac{ee'}{\kappa r^2}$ dynes,	$\dfrac{mm'}{\mu r^2}$ dynes,
where κ is the dielectric constant (or electric inductivity) of the medium in which e, e' are situated r cm apart. The unit of e is chosen to make $\kappa = 1$ in vacuum.	where μ is the permeability (or magnetic inductivity) of the medium in which m, m' are situated r cm apart. The unit of m is chosen to make $\mu = 1$ in vacuum.

(b) *Stress in medium.*

Electric stress F \} (or field strength) at a point is measured by
Magnetic stress H \}
the number of dynes of force per unit charge exerted on a small $\begin{Bmatrix} \text{electric} \\ \text{magnetic} \end{Bmatrix}$ charge if placed there. Thus at r cm from a charge $\begin{Bmatrix} e \\ m \end{Bmatrix}$, the stress is:

$F = \dfrac{e}{\kappa r^2}$ dynes per unit electric charge.	$H = \dfrac{m}{\mu r^2}$ dynes per unit magnetic charge.

(c) *Strain in medium.*

| From every unit electric charge proceeds one indestructible tube of electric flux (Faraday tube). | From every unit magnetic charge proceed 4π indestructible tubes of magnetic flux*. |
| | The 4π may be regarded as an unfortunate departure from uniformity in the definitions of unit tube of electric and magnetic fluxes. |

---

* If these definitions are to embrace magnetic phenomena in non-uniform media, the present mnemonically convenient device of treating magnetic poles

II. ELECTROMAGNETIC RADIATION

ELECTROSTATICS	MAGNETOSTATICS
D = electric flux-density (electric induction) = electric strain.	B = magnetic flux-density (magnetic induction) = magnetic strain.

Thus at r cm from a charge $\begin{Bmatrix} e \\ m \end{Bmatrix}$ the strain is:

$$D = \frac{e}{4\pi r^2} \qquad \qquad B = \frac{m}{r^2}.$$

Although unit electric charge must always have attached to it one end of a unit Faraday tube, it is not to be thought that a Faraday tube must always be terminated on charges. It may be closed upon itself. Similarly with magnetic flux. See (j) below.

(d) *Stress-strain relation in medium.*

From (b) and (c) we have:

$$F = \frac{e}{\kappa r^2} = \frac{4\pi D}{\kappa}, \qquad \qquad H = \frac{m}{\mu r^2} = \frac{B}{\mu}.$$

These relations hold, whether the field is set up by a proximate charge $\begin{Bmatrix} e \\ m \end{Bmatrix}$ or otherwise.

(e) *Potential at point in field.*

The potential V at a point in an $\begin{Bmatrix} \text{electric} \\ \text{magnetic} \end{Bmatrix}$ field = work done on unit $\begin{Bmatrix} \text{electric} \\ \text{magnetic} \end{Bmatrix}$ charge in bringing it there from without. Thus at r cm from charge $\begin{Bmatrix} e \\ m \end{Bmatrix}$, the potential is:

$V = \dfrac{e}{\kappa r}$ ergs per unit electric charge.	$V = \dfrac{m}{\mu r}$ ergs per unit magnetic charge.

---

as analogous to electric charges is not tolerable. The magnetic pole, unlike the electric charge, is purely a mathematical abstraction. It must be visualised only as *part of* an entity, as one end of a pair of equal and opposite poles which are indissolubly united by a thin stalk through which $4\pi m$ lines of flux (in addition to those named in (c)) run from the $-m$ pole to the $+m$ pole. When conditions within as well as without a magnet are under consideration, it is found that tubes of magnetic flux must be endless, whereas tubes of electric flux can end on electric charges.

3. PROPAGATION OF ELECTRIC DISTURBANCE

ELECTROSTATICS	MAGNETOSTATICS

(f) *Energy density of field in medium of constants κ and μ.*

| By considering the uniform electric field F between the plates of an indefinitely extended parallel-plate condenser (where the flux-density D in the dielectric equals the charge density on the plates), it can easily be shown that the work done per unit volume of the dielectric in charging (i.e. stored in the dielectric) is $$\tfrac{1}{2} FD \text{ ergs/cm}^3$$ $$= \frac{\kappa F^2}{8\pi} \text{ ergs/cm}^3.$$ | Since the relations between force, m, H and μ are precisely the same as between force, e, F and κ, we have also for a magnetic field the stored energy density $$\frac{\mu H^2}{8\pi} \text{ ergs/cm}^3$$ $$= \frac{1}{8\pi} HB \text{ ergs/cm}^3.$$ |

ELECTRODYNAMICS

We continue to use the units defined above; and we assume that the stress-strain relations (κ and μ) of (d), there stated for steady conditions, are maintained when the stresses are changing, κ and μ being constant during the process*.

(g) *First law of electrodynamics.*

If electric flux-density D at any point in a medium moves through the medium with velocity v perpendicular to D, it produces a magnetic field at the point perpendicular to D and v of magnitude

$$H = \frac{4\pi}{c} Dv = \frac{\kappa}{c} Fv,$$

where c is a constant of our universe. (See (j) below.) The directions are shown at (i) in Fig. 8, and may be remembered as: If v along *thumb* of right hand and D along *1st finger*, then H along *2nd finger*.

It is easy to deduce from this statement of the law the circuital form for conduction currents (Ampère's law; see p. 14). But the law as here stated embodies Maxwell's extension of Ampère's law.

---

* We are accustomed similarly to assume that Young's Modulus as found for a specimen of steel in the testing machine obtains in dynamical actions, such as a tension wave propagated along a wire of the material.

II. ELECTROMAGNETIC RADIATION

(h) *Second law of electrodynamics.*

If magnetic flux-density B at any point in a medium moves through the medium with velocity v perpendicular to B, it produces an electric field at the point perpendicular to B and v of magnitude

$$F = \frac{1}{c} Bv = \frac{\mu}{c} Hv.$$

The directions are shown at (ii) in Fig. 8, and may be remembered

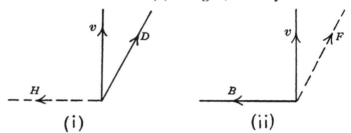

Fig. 8. First and second laws of electrodynamics.

as: If v along *thumb* of right hand and B along *2nd finger*, then F along *1st finger*.

It is easy to deduce from this statement of the law the circuital form (Faraday's Law), viz. the line-integral of the electric field strength round any closed line is $-\frac{1}{c}$ times the rate of change of magnetic flux threading the line.

The second law can be logically deduced from the first with the aid of the principle of the conservation of energy.

(i) *Energy density in space traversed by moving electric field.*

Stationary electric field F at a point has energy density $\frac{\kappa F^2}{8\pi}$ ergs/cm³. (See (f).)

Its motion with perpendicular velocity v sets up at the point magnetic field $H = \frac{\kappa}{c} Fv$. (See (g).)

The energy density of this magnetic field is

$$\frac{\mu H^2}{8\pi} = \frac{\mu \kappa^2 F^2 v^2}{8\pi c^2} = \frac{\kappa F^2}{8\pi} \cdot \frac{\mu \kappa v^2}{c^2} \text{ ergs/cm}^3.$$

Hence the total energy density $= \frac{\kappa F^2}{8\pi} \left(1 + \frac{\mu \kappa}{c^2} v^2\right).$

3. PROPAGATION OF ELECTRIC DISTURBANCE

(j) *Self-supporting field.*

When there are no electric or magnetic charges in proximity, any field can exist only in virtue of its own motion. The symmetry of the foregoing treatment shows that whether we regard the electric field observable at a fixed point as the motion of a magnetic field, or vice versa, the velocity of the motion must be the same. Let this velocity be u. Then from (g) and (h)

$$H = \frac{\kappa}{c} Fu \quad \text{and} \quad F = \frac{\mu}{c} Hu,$$

$$\therefore u = \frac{c}{\sqrt{\kappa\mu}}.$$

When the medium is vacuum, $\kappa = 1$ and $\mu = 1$; therefore $u = c$. Thus the constant c in the two electrodynamic laws is revealed as the velocity of propagation of electric disturbance through vacuum, i.e. the "velocity of light," approximately $3 \cdot 00 \times 10^{10}$ cm/sec.

Hence in a self-supporting field (i.e. radiation) the alternative equivalent moving electric and magnetic fields F and H are perpendicular to each other and to the velocity u through the medium; and

$$\frac{F}{H} = \sqrt{\frac{\mu}{\kappa}} \quad \text{and} \quad u = \frac{c}{\sqrt{\kappa\mu}}.$$

In matter-free space, $F = H$ and $u = c$. Since, whatever the instantaneous spatial distribution of the field may be, the velocity of propagation at every point is the same, it is clear that the field is a wave as defined in I–2.

(k) *Rate of flow of energy in self-supporting field.*

From (i), the total energy density $= \dfrac{\kappa F^2}{8\pi} (1 + 1)$ ergs/cm³.

Hence the energy crossing each sq. cm (perpendicular to direction of flow) per sec.

$$= \frac{\kappa F^2}{4\pi} \times u = \frac{c\sqrt{\kappa} F^2}{4\pi \sqrt{\mu}} = \frac{cFH}{4\pi} = \frac{c\kappa F^2}{4\pi \sqrt{\kappa\mu}} = \frac{c\mu H^2}{4\pi \sqrt{\kappa\mu}} \text{ ergs.}$$

RELATION BETWEEN UNITS IN THE ELECTRO-STATIC AND ELECTROMAGNETIC SYSTEMS

The electrodynamic laws (g) and (h) provide a relation, in terms of the universal constant c, between the electric charge, on which is founded the "electrostatic (E.S.) system," and the magnetic pole, on which is founded the "electromagnetic (E.M.) system." It therefore becomes possible to specify the magnitude of any electric or magnetic or electromagnetic quantity in units belonging to either system. The following example will show how the relation between the two units may be traced out.

In the electrostatic system, unit E.M.F. is given by one unit of electric field strength multiplied by one centimetre; while in the electromagnetic system, unit E.M.F. is given by one tube of magnetic flux cut in one second. Suppose a line 1 cm long is drawn in, and perpendicular to, a magnetic flux of density 1 tube per cm², and is moved in the third perpendicular direction at 1 cm per sec.; the E.M.F. produced in that line is 1 E.M. unit. Now if the second law of electrodynamics (h) is written $Fl = \frac{1}{c} Bvl$, on putting $l = B = v = 1$ it is seen that the E.M.F. whose magnitude is 1 E.M. unit has magnitude F E.S. units, where $F = \frac{1}{c} = \frac{1}{3 \times 10^{10}}$. That is,

1 E.S. unit of E.M.F. $= 3 \times 10^{10}$ E.M. units $= 300$ volts.

In like manner the magnitude of the E.S. and E.M. units of all other quantities may be compared. Thus:

1 E.S. unit of current $= \frac{1}{3 \times 10^{10}}$ E.M. unit $= \frac{1}{3 \times 10^{9}}$ ampere;

1 E.S. unit of resistance or inductance

$= 9 \times 10^{20}$ E.M. unit $= 9 \times 10^{11}$ ohm or henry;

1 E.S. unit of capacitance $= \frac{1}{9 \times 10^{20}}$ E.M. unit $= \frac{1}{9 \times 10^{11}}$ farad.

4. Radiation from a vertical aerial*

In this Section, (a) the field at a large distance from an accelerated electric charge is calculated, using the kinked Faraday tube treatment which was applied by J. J. Thomson to Röntgen rays. The result is then used for the calculation (b) of the remote field set up by the alternating current in a vertical earthed aerial; and (c) of the radiation resistance of the aerial, by performing an integration of rate of energy transference across a large hemisphere described about the aerial as centre. In Subsection 4 (d) the formulas are applied to a numerical example.

4 (a). *The field of an accelerated electric charge at a large distance.* Consider an electric charge e situated at A (Fig. 10). If the charge is stationary, the electric flux-density at P is $\dfrac{e}{4\pi \overline{\mathrm{AP}}^2}$ and is in the direction AP. (See II–3 (c).) If the charge is at B or C or G, the field at P is sensibly the same provided that P is sufficiently remote, i.e. if $\mathrm{AP} \gg \mathrm{AG}$; and it would be sensibly the same if the charge were moving slowly along the line AB with uniform velocity v†. Any electric flux line AP moves parallel to itself, reaching the position CQ after time t, where $\mathrm{AC} = vt$.

Now suppose that for a short time δt after the instant when the charge is at A it has an acceleration f, thus reaching B at time δt with velocity $(v + f . \delta t)$. In the time t it thus reaches G instead of C, where
$$\mathrm{BG} = (v + f . \delta t)(t - \delta t).$$

* Following G. W. O. Howe, "Notes on wireless matters," *Electrician*, Vol. XC (1923), p. 614.

† Since the charge has a direction of motion, the central symmetry of its electric flux lines obtaining in electrostatics is no longer necessary *a priori*. Indeed, it is obvious from the two laws of electrodynamics (II–3 (g) and (h)) that a moving electron cannot carry with it a centrally symmetrical electric field. For if the electric field had the electrostatic distribution, its motion would produce no magnetic field at the point P (Fig. 9) but would produce a magnetic field along OY (perpendicular to the paper) at the point Q. The latter, by virtue of its motion, introduces an additional electric field at Q. Thus the motion has changed the electric field at Q, but not at P. The effect is, however, negligible when v^2 is negligible compared with u^2, where u is the velocity of propagation in the medium.

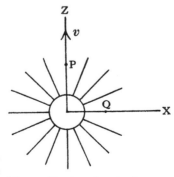

Fig. 9. Moving electric charge.

II. ELECTROMAGNETIC RADIATION

The flux line which, but for the acceleration, would have been CQ emerging from the charge at C, now emerges from the charge at G. But since electric disturbances in the medium surrounding the charge are propagated with the finite speed $u = \dfrac{c}{\sqrt{\kappa\mu}}$ (see II-3 (j)), the occurrence of the acceleration at A must be ineffective at all distances from A exceeding ut. Hence RQ is part of the flux line from the charge at G, where R is the point on CQ given by

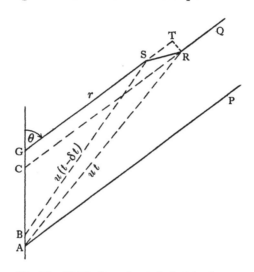

Fig. 10. Field of accelerated electric charge.

$AR = ut$. Similarly, since the acceleration ceased when the charge was at B, GS is also part of the same flux line, where $BS = u(t - \delta t)$. The effect of the acceleration has therefore been to turn the straight flux line CQ into the kinked flux line GSRQ.

Provided that $v \ll u$, AC and BG are very much less than AR and BS; SG is sensibly equal to SB, and RC to RA. Hence if RT be drawn perpendicular to GS,

$$ST = RC - SG = RA - SB = u\,\delta t.$$

Specifying the point S by the polar coordinates r, θ as shown, we have by electrostatics (see II-3 (c)) at S on GS a flux-density $\dfrac{e}{4\pi r^2}$

4. RADIATION FROM A VERTICAL AERIAL

along GS. Since lines of flux are indestructible except at a charge, at S on SR there must be a flux-density $\dfrac{e}{4\pi r^2} \div \cos \text{RST}$ along SR.

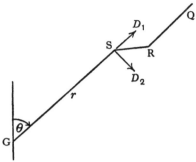

Fig. 11.

Hence between S and R the components (Fig. 11) of flux-density, D_1 along r and D_2 perpendicular to r, are

$$D_1 = \frac{e}{4\pi r^2 \cos \text{RST}} \cdot \cos \text{RST} = \frac{e}{4\pi r^2},$$

$$D_2 = \frac{e}{4\pi r^2 \cos \text{RST}} \cdot \sin \text{RST} = \frac{e}{4\pi r^2} \cdot \frac{\text{RT}}{\text{ST}}.$$

But $\qquad \text{RT} = \text{CG} \sin \theta = \dot{f}t\, \delta t \sin \theta,$
and $\qquad \text{ST} = u\, \delta t.$

$$\therefore \frac{\text{RT}}{\text{ST}} = \frac{\dot{f}t \sin \theta}{u} = \frac{\dot{f}r \sin \theta}{u^2}.$$

$$\therefore D_2 = \frac{e}{4\pi r^2} \frac{\dot{f}r \sin \theta}{u^2},$$

$$= \frac{ef \sin \theta\, \kappa \mu}{4\pi c^2 r}, \quad \because u = \frac{c}{\sqrt{\kappa \mu}} \text{ (see II-3 (j)).}$$

When the acceleration of the charge makes itself felt at S, there thus exists at S the flux-density D_1 moving with small normal speed $v \sin \theta$, and D_2 moving with great normal speed $(u + v \cos \theta) \doteqdot u$. These moving flux-densities are equivalent (see II-3 (g)) to a magnetic field perpendicular to the paper downwards of strength

$$H = \frac{4\pi}{c}[D_1 v \sin \theta + D_2 (u + v \cos \theta)],$$

$$= \frac{4\pi}{c}\left[\frac{ev \sin \theta}{4\pi r^2} + \frac{ef \sin \theta \sqrt{\kappa \mu}}{c^2 r}\right].$$

The Table summarises the results of this investigation; it gives the two components of electric field $\left(\text{radial } F_1 = \dfrac{4\pi}{\kappa} D_1, \text{ and perpendicular thereto } F_2 = \dfrac{4\pi}{\kappa} D_2\right)$, and the perpendicular magnetic field (H) existing at a spot r, θ far distant from a charge e moving with velocity v (if any) when the effect of an acceleration f (if any) arrives $\left(\text{i.e. at a time } \sqrt{\kappa\mu}\,\dfrac{r}{c} \text{ after the occurrence of the acceleration}\right)$.

Motion of charge e	Radial electric field F_1 (dynes per unit el. charge)	Electric field F_2, perp. to F_1 and r (dynes per unit el. charge)	Magnetic field H, perp. to F_1 and to F_2 (dynes per unit mag. pole)
Stationary	$\dfrac{e}{\kappa r^2}$	0	0
Constant veloc. v	$\dfrac{e}{\kappa r^2}$	0	$\dfrac{ev \sin\theta}{cr^2}$
Veloc. v, accel. f	$\dfrac{e}{\kappa r^2}$	$\dfrac{ef \sin\theta\,\mu}{c^2 r}$	$\dfrac{ev\sin\theta}{cr^2} + \dfrac{ef\sin\theta\sqrt{\kappa\mu}}{c^2 r}$

It is to be noticed that the fields due to the acceleration f—termed the *radiated* fields—decrease as the inverse of the distance r, while the other fields decrease as the inverse square of the distance. At large distances, therefore, it is only the former which are of moment. Omitting the subscript from F_2, they are

$$F = \frac{\mu}{c^2} \cdot ef \cdot \frac{\sin\theta}{r},$$

$$H = \frac{\sqrt{\kappa\mu}}{c^2} \cdot ef \cdot \frac{\sin\theta}{r};$$

and in vacuum

$$F = \frac{1}{c^2} \cdot ef \cdot \frac{\sin\theta}{r} = H.$$

4 (b). *Application to an antenna.* Consider a current $i = I \sin pt$ amperes flowing in a straight wire of length l cm. According to the modern view, this alternating current consists of a constant very large number of free electrons, of total charge (say) e E.S. units, drifting slowly* back and forth along the wire, with average instantaneous speed (say) v cm/sec.

---
* Physicists have reason to believe that in any metallic conductor the speed is lower than 1 cm/sec. for all practicable current densities.

4. RADIATION FROM A VERTICAL ANTENNA

If the cross-sectional area of the wire is s cm², the volume density of drifting charge is $\dfrac{e}{ls}$. Hence the current density is $\dfrac{e}{ls} \times v$, and the current is $\dfrac{ev}{l}$ E.S. units or $\dfrac{ev}{l} \times \dfrac{1}{3 \times 10^9}$ amperes. Therefore

$$\frac{ev}{3 \times 10^9 l} = i = I \sin pt,$$

i.e.
$$v = 3 \times 10^9 \frac{l}{e} I \sin pt.$$

Hence if f is the average instantaneous acceleration of the drifting electrons,

$$f = \frac{dv}{dt} = \frac{3 \times 10^9 \, plI \cos pt}{e}.$$

The total fields, at the spot r, θ at any instant t, produced by the acceleration of all the electrons, that is by the changing current in the wire, are then given by the formulas of the last Section on substituting for ef the expression

$$3 \times 10^9 \, plI \cos p \left(t - \sqrt{\kappa \mu} \, \frac{r}{c} \right).$$

The time $\left(t - \sqrt{\kappa \mu} \, \dfrac{r}{c} \right)$ is appropriate because the effect of an acceleration of charge, or rate of change of current, in the wire appears at a distance r only after the lapse of time $\sqrt{\kappa \mu} \, \dfrac{r}{c}$.

Restricting the formulas now to the case of $\kappa = 1 = \mu$ (e.g. free space), the Table on p. 26 shows that the radiation is expressible as:
Moving electric field F

$$= \frac{3 \times 10^9 \, plI \sin \theta}{c^2 r} \cos p \left(t - \frac{r}{c} \right) \text{ dynes/unit el. charge,}$$

$$= \frac{9 \times 10^{11} \, plI \sin \theta}{c^2 r} \cos p \left(t - \frac{r}{c} \right) \text{ V/cm ; or}$$

moving magnetic field H

$$= \frac{3 \times 10^9 \, plI \sin \theta}{c^2 r} \cos p \left(t - \frac{r}{c} \right) \text{dynes/unit mag. pole.}$$

The relative directions are shown in Fig. 12, where OX, OY, OZ are perpendicular axes chosen so that the current is along OZ and the point S⃗ is in the plane XOZ.

These are the field radiated by an isolated Hertz oscillator of length l (Fig. 13 (i)); or, for $\theta \not> 90°$, by the upper half of a vertical Hertz oscillator whose imaginary lower half lies below the horizontal surface of a perfectly conducting flat ground (Fig. 13 (ii)). The sphere conventionally shown at the ends of the Hertz oscillator

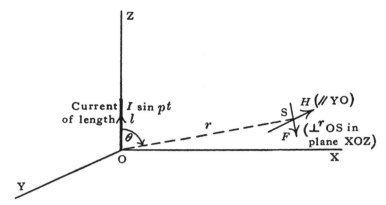

Fig. 12. Directions of electric and magnetic fields.

(Fig. 13), and the substantially flat top of a real vertical earthed antenna (Plate I), are devices for indicating pictorially, and for really approximating to, uniformity of current along the vertical wire; the

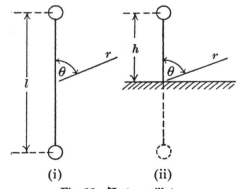

Fig. 13. Hertz oscillator.

capacitance to earth of the flat top being much greater than the distributed capacitance to earth of the vertical wire. Writing

$$h = \frac{l}{2}, \quad p = 2\pi n = \frac{2\pi c}{\lambda}, \quad c = 3 \times 10^{10} \text{ cm/sec.},$$

the formulas for the field radiated (at a large distance) by the

4. RADIATION FROM A VERTICAL ANTENNA

current $I \sin pt$ amperes in the vertical up-lead* of a flat-topped antenna of height h, lengths being in centimetres, are

$$F = 120\pi \frac{h}{\lambda} I \frac{\sin\theta}{r} \cos p\left(t - \frac{r}{c}\right) \text{ V/cm},$$

$$H = 0.400\pi \frac{h}{\lambda} I \frac{\sin\theta}{r} \cos p\left(t - \frac{r}{c}\right) \text{ dynes/unit magnetic pole}.$$

In the horizontal plane, i.e. near the surface of a sensibly flat ground, $\sin\theta = 1$.

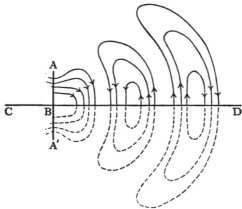

Fig. 14. Lines of electric field near oscillator.

Around the complete Hertz oscillation ABA' (Fig. 14), if isolated in space, the instantaneous distribution of electric field must be such as that portrayed. The figure is symmetrical about the horizontal line CD, and since this line is crossed perpendicularly by the lines of electric field, no change in the field would be caused by introducing a copper sheet CBD perpendicular to the oscillator ABA' (thereby, if copper were of zero resistivity, completely isolating any system above CD from any system below). No currents would flow along the copper sheet.

If the lower half oscillator BA' is removed, no change is observable above CD; but now horizontal currents must flow in the

---

* The currents in the flat top also radiate. But they contribute nothing to the fields at a large distance in the horizontal plane; and in antennas of ordinary proportions they probably contribute very little to the total power radiated. For an exhaustive mathematical treatment of the radiation from a uniform antenna of inverted L form, see G. W. Pierce, *loc. cit.* p. 11, Book II, Chapter ix.

copper sheet, between regions where downward electric lines are growing denser and the corresponding regions where upward lines are growing denser as the whole system of lines sweeps outward radially from the oscillator.

If the ground has sensibly zero resistivity, its surface may take the place of the copper sheet without changing the conditions; but if the resistivity is not negligible, the currents in it develop heat, and there is an absorption of energy from the field above (effected by a slight forward tilt of the wave-front near the ground)*. A more

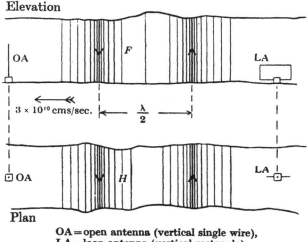

OA = open antenna (vertical single wire),
LA = loop antenna (vertical rectangle).
Fig. 15. Fields at receiving antenna.

rapid attenuation of field strength with distance results. For the present we regard the ground as a perfect conductor.

To stationary observers standing far from the transmitter and endowed with the power of seeing the lines of electric and magnetic field, they would appear instantaneously as shown in Fig. 15. The upper diagram, being a horizontal projection, shows over half a wavelength or so the vertical electric lines near the ground, packed together most densely and directed downwards at one distance from the transmitter, distributed sinoidally on either side, and so packed together most densely again but directed upwards at a distance greater or less by $\frac{\lambda}{2}$. The lower diagram, being a vertical projection,

* The peculiar Beverage antenna depends upon the existence of this tilt. See XIV-4.

shows the horizontal magnetic lines similarly distributed. Either (but not both) collection of lines may be regarded as sweeping past the observer, in the direction away from the transmitter, with the velocity of light. If the observer were provided with a receiving antenna, such as a vertical wire OA or vertical loop LA, he would find in it an alternating electromotive force. In the one case (OA), this is most easily thought of as the result of the changing electric field along the wire; in the other case (LA), it is the result of the difference of the changing electric fields along the two vertical sides of the loop, or alternatively* the result of the changing magnetic flux through the loop.

It is to be noted that, whereas close to the transmitting antenna the electric and magnetic fields are in time quadrature, in pure radiation—i.e. in self-supporting fields, remote from charges and poles—they are synphased in time.

4 (c). *Radiation resistance of antenna.* If it is assumed, as hitherto, that no power is absorbed by the ground, all the power radiated by

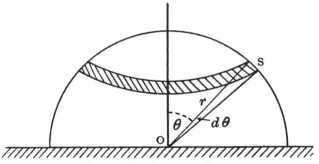

Fig. 16. Total power radiated.

the antenna at O passes across the surface of a hemisphere of radius r, described about O as centre, through the distant point S (Fig. 16). The field at S has been calculated; hence the rate of flow of energy

---

* The reader is cautioned against the temptation he may feel to take the E.M.F. generated in the loop as the *sum* of the E.M.Fs. calculated in these alternative ways. He must discipline himself to realise that the electric field *is* the moving magnetic flux, and the magnetic field *is* the moving electric flux. The observer, stationary at his post, may attribute his experience *either* to a stationary vertical alternating electric field, plus a stationary horizontal alternating magnetic field; *or* to the motion past him of the constant vertical electric system portrayed in the upper part of Fig. 14; *or* to the motion past him of the constant horizontal magnetic system portrayed in the lower part of Fig. 14. He cannot have his cake and eat it.

per unit area of the sphere at S is known. On integrating over the surface of the hemisphere, the total power radiated is found.

Thus (see II–3 (k)), time-mean of power crossing unit area at S

$$= \frac{c}{4\pi} \text{ (mean } H^2) \text{ ergs/sec.,}$$

$$= \frac{c}{4\pi} \cdot \frac{1}{2} \left(0 \cdot 400\pi \frac{h}{\lambda} I \frac{\sin \theta}{r}\right)^2 \text{ ergs/sec.}$$

∴ Total power radiated

$$= \frac{6 \times 10^8 \pi\, h^2 I^2}{\lambda^2 r^2} \int_0^{\frac{\pi}{2}} \sin^2 \theta \,.\, 2\pi r \sin \theta \,.\, r \, d\theta \text{ ergs/sec.,}$$

$$= 80\pi^2 \frac{h^2}{\lambda^2} I^2 \text{ watts.}$$

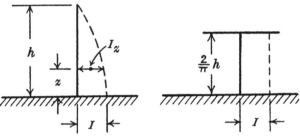

Fig. 17. Antennas without and with flat top.

But if R ohms is the *radiation resistance*, the total power radiated is $\tfrac{1}{2} I^2 R$ watts. Hence

$$R = 160\pi^2 \frac{h^2}{\lambda^2} = 1580 \frac{h^2}{\lambda^2} \text{ ohms.}$$

If the antenna had no flat top, but were a uniform vertical conductor of length h, the current would decrease from (say) I at the bottom to zero at the top. Such an antenna loaded with inductance (as at C in Fig. 40) sufficiently to make λ much exceed the fundamental wavelength—say if λ exceeds 2 or 3 times $4h$—has a current amplitude falling approximately linearly along the antenna from I at the bottom to zero at the top. The antenna is then equivalent to one with uniform current I (i.e. the ideal flat-top antenna) of height approximately $\tfrac{1}{2}h$. Its radiation resistance is therefore

$$\frac{1580 \,(\tfrac{1}{2}h)^2}{\lambda^2} = 395 \frac{h^2}{\lambda^2} \text{ ohms.}$$

4. RADIATION FROM A VERTICAL ANTENNA

If, on the other hand, the straight antenna is unloaded, so that λ is its natural wavelength, the distribution of current is sensibly

$$I_z = I \cos\left(\frac{\pi}{2} \cdot \frac{z}{h}\right) * \text{ (see Fig. 17)};$$

the equivalent flat-top height is $\frac{2}{\pi} h$; and the radiation resistance is

$$\frac{1580 \left(\frac{2}{\pi} h\right)^2}{\lambda^2} = 640 \frac{h^2}{\lambda^2} \fallingdotseq 640 \frac{h^2}{(4h)^2} = 40 \text{ ohms}\dagger.$$

4 (d). *Numerical example of antenna radiation.*

Transmitter: The large Government Station at Rugby (Hillmorton), using its complete antenna supported on 12 masts of height 250 m (Fig. 44, p. 81).
$\lambda = 18,800$ m   ($n = 16,000$ c/s),
$h = 175$ m$\ddagger$,
$I = 725$ A (R.M.S. value).

Receiver: Several wavelengths distant, so that the analysis of Section 3 (b) may be applicable, but not far enough for the Earth's curvature to have much effect; say at Cambridge, where $r = 97$ km.

---

* This would be the correct expression for the distribution at the natural wavelength if the capacitance and inductance per unit length were uniform from top to bottom.

† Cf. 36·57 ohms calculated by M. Abraham and by G. W. Pierce. (See the latter's book, *loc. cit.* p. 11, p. 481.) Pierce criticises the derivation of the antenna resistance from that of the Hertzian doublet (Rüdenberg), and quotes 98·7 ohms as the resistance in the above case "calculated on the doublet theory." In this Section, however, the double theory has led us to 40 ohms, which agrees well enough with Pierce's own value. It seems that Pierce has omitted the factor $\frac{2}{\pi}$ in applying Rüdenberg's formula for the Hertz doublet, in which the current is I throughout the length, to the uniform vertical wire, in which the current is I at the bottom and zero at the top. The fair deduction from Rüdenberg's formula appears to be

$$98\cdot 7 \times \left(\frac{2}{\pi}\right)^2 = 39\cdot 8 \text{ ohms.}$$

‡ The effective height is less than the height of the masts supporting the flat top on account of sag, etc.; and because the presence of conducting masts (of steel lattice) extending from near the flat top to near the ground reduces their effective mean separation. The effective height of large aerials supported by steel masts or towers is usually about 70°/₀ of the mast height; and in this instance 175 m is the mean of a number of measured values. See R. V. Hansford and H. Faulkner, "Design Details of a High-Power Radio-Telegraphic Transmitter using Thermionic Valves," *Journ. I.E.E.* Vol. 65 (1927), p. 297.

The formulas on p. 29 give for the field strengths at Cambridge:

$$F = 120\pi \times \frac{175}{18,800} \times 725 \times \frac{1}{97 \times 10^5} \text{ V/cm},$$

$$= 2\cdot 62 \times 10^{-4} \text{ V/cm}.$$

$$H = 8\cdot 7 \times 10^{-7} \text{ dynes/unit magnetic pole}.$$

The formula on p. 33 gives for the radiation resistance at Rugby

$$R = 640 \frac{175^2}{18,800^2} \Omega,$$

$$= 0\cdot 055 \ \Omega.$$

The power radiated is therefore

$$725^2 \times 0\cdot 055 \text{ W} = 29 \text{ kW}.$$

The low value, $0\cdot 055\Omega$, of the radiation resistance illustrates the difficulty of providing a good radiator for very long waves, using masts of practicable height. When the radiation resistance is small, it is of course important to keep the total resistance very low, i.e. to reduce the heat losses in and around the antenna circuit. A large proportion of the loss occurs in the ground near the antenna, where the current densities are large; and this loss is usually moderated by laying in the ground in appropriate directions a great quantity of copper wire. At Rugby this was done very thoroughly*, and the total resistance of the antenna is exceptionally low, viz. $0\cdot 5$ ohm†. Even so, it is clear that the efficiency of the antenna is very poor, viz.:

$$\text{Efficiency} = \frac{\text{power radiated}}{\text{power supplied}} = \frac{0\cdot 055}{0\cdot 5} = 11 \%.$$

With short waves it is, of course, feasible to obtain much higher radiation resistances, and higher antenna efficiencies.

* Below the portion of the antenna carried by eight masts, about 120 miles of copper wire is laid a few inches deep in the ground.
† See Hansford and Faulkner, *loc. cit.* p. 33.

CHAPTER III

PROPAGATION OF WIRELESS WAVES ROUND THE EARTH

1. Curvature of the ground; the Heaviside layer

In the last Chapter an expression was derived for the field radiated by the current in an antenna situated on, and perpendicular to the plane surface separating a perfect conductor from a perfect dielectric. At places near that surface and distant r from the transmitter, the formula on p. 29 for the electric field becomes

$$F = 0{\cdot}377 \times 10^6 \frac{hI}{\lambda}\frac{1}{r} \ \mu\text{V/m},$$

if lengths are in kilometres and current in amperes*.

Actual terrestrial conditions differ from these ideal conditions in the following respects:

(i) The surface of the ground, being roughly spherical, can be treated as sensibly plane only over short distances.

(ii) The atmosphere is not *throughout* a sensibly perfect dielectric, conduction phenomena playing an important rôle in the auroral regions at altitudes of the order of 100 km above ground.

(iii) The ground is not formed of materials of zero resistivity; and its geometrical irregularities, constituted by mountains, trees and the like, cannot always be ignored.

Of these real departures from the ideal conditions, the third is relatively trivial, but the first and second are of immense significance. The solid Earth being substantially opaque to wireless waves, as it is to light waves, how shall wireless rays reach points below the horizon when the rays from a searchlight will not do so? This question would be one of rhetorical despair were it not for the second departure from the ideal conditions, the conducting properties of the upper atmosphere.

A conductor is a material which contains electric charges capable of moving through the material under the forces exerted on them by an electric field. Near the ground the air is substantially a non-conductor; its atoms are electrically neutral, and interchange of

* These units are adhered to throughout this Chapter.

III. WIRELESS WAVES ROUND THE EARTH

electrons between the atoms (to which a wire owes its conductivity) does not take place. At the other extreme, far above the ground, there is substantially perfect vacuum; there are no charges upon which a field can act. But in an intermediate region, commonly referred to as the *Heaviside Layer*, probably especially between 40 km and 100 km high, the air is strongly ionised by the ultraviolet rays of solar light; and at altitudes where the atoms are few and far between, so that recombination is slow, it remains ionised during the hours of darkness.

A radiated field arriving in such a region therefore finds, mingled with electrically inert atoms, electrically charged particles capable of movement under the forces produced upon them—negative ions (electrons) of small mass and positive ions (atoms deficient of electrons) of large mass. Qualitatively such a region simulates a conducting spherical ceiling to the sky which, if it were perfectly conducting, would confine the radiation entirely to the space between ground and ceiling, as the walls of a speaking tube confine the sound waves. Although, as explained later, the properties of the Heaviside layer cannot be fully expressed in any such simple terms, its effect does seem sometimes to be much the same as that of a concentric copper shell some 40 to 100 kilometres up, between which shell and the ground the radiation is confined. It is instructive to obtain a quantitative estimate of the enormous attenuation of field strength at great distances for which, in the absence of the Heaviside layer, the curvature of the ground would be responsible; and of the great service a well conducting Heaviside layer would render.

Let us compare the propagation in the following four ideal cases, viz.:

(a) A flat perfectly conducting ground bounded by a perfect vacuum (or the unionised air we breathe).

(b) Case (a) modified by the addition of a perfectly conducting (reflecting) layer at a uniform height H above the ground.

(c) Case (b) modified by making the ground spherical of radius R, R being very large compared with H.

(d) Case (c) modified by removal of the reflecting layer.

Case (a). The field at a large distance r has been found to be

$$F = 0.377 \times 10^6 \frac{hI}{\lambda} \frac{1}{r}.$$

1. CURVATURE OF THE GROUND; THE HEAVISIDE LAYER

An attempt to show roughly what is happening is made at (a) in Fig. 18*, where the energy at one instant in the shaded volume at P is spread at a later instant over the larger volume at P'. The shaded ring at P' is of the same thickness δr as at P, but its perimeter has increased from $2\pi r$ to $2\pi r'$, and the height too has

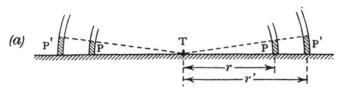

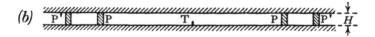

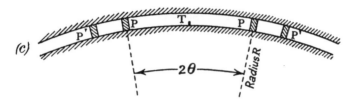

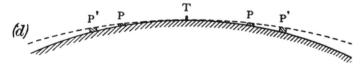

Fig. 18. Dispersion of radiation.

increased in the ratio $r':r$. The energy at P therefore now occupies a volume increased in the ratio $r'^2:r^2$. This implies that the field strength must vary inversely as r, which agrees with the formula.

As a numerical example exposing the significance of the several formulas to be compared, let the dimensions of the transmitter be:

$$h = 0.075 \text{ km}; \quad \lambda = 14.1 \text{ km}; \quad \mathscr{I} = 300 \text{ A}.$$

* This Figure and the Table on p. 38 are taken from an article by the author in *The Electrician*, Vol. XCVII, p. 42, which was supplemented—with reference to some criticism by T. L. Eckersley—on p. 176.

III. WIRELESS WAVES ROUND THE EARTH

Line (a) of the Table then gives the electric field strengths at distances r between 1000 km and 15,000 km, according to the above formula.

	Field strengths in microvolts per metre					
Case	Conditions	Distance from transmitter in kilometres				
		1000	2000	5000	10,000	15,000
(a)	Flat ground; no reflecting layer	600	300	120	60	40
(b)	Flat ground; reflecting layer 40 km above it	275	190	120	86	70
(c)	Spherical earth; reflecting layer 40 km above it	275	195	130	110	130
(d)	Spherical earth; no reflecting layer	296	43	0·28	1×10^{-4}	5×10^{-8}
(e)	Real (measured strengths)	600	200	40	6	2

NOTE. Great-circle semi-circumference of Earth = 20,000 km.

Case (b). The perfectly conducting layer now introduced prevents the spread of energy in the vertical direction, as indicated at (b) in Fig. 18. The volume of the shaded ring at P is now $H \cdot \delta r \cdot 2\pi r$; it increases linearly with r. If the power radiated remains

$$I^2 \times 1580 \frac{h^{2*}}{\lambda^2},$$

it is easily found that

$$F = 3 \cdot 09 \times 10^5 \frac{1}{\sqrt{H}} \frac{hI}{\lambda} \frac{1}{\sqrt{r}}.$$

T. L. Eckersley has calculated, however, that the power radiated in the presence of the reflecting layer is not the same as in its absence, and his analysis leads to the result

$$F = 1 \cdot 88 \times 10^5 \frac{1}{H} \frac{hI}{\sqrt{\lambda}} \frac{1}{\sqrt{r}}.$$

Line (b) gives the values in the numerical example obtained from

---

* I is here the R.M.S. value of the current, not the peak value as in II-4 (c), where the radiation resistance is shown to be $1580 \frac{h^2}{\lambda^2}$.

1. CURVATURE OF THE GROUND; THE HEAVISIDE LAYER

this formula, on the assumption that $H = 40$ km, which various researches have shown to be a likely value during daylight.

Case (c) is shown at (c) in Fig. 18. If the distance $r = R\theta$ is still measured along the ground—i.e. along great circles through the transmitter instead of straight lines—the volume of the shaded ring at P is now

$$H \cdot \delta r \cdot 2\pi R \sin \theta = H \cdot \delta r \cdot 2\pi \frac{r \sin \theta}{\theta}$$

instead of $H \cdot \delta r \cdot 2\pi r$; and the only modification required in the last formula is a geometrical factor $\sqrt{\dfrac{\theta}{\sin \theta}}$ to cover the curvature of the ground. Hence

$$F = 1 \cdot 88 \times 10^5 \frac{1}{H} \frac{hI}{\sqrt{\lambda}} \frac{1}{\sqrt{r}} \sqrt{\frac{\theta}{\sin \theta}}.$$

It is interesting to compare the attenuations of field strength with increasing distance in cases (b) and (c). The curvature factor $\sqrt{\theta/\sin \theta}$ is always greater than unity; it is $1 \cdot 25$ at $\theta = \pi/2$ (i.e. halfway to the antipodes), and becomes infinity at the antipodes. In words: if the transmitter were situated at the North Pole, the length perpendicular to the paper of the element P would be the length of the circumference of the parallel of latitude through P. This increases to about 40,000 km when P is on the equator, and dwindles to zero again when P is at the South Pole. The field strength therefore falls towards a minimum as r rises towards about 10,000 km, and thereafter rises again towards the antipodes. In our imaginary case, since there is no dissipation of energy, we ought to take into account the waves travelling beyond the antipodes and back towards the transmitter; but this is unnecessary with dissipation even much less than that usually experienced*, unless a point is under consideration which is much nearer to the antipodes than to the transmitter.

The effect of the sphericity of the Earth and the Heaviside layer in re-concentrating the power when once a distance of 10,000 km is passed is illustrated in the numerical example by a comparison of lines (b) and (c) in the table. It is clear that if anything like a perfectly conducting earth and a perfectly conducting Heaviside layer exist, a range of (say) 15,000 km is little more remarkable than a range of 5000 km. The occurrence from time to time of "freak" ranges, when a listener hears signals from some remote transmitting

* Except sometimes with short waves. See p. 58.

station ordinarily far out of range, has sometimes been attributed to the occurrence of an occasional channel of exceptionally good propagating quality called into being by meteorological conditions, like a rift in a fog. But the recognition of 10,000 km as—apart from absorption—the hardest range to obtain in terrestrial communication relieves us from the necessity of resorting to so unsatisfactory an explanation.

Experience gained during the last few years, especially with very short waves, seems to indicate that sometimes and for some wavelengths the Heaviside layer really does behave as a good reflector. This confers a large measure of reality on the ideal we have been examining as Case (c).

Case (d). The fourth case to be considered is the spherical earth without an atmosphere. This is Case (a) modified from plane to sphere. Long distance propagation now depends wholly on diffraction. It is portrayed, very inadequately, at (d) in the series of diagrams. The formula for the field strength is due to G. N. Watson and van der Pol*, and with the symbols and units here employed is

$$F = 5\cdot 36 \times 10^2 \cdot \frac{hI}{\lambda^{\frac{7}{6}}} \cdot \frac{\epsilon^{-\frac{23\cdot 9\theta}{\lambda^{\frac{1}{3}}}}}{(\sin\theta)^{\frac{1}{2}}} \dagger.$$

The numerical significance of this formula is shown by line (d) of the Table. Comparison of lines (a) and (d) makes it clear how essential to the wireless engineer a flat Earth would be, had he no Heaviside layer in support; and comparison of lines (c) and (d) shows how immense the effect of a really good Heaviside layer may be.

It is of interest to tabulate, with the four ideal limiting cases considered, field strengths as actually observed. The transmitter taken for the numerical examples was chosen to enable this to be done. It is Carnarvon ("M.U.U."); and in line (e) of the Table are extracted ‡ some observations of that station which were made by the Marconi Co. in the course of an expedition undertaken to measure signal strengths up to great distances.

* Van der Pol, "On the propagation of electromagnetic waves round the earth," *Phil. Mag.* Vol. xxxviii (1919), p. 365.

† Valid for distances $r > 140\,\lambda^{\frac{1}{3}}$, and small enough to permit neglect of the radiation arriving by the longer great-circle path.

‡ From Fig. 19, *Journal I.E.E.* Vol. 63 (1925), p. 946.

2. Degradation of Radiated Energy

In the foregoing imaginary case (b), and in the less unreal case (c), it was assumed that at large distances from the transmitter the wave-front stretched from the ground to the Heaviside layer, but that no loss of energy from the wave occurred in either. The attenuation was due solely to the spread of the wave-front. Actually there must be some absorption—a conversion from radiation to heat—both in the ground because its conductivity is finite, and in the upper atmosphere because collisions occur between the ions and the uncharged atoms*. Any field strength—distance formula based upon actual conditions may therefore be expected to contain an absorption factor of the form ϵ^{-kr}, where k is a positive quantity independent of r but probably a function of the wavelength and certainly a function of the condition of the ground and of the upper atmosphere. These conditions vary from place to place—for example, the conductivity of sea water is greater than that of dry land; and from time to time—for example, from light to dark in the upper atmosphere. Moreover, hills and valleys mar the sphericity of the ground; and doubtless like irregularities (less constant in time) occur in the Heaviside layer. It is therefore not to be expected that any single attenuation formula could do more than show rather rough agreement with an average of the observed results. It would be unreasonable to expect to find wireless attenuations as steady and as calculable as attenuations in a telephone cable.

The famous Austin-Cohen formula was for long the most generally respected attenuation formula. In our same units it is

$$F = 0.377 \times 10^6 \frac{hI}{\lambda} \frac{1}{r} \epsilon^{-\frac{0.0015r}{\sqrt{\lambda}}} \dagger.$$

This formula was developed mainly with reference to transatlantic

* See Section 4 of this Chapter.
† Recently, in order to give better agreement with observations at longer ranges, Austin has adopted, on purely empirical grounds, a slightly changed exponential factor. See L. W. Austin, *Proc. Inst. Radio Engineers*, Vol. 14 (1926), p. 377. The factor $\sqrt{\frac{\theta}{\sin \theta}}$ has also come to find a place in the formula, although it appears to have no logical basis there, since the distance function $\frac{1}{r}\sqrt{\frac{\theta}{\sin \theta}}$ is not appropriate to a spherical Earth either with or without a reflecting layer. The modern Austin formula is accordingly

$$F = 0.377 \times 10^6 \frac{hI}{\lambda} \frac{1}{r} \sqrt{\frac{\theta}{\sin \theta}} \epsilon^{-\frac{0.0014r}{\lambda^{0.6}}}.$$

measurements with fairly long waves many years ago, before much diversity of experimental data was available, and before any reasonable theory of propagation had been framed. To the theoretical expression for propagation over a plane perfectly conducting ground without a Heaviside layer was attached an empirical factor, the exponential term, of a form appropriate to allow for actual losses in the ground. Now that working ranges are so great that the flat ground approximation becomes an absurdity, and now that all evidence points to the existence of an operative Heaviside layer, efforts to strain the Austin formula into agreement with observed field strengths seem to be misdirected.

A truer formula is likely to be given by attaching an appropriate absorption factor to Case (c); and the work of G. N. Watson and T. L. Eckersley indicates that this factor should be of the form $\epsilon^{-\frac{ar}{\sqrt{\lambda}}}$, where a is a function only of the qualities of the ground and the Heaviside layer, and would be calculable if these were uniform. The formula accordingly becomes

$$F = 1\cdot 88 \times 10^5 \frac{1}{H} \frac{hI}{\sqrt{\lambda}} \frac{1}{\sqrt{r}} \sqrt{\frac{\theta}{\sin\theta}} \, \epsilon^{-\frac{ar}{\sqrt{\lambda}}}.$$

Although a in the Watson-Eckersley formula is not a function of distance r or (for long waves) of wavelength λ, it is not likely to be a constant all over the world, and it is likely to vary somewhat from winter to summer and from light to dark. Fig. 19* gives some idea of the closeness of agreement between observed field strengths and those calculated from the above formula. The transmitting stations were those named in the figure, and the receiving station at which the field strengths were measured was on a ship travelling from Liverpool via Newport News and Panama to New Zealand. The full-line curves are prepared by taking $a = 0\cdot 0016$ from Liverpool to Panama (7700 km) and thereafter $a = 0\cdot 00095$; whereas the dotted "uncorrected" curves show the calculated strengths if $a = 0\cdot 0016$ throughout. Other evidence supports the conclusion that the attenuation over the Pacific Ocean is much less than elsewhere; but the Figure shows that, with only two values of a along the whole journey, a good agreement with the observed strengths is given.

* Taken, by permission of the Institution, from Fig. 89 of the report on the Marconi expedition referred to on p. 40: Round, Eckersley, Tremellen and Lunnon, "Report on measurements made on signal strength at great distances," *Journ. I.E.E.* Vol. 63 (1925), p. 978.

The formula, however, fails altogether to express the facts when short waves are used.

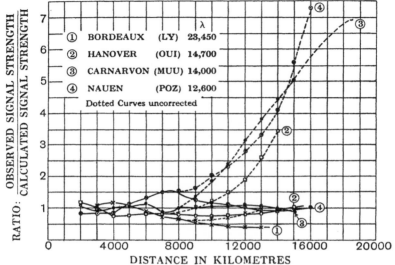

Fig. 19. Field-strength observations with long waves.

3. COMPLICATIONS IN PROPAGATION PHENOMENA

Any ultimate theory of the manner in which wireless signals are propagated round the Earth must correlate a great variety of observed phenomena, only now beginning to fall into order. Especially the behaviour of short waves—which until about 1923 were believed to be incapable of propagation over any but short distances—has shown that the process of propagation is much more complicated than was suspected. Only a sketch of some of the more salient points is here attempted, but this will suffice to show the directions along which research is proceeding.

(i) *The skip.* Both the Austin-Cohen and the Watson-Eckersley formulas imply that with a given transmitter (h, I and λ) the field strength decreases continuously as the distance r increases; and that the distance-rate of attenuation increases continuously with reduction of the wavelength λ. Experience has proved, however, that while these relations do hold with wavelengths above, say, 1000 m, they are flagrantly untrue with wavelengths below, say, 300 m. With short waves the signals fall off rapidly as the distance from the transmitter increases—indeed much more rapidly than the long-wave attenuation formulas would suggest—only to rise again as the distance is further increased. This is illustrated in Fig. 20*, which

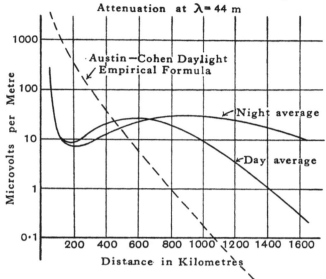

Fig. 20. Field-strength observations with short waves.

is based on absolute field-strength measurements of signals from a transmitter radiating some 2 kW at a wavelength of 44 m. Other observers, particularly with still shorter waves, have reported zones apparently wholly jumped by the signals extending up to 8000 km, the signals reappearing in strength at greater distances. The phenomenon is often called the *skip*, and the no-signal range, the *skip distance*. It is well established that the skip distance is greater by night than by day.

(ii) *Fading.* A well-known phenomenon, experienced even with broadcasting signals of the shorter wavelengths when received at long range (e.g. wavelengths around 300 m at upwards of 100 miles) is that of *fading*. The field strength fades intermittently to a small fraction of its normal intensity—often to zero, as far as can be told—and sometimes fades and recurs fairly rapidly and regularly, such as once a minute. Diurnal strength variation, familiar in a relatively mild form with long waves as a strengthening during the

* Replotted, by permission of the Institute, from Fig. 14 of Heising, Schelleng and Southworth, "Measurements of short-wave transmission," *Proc. Inst. Radio Engineers*, Vol. 14 (1926), p. 613.

3. COMPLICATIONS IN PROPAGATION PHENOMENA 45

dark hours, is in general much more marked with short waves and is often reversed, signals audible in the day becoming inaudible at night.

(iii) *Direction finder aberrations.* Much significant information has come from observations with directive receivers capable of ascertaining what is the single direction, if any, in which the signals arrive. It is a common experience that such receivers, while giving good indications at most times during the daylight, give different indications, or seem to show that there is no single horizontal direction, at other times, especially during the night*.

(iv) *Polarisation phenomena.* Near the ground the electric field radiated by a vertical antenna, as investigated in II-4, is vertical; the wave is said to be plane-polarised vertically, because a vertical plane can be drawn in which at every point both the direction of propagation and the electric field lie. In the most usual circumstances, the field arriving at the receiver is likewise vertically polarised; but sometimes the incoming radiation has a component whose electric field lies in a plane which is horizontal, or which even rotates (circular polarisation—see footnote to p. 55).

(v) *Anomalous effects near* $\lambda = 200$ m. The attenuation, at any rate in the daytime, appears to be especially great at wavelengths near 200 m—greater than with either longer or shorter waves.

(vi) *The effect of direction on propagation.* There is strong evidence that propagation around any great circle ABCD (Fig. 21) depends to some extent upon its inclination to the magnetic axis NS, and this apart from the light-dark conditions along the path between the stations. It seems even that the attenuation between AC when signalling in the sense ABC may differ from the attenuation when signalling between the same points in the sense CBA.

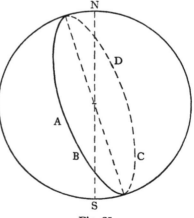

Fig. 21.

(vii) *Lower wavelength limit.* With very short waves, say between

* See XIV-6 (c).

100 m and 10 m, the attenuation over great distances is so slight that signals are received at C (Fig. 21) from both directions ABC and ADC; it may be from ABC at one time of day and from ADC at another; it may be simultaneously, so that interference and echo effects are produced. This is even so when the two stations A and C are not far apart, the length of one route exceeding the other by nearly 40,000 km. But although, as wavelengths have been reduced, signals (in daylight) have seemed to span the Earth more and more readily, there appears to be a limit, in the neighbourhood of 10 m, below which no long-range signalling is possible.

These are some of the divers phenomena which must be embraced by a propagation theory. A deeper study of the influence of the upper atmosphere on wireless waves originating on or near the ground is leading towards the establishment of such a theory. Some fundamental considerations involved are sketched in the next Section.

4. Theory of the Heaviside layer

4 (a). *Free electrons in the upper atmosphere.* Consider an alternating electric field,

$$f = F \sin pt,$$

existing in a region occupied by a cloud of N ions per unit volume. If the charge and mass of each ion are e and m, between collisions the equation of motion of each ion must be

$$m \frac{dv}{dt} = fe = Fe \sin pt,$$

where the velocity v is measured in the direction of the field. The velocity produced by the field is accordingly

$$v = -\frac{Fe}{pm} \cos pt.$$

There is therefore a convection current density

$$Nev = -\frac{FNe^2}{pm} \cos pt.$$

The Maxwell displacement current* density, which in the absence of the ions would exist alone, is

$$\frac{\kappa}{4\pi} \frac{df}{dt} = \frac{\kappa}{4\pi} Fp \cos pt,$$

* See II–2.

4. THEORY OF THE HEAVISIDE LAYER

where κ is the dielectric constant of the medium (unity in the case of vacuum). The presence of the ions has thus changed the total current density produced by the field from

$$\kappa \frac{Fp}{4\pi} \cos pt$$

to
$$\left(\kappa - \frac{4\pi Ne^2}{p^2 m}\right) \frac{Fp}{4\pi} \cos pt;$$

in other words, the presence of the ions has reduced the effective dielectric constant of the medium by $\dfrac{4\pi Ne^2}{p^2 m}$.

In the unionised atmosphere, κ is sensibly unity, so that the presence of the ions produces an apparent dielectric constant of $\left(1 - \dfrac{4\pi Ne^2}{p^2 m}\right)$*; and therefore the velocity of propagation $u = \dfrac{c}{\sqrt{\kappa\mu}}$ exceeds the velocity in vacuum, and is

$$u = \frac{c}{\sqrt{1 - \dfrac{4\pi Ne^2}{p^2 m}}}.$$

In the atmosphere it may be supposed that the negative ions (electrons) equal or exceed in number the positive ions (probably at great heights hydrogen nuclei); and as their charges are the same while the mass of the electron is only $\frac{1}{1850}$ that of the hydrogen

---

* A reduction of $\dfrac{4\pi Ne^2}{p^2 m}$ below unity suggests the intriguing possibility of the occurrence of a *negative* dielectric constant. For any specified frequency $n \equiv \dfrac{p}{2\pi}$, it seems necessary only to have sufficiently dense an atmosphere of electrons, viz.
$$N > \frac{p^2 m}{4\pi e^2},$$
i.e.
$$N > \left(\frac{n}{9100}\right)^2 \text{ approx.,}$$
in order to realise this topsy-turvy condition. E. V. Appleton points out that no radiation of terrestrial origin can penetrate into such a region, if it exists, since the intermediate region where $\kappa = 0$ $\left(\text{i.e. where } N = \left(\dfrac{n}{9100}\right)^2\right)$ behaves as a perfect reflector to radiation of that frequency. Most of us dwellers in a world where dielectric constants are positive must contemplate this impenetrable layer separating us from the region of negative dielectric constants much as Alice stared at the looking-glass before she managed to pass through to her surprising experiences on the other side.

atom, it is only the electrons which need be considered*. N then stands for the number of free electrons per cm³, and

$$\frac{e^2}{m} = \frac{(4{\cdot}77 \times 10^{-10} \text{ E.S. units})^2}{8{\cdot}8 \times 10^{-28} \text{ gm}} = 2{\cdot}59 \times 10^8.$$

Writing $n \equiv \dfrac{p}{2\pi}$, we then have

$$u = \frac{c}{\sqrt{1 - \dfrac{8{\cdot}27 \times 10^7 N}{n^2}}}.$$

Ionisation in the upper atmosphere increases the speed of wave propagation above the "velocity of light," and the more so the smaller the frequency, i.e. the longer the wavelength. There is evidence† that at greater altitudes N reaches values of the order 10^5. Accordingly, even for very short wavelengths, there is a region in which the propagation velocity u very sensibly exceeds the velocity in free space and at smaller altitudes. Thus with $N = 10^5$ and $n = 10^7$ ($\lambda = 30$ m),

$$u = \frac{c}{\sqrt{1 - 0{\cdot}0827}} = c\,(1 + 0{\cdot}04).$$

Now near the ground N is sensibly zero, but at high altitude h, where the ultra-violet light is strong and the rate of ionic recombination small, N rises to large values such as those named above. Higher still N presumably falls again towards zero in sensibly empty space. Radiation from a source on the ground, therefore, on reaching heights such as H, passes into an ionised region called the Heaviside layer, where there is a positive ionisation gradient $\dfrac{dN}{dh}$. The higher the wave-front penetrates into the layer the greater is the velocity of propagation, so that the wave-front is progressively tilted downwards. This is the same phenomenon as optical refraction in a transparent medium of not uniform refractive index; it becomes

---

* W. H. Eccles appears to have been the first (1912) to point out the effect on the dielectric constant of a medium of electrically charged particles therein. The full significance of this discovery as a vital contribution to a theory of wireless propagation round the Earth was not appreciated until J. Larmor applied it (1924) to media in which the massive ions envisaged by Eccles are replaced by electrons, of equal charge but much smaller mass.

† See E. V. Appleton, "On the diurnal variation of ultra-short wave wireless transmission," *Camb. Phil. Soc. Proc.* Vol. XXIII (1926), p. 155. Later measurements by Appleton (*Nature*, 7 Feb. 1931) throughout a mid-winter day in England showed that N varied between a maximum of 2×10^5 electrons per cm³ at noon to a minimum of 3×10^4 lasting most of the night.

4. THEORY OF THE HEAVISIDE LAYER

reflection if the refractive index gradient is sufficiently sharp to make the penetration of the ray small compared with the wavelength. It seems probable that both reflection (with long waves) and gradual refraction (with short waves) are to be found in wireless telegraphy.

An upward ray AB (Fig. 22) entering such an ionised medium at BD is bent along a path BC and will, if the bending is sufficient, leave the medium again at D in the downward direction DE. It may be seen from elementary optics that $M \sin \theta$ is constant along the ray, where M is the absolute refractive index* of the medium

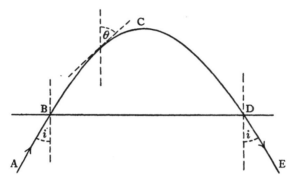

Fig. 22. Refraction in ionised medium.

at the point where the ray inclination is θ. Now the wave velocity u and the absolute refractive index M have the relation $u = \dfrac{c}{M}$; hence if u_0 is the velocity at the highest point of the ray, where $\theta = 90°$, the angle i of incidence and emergence is given by

$$\sin i = \frac{c}{u_0} = \sqrt{1 - \frac{8 \cdot 27 \times 10^7 N_0}{n^2}}, \text{ i.e. } \cos i = \frac{9100 \sqrt{N_0}}{n},$$

where N_0 is the electron density at the highest point. For grazing incidence ($i = 90°$) on the Heaviside layer, an infinitesimal N_0 is sufficient to bend the ray down (through an infinitesimal angle); but for normal incidence ($i = 0$), the ray of frequency n is returned only if the maximum electron density $N_{\max}$ in the layer exceeds

$$\frac{n^2}{8 \cdot 27 \times 10^7} \dagger.$$

---

* M is used here, instead of the symbol μ customary in optics, to avoid confusion with permeability.

† Cf. footnote on p. 47.

III. WIRELESS WAVES ROUND THE EARTH

With a limited electron density in the upper atmosphere, therefore, all rays of small frequency $n < 9100 \sqrt{N_{max.}}$ projected from the ground are bent back towards the ground; but rays of frequency above the critical value $9100 \sqrt{N_{max.}}$ return from the layer only if they strike it with sufficiently glancing incidence. Let us for simplicity take $N_{max.} = 1{\cdot}09 \times 10^5$ electrons per cm^3; then for normal incidence the critical frequency is 3×10^6 c/s and wavelength 100 m.

In Fig. 23 (which is not to scale), O is the centre of the Earth of radius R; AEE$'$E$''$ is the surface of the ground, and CC$'$C$''$ is the Heaviside layer at a height H above. With long wave radiation it is possible to call upon the Heaviside layer to assist in communication between two spots A and E, however close they may be and whatever the height H of the layer; but with radiation from A of wavelength less than 100 m, a ray reaching the Heaviside layer can be deflected to E$'$ only if E$'$ is far enough from A, and/or the layer is low enough, to give the requisitely large angle of incidence i. Since it is geometrically impossible for i to exceed the value i_l given by OAC$''= 90°$, an upper limit to the frequency of rays which can be returned to the ground at all (viz. at E$''$) is given by

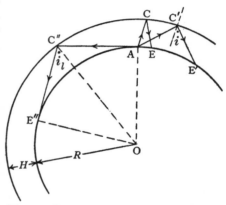

Fig. 23. Reflection from Heaviside layer.

$$\sin i_l = \frac{R}{R+H}*.$$

$$\therefore \cos^2 i_l = \frac{2RH + H^2}{(R+H)^2},$$

$$\fallingdotseq \frac{2H}{R}.$$

---

* We are here adopting the geometrical simplification of treating the points B, C, D in Fig. 22 as coincident; i.e. in physical terms, of assuming that the depth of penetration into the refracting layer is small compared with the mean height above ground of the curved portion of the ray.

4. THEORY OF THE HEAVISIDE LAYER

The limiting frequency is thus n_l, where

$$\frac{8 \cdot 27 \times 10^7 N_{max.}}{n_l^2} = \cos^2 i_l = \frac{2H}{R}.$$

This gives $n_l = 2 \cdot 69 \times 10^7$ c/s, i.e. $\lambda_l = 11 \cdot 2$ m.

To examine where a ray, re-emergent from the Heaviside layer, meets the ground, it is necessary to consider its path BCD (Fig. 22) within the layer. It is easy to see that the curvature of the ray at any point is

$$\frac{1}{\rho} = \frac{1}{u}\frac{du}{ds},$$

where distance s is measured in the direction perpendicular to the ray. Since

$$u = \frac{c}{\sqrt{1 - \frac{8 \cdot 27 \times 10^7 N}{n^2}}},$$

this is

$$\frac{1}{\rho} = \frac{4 \cdot 13 \times 10^7 \frac{dN}{ds}}{n^2 - 8 \cdot 27 \times 10^7 N}.$$

If, where the ray becomes horizontal at C, the electron density N and the gradient $\frac{dN}{ds}$—which is here $\frac{dN}{dh}$, where h is the vertical height—are such that ρ equals the radius of the Earth (more pre-

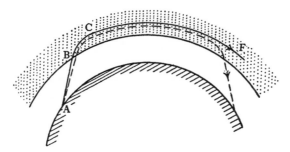

Fig. 24. Critical curvature of refracted ray.

cisely, $R + H$), the ray will thenceforth proceed at uniform height, as indicated by CF in Fig. 24. It seems that for each wavelength there must be one such height and one such angle of incidence which will take the ray round and round the Earth within the Heaviside layer. An appropriate minute increase of the angle of incidence would cause the ray to descend to ground at any specified spot sufficiently remote, as indicated by the ray shown dotted in Fig. 24. This spot could be the antipodes of A. [Cf. (i) and (vii) of Section 3, pp. 43 and 46.]

Even with the assumption of uniform spherical distribution, direct test of such deductions as these is impeded on the theoretical side by our ignorance of the actual values of N and $\frac{dN}{dh}$ in the upper atmosphere, and on the experimental side by the difficulty of emitting or earmarking a sharply defined bundle of rays at the transmitter. Further, there are always two great-circle routes round the world between any two stations, and in general at least one of these routes is neither all dark nor all light; and theory and experiment unite in assuring us that the change from light to dark profoundly alters the ionisation conditions, the effective layer being changed in height and probably in ionisation gradient when the ionising agent is removed. There is also the question of the absorption experienced by the radiation at different altitudes along the ray, which has so far been ignored in this examination.

4(b). *Absorption in the Heaviside layer.* Absorption in the atmosphere is due to collisions between the vibrating electrons and the atoms (or massive ions); it is negligible at any wavelength at small altitudes, for although there the atoms are densely distributed free electrons are sensibly absent; and it is negligible at great altitudes, for there atoms are very rare. The lower regions of the ionised atmosphere are therefore probably the seat of most of the absorption experienced. But the absorption depends on the lack of freedom of the electrons to execute vibrations under the alternating force imposed by the radiated field. It is measured by the product of the number of collisions occurring per second and the energy lost per collision; consequently if long waves and short waves followed the same track, the former would be more attenuated than the latter. But in general they do not follow the same track, since at the same point in the layer long waves suffer bending more than short. The tracks therefore differ both in length and in the intrinsic absorptive quality of the medium in which they chiefly lie. The refracting

4. THEORY OF THE HEAVISIDE LAYER

and absorbing properties of the upper atmosphere for long and short waves are roughly summarised in the following Table:

Feature of the Heaviside layer	For long waves	For short waves
Depth of penetration	Small	Great
Length of track in layer	Small	Great
Intrinsic absorptive quality of medium traversed	Great, and greater for long than for short waves in the same region	Small, and smaller for short than for long waves in the same region
Power to refract ray to ground again	Complete for any incidence	The shorter the wave the less steep must be the ray inclination. For waves shorter than some limiting value (such as 11 m), no return is possible

[Cf. (ii) of Section 3, p. 44.]

4 (c). *The Earth's magnetic field.* There is another agent in the upper atmosphere to which no reference has yet been made, viz. the Earth's magnetic field. If the Heaviside layer is effective by virtue of the motion of its free electrons, this must be taken into account; for an electron moving across an independent magnetic field experiences a force in the third perpendicular direction.

At time $t = 0$ let an electron of charge e and mass m have velocities $\dot{x}$ and $\dot{y}$, and let it be situate at O in a steady magnetic flux-density B along the z-axis and an electric field $F \sin pt$ along the x-axis (Fig. 25). Then the forces applied to the electron are:

along OX, $e(F \sin pt - B\dot{y})$;
along OY, $eB\dot{x}$;
along OZ, 0.

Therefore
$$m\ddot{x} = e(F \sin pt - B\dot{y}) \quad ...(1);$$
$$m\ddot{y} = eB\dot{x} \quad ....(2).$$

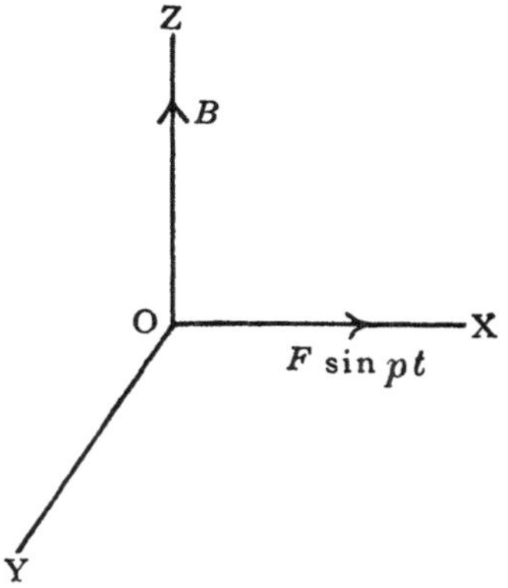

Eliminating $\ddot{x}$ in (1) by use of (2),

$$\frac{m^2}{eB} \dddot{y} + eB\dot{y} = eF \sin pt.$$

Fig. 25. Effect of magnetic field on motion of electron.

III. WIRELESS WAVES ROUND THE EARTH

The final motion along OY is therefore given by

$$m\ddot{y} + \frac{e^2 B^2}{m} y = \frac{e^2 BF}{m} \int \sin pt \cdot dt,$$

$$= \frac{e^2 BF}{pm} \cos pt + \text{(a constant)}.$$

$$\therefore y = \frac{e^2 BF/p}{p^2 m^2 - e^2 B^2} (\cos pt - 1).$$

The motion along OY is thus a harmonic motion of amplitude $A_y \equiv \frac{e^2 BF}{p(p^2 m^2 - e^2 B^2)}$. The addition of the magnetic field has introduced resonance features, of which there was no trace without it. By considering also the motion along OX, it is seen that the electron moves with frequency $\frac{p}{2\pi}$ round the ellipse shown in Fig. 26. As the frequency tends to the resonant value $\frac{1}{2\pi} \frac{eB}{m}$, the ellipse tends to enlarge into a circle of infinite radius. If $B = 0.5$ C.G.S. E.M. units, the resonant frequency is 1.4×10^6 c/s, corresponding to the wavelength 214 m.

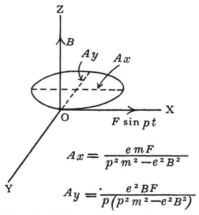

Fig. 26. Elliptical motion of electron.

In the Heaviside layer, electrons more or less free, and the Earth's field of about 0·5 C.G.S. units, coexist; so that the above analysis applies to a ray polarised with its electric field perpendicular to the Earth's magnetic field—e.g. to the signals from an ordinary transmitter near the equator propagated in the E-W direction (along OY, Fig. 26), or horizontally in the N-S direction (along OZ, Fig. 26). In both cases it is clear that the absorption from collisions between

4. THEORY OF THE HEAVISIDE LAYER

atoms in the Heaviside layer and electrons attempting to execute the motions we have calculated will shoot up towards a maximum at wavelengths near 214 m. [Cf. (v) of Section 3, p. 45.]

It is interesting that for waves much longer than 214 m, the amplitude A_x is much smaller than it would be in the absence of the magnetic field, indicating that for long waves propagated in the direction perpendicular to the Earth's magnetic field the presence of this field may greatly reduce the attenuation. [Cf. (vi) of Section 3, p. 45.]

4 (d). *Polarisation effects.* The very incomplete treatment already given strongly suggests on physical grounds that the magnetic field may produce polarisation effects on the waves passing through the layer. Fuller analysis* shows that such effects do occur, in complicated variety.

If the ray (assumed plane-polarised; see p. 45) has its electric vector (F) parallel to the Earth's magnetic flux (B), the latter has no effect at all.

If, however, F is normal to B, different effects are produced according as the ray direction—not specified in our introductory analysis—is along or normal to B. If the ray is along B, it is split into two oppositely circularly polarised † components, which undergo unequal

* H. W. Nichols and J. C. Schelleng, "Propagation of electric waves over the Earth," *Bell System Technical Journal*, Vol. IV (1925), p. 215.

† A ray is said to be circularly polarised if the wave-front contains two mutually perpendicular equal electric fields in time quadrature. Thus in

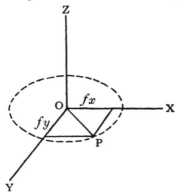

Fig. 27. Circular polarisation.

Fig. 27, with reference to the fixed axes XYZ, if the ray direction is OZ, the wave-front is in the plane XY, and OP is the resultant electric vector at any instant, where the coordinates of P are

$$f_x = F \sin pt,$$
$$f_y = \pm F \cos pt.$$

The point P therefore rotates in a circle of radius F in the plane XY.

56   III. WIRELESS WAVES ROUND THE EARTH

curvatures and unequal absorptions. If the ray is normal to B, it is split into two component plane-polarised rays, with other unequal curvatures and absorptions. [Cf. (ii), (iii), (iv) and (vi) of Section 3, pp. 44 and 45.]

It seems, therefore, that the Earth's magnetic field may be violently and anomalously in evidence at wavelengths near 214 m, and may produce important divers effects at great wavelengths. Fortunately, at very small wavelengths its effects tend to vanish.

5. Experimental knowledge of the Heaviside layer

5(a). *Phase and group velocities.* That refraction or reflection of rays in the upper atmosphere does occur, and with the divers effects outlined in the preceding Sections, is no longer open to doubt. The height of the effective Heaviside layer, for various conditions of wavelength, season and range of transmission, has been the subject of much research during recent years. The latest experiences have indicated that more than one layer is involved; and a still more startling discovery is that a terrestrially transmitted signal may sometimes reach its terrestrial destination after a journey whose length appears to be scores or hundreds of times the great-circle distance round our globe.

Since the return of the ray to Earth from the upper atmosphere is effected rather in a refracting region than at a reflecting surface, the meaning of the effective height of the Heaviside layer has not always been clear. Fundamentally, the information sought is the track of the ray between transmitter and receiver. Attempts to deduce the track have been made in four different ways, briefly set out below; but before these can be understood it is necessary to scrutinise a little more closely than in Chapters I and II the process by which signals are carried by waves.

The wave of χ-ishness described in I–2 and portrayed in Fig. 1 is transmitted without change of shape only (unless it be of sine shape) if the medium is *non-dispersive*; or—expressed in terms of persistent waves—only if the velocity of propagation v is independent of the frequency n (or wavelength λ). It is obvious that for electric waves a medium with constant μ and κ (see II–3 (j)) must be non-dispersive. But an elastic rod, for example, is a dispersive medium for the propagation of flexural waves (the velocity of propagation being here inversely proportional to the wavelength); and a cloud

5. EXPERIMENTAL KNOWLEDGE OF HEAVISIDE LAYER

of free ions is a dispersive medium for electric waves, since the velocity of propagation,

$$v = \frac{c}{\sqrt{1 - \dfrac{4\pi N e^2}{p^2 m}}} \quad \text{(p. 47)},$$

contains a term in $p \equiv 2\pi n$. The velocity of propagation of a persistent wave is termed the *phase velocity*; if the medium is dispersive, the phase velocity varies with the frequency; if it is non-dispersive, the phase velocity is the same for all frequencies. The phase change from point to point along the track of a ray depends upon the values of phase velocity in the medium; and the bending of the ray (refraction) in a medium which is not homogeneous in phase velocity depends upon the changing value of phase velocity along the track.

Now a signal cannot be conveyed by a persistent wave, but only by some sort of modulation therein, some break in the monotony. A modulated persistent wave can be analysed as a set of persistent waves of different frequencies*, and if these components have different velocities of wave propagation, i.e. phase velocities, we need not feel surprised to find that the velocity of propagation of the signal, i.e. the *group velocity*, differs from the velocity of propagation of the unmodulated or persistent wave. Analysis shows that the group velocity v_g in a medium whose phase velocity is v at a frequency n is

$$v_g = \frac{v}{1 - \dfrac{n}{v} \cdot \dfrac{dv}{dn}}.$$

When v depends upon n in the way that we have seen it does in an ionised atmosphere (in the absence of a magnetic field), this expression yields the relation

$$v \cdot v_g = c^2,$$

where c is the velocity in empty space. If M is the absolute refractive index of the medium at that frequency,

$$v = \frac{1}{M} \cdot c \quad \text{and} \quad v_g = M \cdot c.$$

---

* Perhaps the simplest example is the acoustically modulated carrier wave from a telephony transmitter emitting a single tone with a modulation ratio of $100°/_\circ$ (see p. 390). A derivation of the group velocity in this case is given by P. O. Pedersen, *The Propagation of Radio Waves*, p. 169 (1927, Copenhagen). This valuable treatise contains a comprehensive treatment of the whole rapidly developing subject of propagation, carried right up to the date of its publication.

Thus in the ionised upper atmosphere, while the phase velocity exceeds the velocity in empty space, any sort of transient by which a signal is conveyed travels at less than $c = 3 \times 10^{10}$ cm/sec.

5 (b). *Methods of estimating the height of the Heaviside layer.*
(i) Quäck and Wagner: transmission round the Earth. With short waves a signal from a transmitter T (Fig. 28) has on many occasions been observed as a dual signal at a receiver R, being carried apparently by rays TAR and TBR along both the shorter and the longer great-circle paths between T and R. If the velocity of propagation of the signal were known, measurement of the time interval between the arrivals via the A-route and the B-route would give the difference between the lengths of the tracks TAR and TBR, and thus the radius $(R+H)$ of the track; whence the height H of the track above the surface of the Earth would be found.

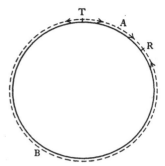

Fig. 28. Two tracks round the Earth.

Thus, in one experiment (using a wavelength of 15 m and a power of 8 kW), T and R were close together (at Nauen and Geltow, near Berlin), and the observed time interval was 0·141 sec.* If the signal velocity had been the velocity in empty space, this would have indicated a track of length 42,180 km and consequently a height H of 350 km. But the appropriate velocity is the group velocity at the height H, and this is not known unless the electron density at various heights is known. In the absence of a knowledge of electron density, the experiment proves only that the height is less than 350 km; and, as Howe points out, observations of this character are better adapted to provide estimates of electron density than of height. Probable values of electron density suggest that the group velocity might be some 3°/₀ or 4°/₀ below the velocity in empty space, and the height H of the order of 100 km.

It is interesting to note, in passing, that the Quäck, or round-the-globe, echoes are of practical importance—i.e. apart from their bearing on propagation theory; for they may seriously impede good telegraphic, and still more telephonic, communication. The

* See Pedersen (*loc. cit.* p. 57), p. 202. Also G. W. O. Howe, "Phase and group velocities in an ionised medium," *Experimental Wireless*, Vol. IV (1927), p. 259.

5. EXPERIMENTAL KNOWLEDGE OF HEAVISIDE LAYER 59

avoidance of interference from such echoes is a current engineering problem*.

The other methods to be described† depend upon comparisons between the radiation received at R (Fig. 29) from a fairly near transmitter T along the direct route TAR and along a ray TBR deflected from the upper atmosphere. These comparisons constitute methods (ii), (iii) and (iv) below.

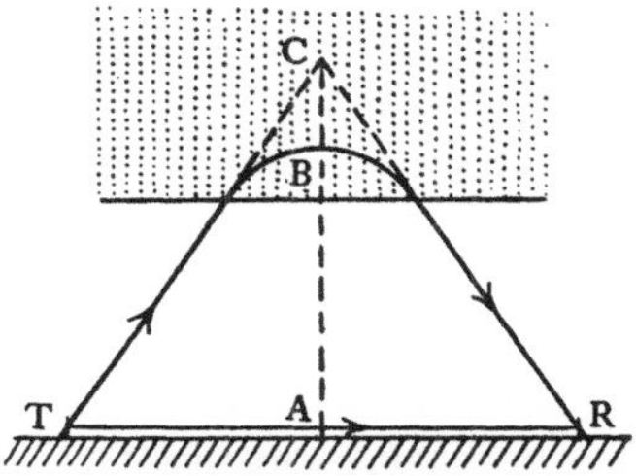

(ii) Appleton and Barnett: observation of the inclination TRC of the downcoming ray.

(iii) Appleton and Barnett: observation of the number of changes between maxima and minima of intensity at R, due to interference between the two rays arriving at R, as the wavelength of a steady oscillator at T is slowly and continuously varied.

Fig. 29. Direct and refracted rays.

(iv) Breit and Tuve: observation of the time interval between the arrivals at R of a signal (modulation) occurring at T as received along the track TAR and along the track TBR.

In methods (ii) and (iii), the observations depend upon the values of phase velocity in the ionised region; and in method (iv) upon the values of group velocity. Method (iv) differs from method (i) in that in (iv) a large proportion of the length of track TBR is outside the ionised region, whereas in (i) the track TBR was almost wholly within. Uncertainties as to the ionised region B are, therefore, in (iv) relatively unimportant for a determination of the height of that region.

A mathematical investigation by E. V. Appleton‡ has yielded the surprising result that in each of the methods (ii), (iii) and (iv) the several observed quantities are the same as would have been found if the ray, instead of experiencing the actual variation of phase and group velocities along the track in the ionised region, had traversed no ionised region at all, but had followed the rectilinear path TCR

* See, for example, E. Quäck and H. Mögel, "Double and multiple signals with short waves," *Proc. Inst. Radio Engineers*, Vol. XVII (1929), p. 791.

† See *Experimental Wireless*, Vol. V (1928), p. 657; or Pedersen, *loc. cit.* p. 57.

‡ "Some notes on wireless methods of investigating the electrical structure of the upper atmosphere," *Proc. Phys. Soc.* Vol. XLI (1928), p. 43. Any effect of the Earth's magnetic field is neglected.

(Fig. 29), undergoing reflection at C without change of phase. The height AC disclosed by these observations, while somewhat exceeding the greatest height of the track, is the same in all three methods, so that results are strictly comparable.

5 (c). *Beyond the Heaviside layer.* Apart from the moderate diurnal and seasonal changes found in the height H of the effective layer—more precisely, in the length of AC (Fig. 29)—which changes are attributable to continuous processes of ionisation and recombination, evidence is now forthcoming for the occurrence of at least two separate effective Heaviside layers of widely differing heights. These appear to be not merely different locations from time to time of an ionised region transferring itself more or less continuously from place to place, but—at least sometimes—actually coexistent. This was discovered first with waves of medium length during the dark hours when alone such waves could penetrate the lower layer; but more recently, by the use of shorter waves, intermittent penetration of the lower layer and reflection at a much more strongly ionised upper layer has been observed even at mid-day. E. V. Appleton quotes* the following determinations of height AC (Fig. 29), obtained by method (i) at 10-minute intervals between 10 a.m. and 2 p.m. (G.M.T.) on 13 Jan. 1929:

229, 229, 236, 244, 217, 229, 229,   230,   204, 196, 229, 100,
 99, 93, 98, 99, 96, 98, 232, 99 & 220, 99, 229, 229, 99 km.

"It will be seen that these heights fall into two definite series of mean values 226 km and 98 km."

On the same hypothesis of two layers, he is able to interpret observations made by method (iv) under similar conditions in America† as giving heights 225 km and 105 km.

In the autumn of 1928, G. Störmer and J. Hals, observing signals of terrestrial origin on a wavelength of about 31 m, detected repetitions or echoes‡, with delays of several seconds. Further observations have since been made by the same and other experimenters, and have disclosed echo intervals of the order of 10 and even 30 seconds

---

* "The equivalent heights of the atmospheric ionised regions in England and America," *Nature,* Vol. cxxiii (1929), p. 445. In this series, T = Teddington; R = King's College, London; λ = 98·8 m.

† By Breit, Tuve and Dahl; wavelength 75 m; during daylight in December 1927 and January 1928.

‡ It is customary, and etymologically correct, to call repeated or multiple signals *echoes,* without prejudice to the question whether they are caused by reflection.

5. EXPERIMENTAL KNOWLEDGE OF HEAVISIDE LAYER

with tolerable certainty; and Hals reports* that in Feb. 1929 he observed delays up to 260 seconds—a time interval sufficient for a journey through empty space of over half the distance between the Sun and the Earth. These astonishing observations—recalling the romances of Jules Verne and H. G. Wells—while hardly likely to have a direct bearing on the art of wireless communication between inhabitants of this globe, are clearly very relevant to theories of wireless propagation.

The distance travelled in (say) 30 seconds in empty space is longer than 200 great-circle perimeters of the Earth. In its track from transmitter to receiver, did the signal circumscribe the Earth 200

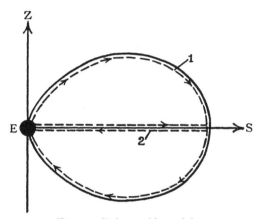

Fig. 30. Echoes of long delay.

times at a speed near $c = 3 \times 10^{10}$ cm/sec.? Or much fewer times with a group velocity much lower than c? Or did it penetrate the Heaviside layer, find a reflector far out in space, and return to Earth? No generally accepted answer has yet been reached, but the reflector far out in space seems the most probable. Pedersen, after examining the alternatives, finds the circumscription hypothesis untenable, not because tracks in the upper atmosphere of very low group velocity are improbable, but because the attenuation along any such probable track would be much too great. He supports Störmer's own view that, far from the Earth's surface, but within its sensible magnetic field, there exist reflecting electron cloud formations,

* See P. O. Pedersen, "Wireless echoes of long delay," *Vid. Selsk. Mat.-fys. Medd.* IX, 5, 1929 (Copenhagen; in English), p. 48.

shaped by the interaction of the Earth's magnetic field and flying streams of charged particles ejected by the Sun. Fig. 30, reproduced from his paper, illustrates the suggested mode of production of echoes with delays up to 30, or perhaps 60 seconds. The full-line curve indicates the inner limits of the paths of charged particles. E represents the Earth, with magnetic axis along EZ; and ES points towards the Sun. The dotted curves 1 and 2 indicate the two echo tracks possible.

CHAPTER IV
OSCILLATORY CIRCUITS

1. Reactances and resistances

When we leave the subject of radiation and come to consider the sensibly non-radiating circuits in which high-frequency currents are generated, transformed and controlled, we are amongst the phenomena of more ordinary alternating current engineering, with only such quantitative differences as derive from the much higher frequency of the currents. These circuits are used at the transmitter in conveying power to the antenna for producing electromagnetic waves in space, and at the receiver in withdrawing power from the antenna and passing it to some form of detector.

From a variety of circumstances, in wireless circuits the reactances are usually very large compared with the resistances. Sometimes this condition cannot be avoided, as in antennas; sometimes it is convenient for the production with small power of the large E.M.F. required by accessory apparatus such as crystal detectors and triodes; sometimes, too, it is cultivated with the object of filtering out one frequency from another, an operation requiring oscillatory circuits of low damping. Since it is only the resistance component of the E.M.F. which is involved in the work done with which we are ultimately concerned, it is often necessary to counter an inductive E.M.F. by an equal and opposite capacitative E.M.F. In other words, wireless circuits are generally—not invariably—tuned oscillatory circuits, comprising inductance and capacity so adjusted that their reactances are approximately numerically equal at the particular frequency employed, each of these reactances alone very greatly exceeding the resistance.

The inductance coils of wireless circuits, unlike those used in low-frequency engineering, are seldom provided with an iron core*, for two reasons. Owing to the high frequency, the eddy currents within the iron, which are kept moderate at 50 c/s by fairly coarse lamination of the iron, would be too great even with fine lamination; and in sharply tuned circuits it is essential that the inductance $L = \dfrac{d\phi}{di}$ (where ϕ is the flux-turns and i the current) shall be both

* See, however, V–5 (c), (f) and (g).

constant throughout the cycle and be independent of the amplitude. They are therefore usually air-core coils, of shape chosen with two ends in view, viz. to obtain the greatest inductance with a given length of wire of given thickness, and to avoid excessive self-capacitance. Small coils may be single-layer solenoids, of length preferably about equal to the radius; and large coils are wound with several layers well spaced apart. They may be of all sizes, ranging from the small coils wound with thin solid copper wire familiar in broadcast receivers, to great transmitting coils such as those of the Rugby station shown in Plate II. The latter are 16 ft. across, and are wound of cable about $1\frac{3}{4}$ in. in diameter containing 6561 separately insulated strands*. The inductance of these coils is about 500 μH, and the power factor at the working frequency of about 17 kc/s is about 0·0005† (i.e. the reactance of the coil is 2000 times its total resistance); but power factors some 10 or 20 times greater are more common.

The inductance of a coil can be adjusted in steps by the use of tapping points; or continuously by arranging that it shall comprise two parts L_1, L_2, whose mutual inductance M is adjustable by altering their relative position. In Fig. 31, since $i_1 = i_2 = i$, the inductance of the whole is L, where

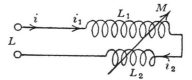

Fig. 31. Smooth variation of inductance.

$$L \frac{di}{dt} = \left(L_1 \frac{di_1}{dt} + M \frac{di_2}{dt} \right) + \left(L_2 \frac{di_2}{dt} + M \frac{di_1}{dt} \right);$$

i.e. $L = L_1 + L_2 + 2M$.

The condensers used to provide the negative reactances for high-frequency currents commonly have mica or air as the dielectric. Mica condensers can be varied in steps by switching; air condensers can be constructed to be continuously variable by making one set of plates movable in relation to the other. The ordinary variable air condenser used in wireless receivers has a capacitance adjustable between perhaps 80 and 500 $\mu\mu F$ and is capable of withstanding a P.D. of only a few hundred volts. Such a condenser is shown in

---

* For reduction of eddy-currents within the copper. See XV.
† Many interesting details of the apparatus in the Rugby station are published in R. V. Hansford and H. Faulkner's paper, *loc. cit.* p. 33.

Plate III. Final tuning adjustments are usually made by means of a variable air condenser—small as in Plate III for low-pressure circuits, or with larger spacing for transmitting circuits. Two large banks of mica condensers* in oil-filled steel tanks are shown in Plate IV; one of these has a capacitance of $\frac{1}{4}\,\mu$F for working at 34 kV, and the other 1μF for 10 kV. The power lost in an oscillatory circuit is usually more in the coil than the condenser. The power factor of the mica condensers of Plate IV is about 0·0002; but again, figures some 5 times greater are more common.

2. Undamped impressed E.M.F.

2 (a). *General case.* The equation of E.M.Fs. in the circuit of Fig. 32 is

$$L\frac{d^2q}{dt^2} + R\frac{dq}{dt} + \frac{1}{C}q = e = E\sin 2\pi nt,$$

where q is the charge in the condenser, and $\frac{dq}{dt}$ is the current i. Writing for brevity $p \equiv 2\pi n$, the solution of this equation is

$$q = Q\cos(pt + \theta) + Q'\epsilon^{-bt}\cos(p't + \phi);$$
$$i = -pQ\sin(pt + \theta) - Q'\epsilon^{-bt}[p'\sin(p't + \phi) + b\cos(p't + \phi)].$$

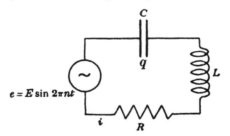

Fig. 32. E.M.F. impressed in oscillatory circuit.

The first term is the forced oscillation (the particular integral), of amplitude

$$Q = -\frac{E}{p\sqrt{\left(pL - \frac{1}{pC}\right)^2 + R^2}},$$

and leading phase angle

$$\theta = \tan^{-1}\frac{pL - \frac{1}{pC}}{R}.$$

* At the Rugby station. The coil of Plate II is visible at the top of the photograph.

The second term is the free oscillation (the complementary function), of frequency

$$n' \equiv \frac{p'}{2\pi} = \frac{1}{2\pi}\sqrt{\frac{1}{CL} - \frac{R^2}{4L^2}},$$

and attenuation constant $b = \dfrac{R}{2L}$.

Q' and ϕ' depend upon initial conditions, which have not been specified.

2 (b). *The decrement.* The ratio between the attenuation constant b and the natural frequency n' is called the *decrement**, and is here written δ.

$$\delta = \frac{b}{n'} = \frac{R}{2n'L}.$$

We shall see in Subsection 2 (e) below that the decrement also specifies completely the tuning properties of the circuit in the steady state, i.e. for the forced oscillation after the transient free oscillation has died away. The decrement is, therefore, a most important characteristic of the circuit.

Mechanical oscillators, e.g. the pendulums of astronomical clocks, can be made with extremely low decrements†; such decrements could not be approached in plain electrical circuits unless some material of vastly better conductivity than copper were available. Actually wireless circuits in which low damping is desired ordinarily have decrements of the order $0{\cdot}005 - 0{\cdot}05$.

Since
$$n'^2 = \frac{1}{4\pi^2}\left(\frac{1}{LC} - \frac{R^2}{4L^2}\right) = n_0^2 - \frac{n'^2\delta^2}{4\pi^2},$$

where n_0 is the resonance frequency (see Subsection 2 (d) of this Chapter),

$$\frac{\text{resonance frequency}}{\text{natural frequency}} \equiv \frac{n_0}{n'} = \sqrt{1 + \frac{\delta^2}{4\pi^2}}.$$

* An abbreviation of *logarithmic decrement*, so called because δ is the logarithm of the ratio between any two successive maxima of the free oscillation. The decrement δ is generally a more convenient measure of damping than is the attenuation constant b because it can be used without reference to frequency. Thus, the amplitude of an oscillatory system, when disturbed and left to oscillate freely, is divided by $\epsilon \doteq 2{\cdot}7$ during the course of every $\dfrac{1}{\delta}$ cycles, and is reduced to $0{\cdot}7°/_\circ$ of its initial value in $\dfrac{5}{\delta}$ cycles.

† The pendulums made by the Cambridge Instrument Co. for measuring the local variations in the Earth's gravitational field have a decrement of about 2×10^{-6}.

2. UNDAMPED IMPRESSED E.M.F.

Hence if $\delta = 0{\cdot}009$, $\dfrac{n_0}{n'} = 1 + 10^{-6}$;

$\delta = 0{\cdot}09$, $\dfrac{n_0}{n'} = 1 + 10^{-4}$;

$\delta = 0{\cdot}9$, $\dfrac{n_0}{n'} = 1 + 10^{-2}$.

With decrements lower than $0{\cdot}09$, therefore, the resonance frequency n_0 differs from the synchronous frequency n' by less than 1 in 10,000. In dealing with the decrement δ it is in practice never necessary to distinguish between

$$\frac{R}{2n'L} \quad \text{and} \quad \frac{R}{2n_0 L} = 2\pi^2 n_0 CR = \pi\sqrt{\frac{C}{L}}\, R.$$

2 (c). *Certain initial conditions.* Let us suppose that the alternator of Fig. 32 was switched on when its E.M.F. was zero, and the circuit was electrically dead; i.e. at

$$t = 0, \quad q = 0, \quad i \equiv \frac{dq}{dt} = 0.$$

Inserting these initial conditions in the complete solution and its first differential (p. 65), we find

$$Q' = -\frac{Q\cos\theta}{\cos\phi} \quad \text{and} \quad \tan\phi = \frac{p\tan\theta - b}{p'}.$$

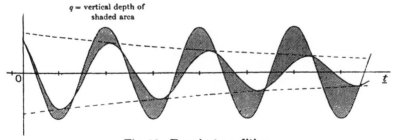

Fig. 33. Transient condition.

Graphically exhibited, the solution is given by the vertical depth of the shaded area in Fig. 33, which is drawn approximately to scale for the case

$$\frac{b}{p'} = \frac{0{\cdot}31}{2\pi}, \quad \frac{p}{p'} = 0{\cdot}85.$$

The sign of the forced oscillation is reversed in the Figure in order

to show q as the *difference* of the two oscillations. In this numerical case,
$$\theta = 45°, \quad \phi = 40\cdot 3°, \quad \frac{Q'}{Q} = -0\cdot 93.$$
After a few more periods the solution would settle down to the forced oscillation alone, viz.
$$q = Q\cos(pt + 45°).$$

2(d). *Resonance, with the same initial conditions.* The eventual amplitude of the current,
$$\frac{E}{\sqrt{\left(pL - \frac{1}{pC}\right)^2 + R^2}},$$
is a maximum for choice of p or L or C when $pL - \frac{1}{pC} = 0$. This is called the condition of resonance. Writing the resonant frequency $\frac{p_0}{2\pi}$, at resonance
$$p^2 = p_0^2 = \frac{1}{LC},$$
$$p'^2 = p_0^2 - b^2.$$
Then, with the same initial conditions as in 2(c),
$$Q = -\frac{E}{pR}, \quad \theta = 0,$$
$$\tan\phi = -\frac{b}{p'}, \quad \therefore \cos\phi = \frac{p'}{p_0},$$
$$Q' = -\frac{Q}{\cos\phi} = \frac{E}{p'R}.$$
$$\therefore q = -\frac{E}{pR}\left[\cos pt - \frac{p}{p'}\epsilon^{-bt}\cos(p't + \phi)\right],$$
$$i = \frac{E}{R}\left[\sin pt - \frac{p}{p'}\epsilon^{-bt}\sin(p't + \psi)\right],$$
where $\psi \equiv 2\tan^{-1}\frac{b}{p'}$.

If the damping is not very heavy, viz. if $\frac{b^2}{2p^2} \ll 1$, this reduces sensibly to
$$i = \frac{E}{R}(1 - \epsilon^{-bt})\sin pt^*.$$

* To expose better this function of t while bt is small, ϵ^{-bt} may be expanded, giving $i = \frac{E}{R}bt\sin pt$ for $bt \ll 2$.

2. UNDAMPED IMPRESSED E.M.F.

Fig. 34 shows this curve drawn to scale for the same case as Fig. 33, viz. $\dfrac{b}{p} = \dfrac{0\cdot 31}{2\pi}$.

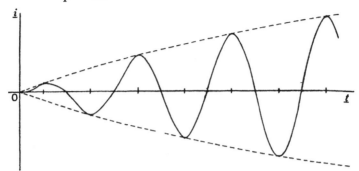

Fig. 34. Transient condition with resonance.

2 (e). *The forced oscillation alone.* After the lapse of sufficient time, the transient free oscillation has sensibly vanished, and there remains only the forced oscillation

$$i = I \sin (pt + \theta),$$

where $I = \dfrac{E}{\sqrt{\left(pL - \dfrac{1}{pC}\right)^2 + R^2}}$ and $\theta = -\tan^{-1} \dfrac{pL - \dfrac{1}{pC}}{R}$.

At the resonant frequency $\dfrac{p_0}{2\pi} = \sqrt{\dfrac{1}{CL}}$, the amplitude is

$$I_0 = \dfrac{E}{R}.$$

If pL and $\dfrac{1}{pC}$ are, as usual, very large compared with R, slight inequalities in pL and $\dfrac{1}{pC}$, i.e. slight departures from resonance, make great changes in the values of I. As an example, in Fig 35 is plotted a curve, commonly called a *resonance curve*, between I and n for a lightly damped circuit such as might occur in practice, where (Fig. 32)

$R = 10\,\Omega,$ $L = 5000\,\mu\text{H},$ $C = .508\,\mu\mu\text{F},$
$\lambda = 3000\,\text{m},$ $n' = 100\,\text{kc/s},$ $\delta = 0\cdot 01.$

If $E = 10$ volts, the ordinate scale is in amperes.

The "peakiness" of the resonance curve is clearly a measure of the "sharpness of tuning" obtainable with the circuit. If it is drawn with abscissas $\frac{n}{n_0}$ and ordinates $\frac{I}{I_0}$, where I_0 is the value of I at the resonant frequency n_0, its shape is a function of the decrement only.

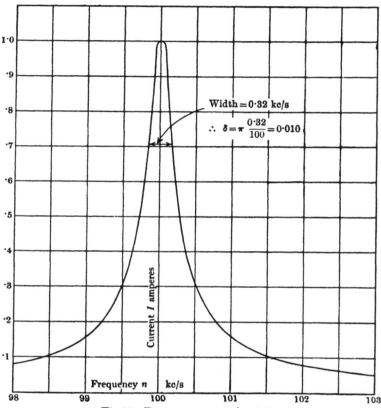

Fig 35. Resonance curve, $\delta = 0.01$.

For
$$\left(\frac{I_0}{I}\right)^2 = \frac{R^2 + \left(pL - \frac{1}{pC}\right)^2}{R^2},$$

$$= 1 + \left(\frac{pL - \frac{1}{pC}}{R}\right)^2.$$

$$\therefore \frac{pL - \frac{1}{pC}}{R} = \pm \sqrt{\left(\frac{I_0}{I}\right)^2 - 1} \equiv \pm m \text{ (say)}.$$

2. UNDAMPED IMPRESSED E.M.F.

The same values of curve ordinate, and therefore of m, are given by the two values of p, one on each side of p_0, satisfying the equation

$$pL - \frac{1}{pC} = \pm mR,$$

i.e.
$$p^2 \mp \frac{mR}{L}p - \frac{1}{LC} = 0;$$

$$\therefore p_1, p_2 = \frac{\pm \frac{mR}{L} + \sqrt{\frac{m^2 R^2}{L^2} + \frac{4}{LC}}}{2},$$

$$\therefore p_1 - p_2 = \frac{mR}{L},$$

$$\therefore \frac{n_1 - n_2}{n_0} = \frac{p_1 - p_2}{p_0} = \frac{m}{\pi} \cdot \frac{R}{2n_0 L} = \frac{m}{\pi} \cdot \delta,$$

i.e.
$$\delta = \frac{\pi}{m}\left(\frac{n_1}{n_0} - \frac{n_2}{n_0}\right).$$

In words, the decrement is $\frac{\pi}{m}$ times the width, $\left(\frac{n_1}{n_0} - \frac{n_2}{n_0}\right)$, of the peak at any height $\frac{I_0}{I}$, where $m \equiv \sqrt{\left(\frac{I_0}{I}\right)^2 - 1}$.

In using this relation for determining the decrement of a circuit from its observed resonance curve, a convenient height to take is $\frac{1}{\sqrt{2}}$ of the height of the summit, making $m = 1$. This is illustrated in Fig. 35.

At resonance, the current is in phase with the impressed E.M.F.; but at frequencies below and above the resonance value the current leads or lags respectively by the angle θ, where

$$\tan \theta = \frac{pL - \frac{1}{pC}}{R} = \frac{\pi}{\delta}\left(\frac{n}{n_0} - \frac{n_0}{n}\right).$$

A cognate expression for the amplitude is

$$\frac{I}{I_0} = \sqrt{\frac{1}{1 + \left(\frac{pL - \frac{1}{pC}}{R}\right)^2}} = \cos \theta.$$

Circuits are most commonly used in the condition of resonance, when $\theta = 0$; but phase displacement by slight distuning is sometimes intentionally effected*.

For small distuning, say $x \equiv \dfrac{n}{n_0} - 1$, the frequency function $\left(\dfrac{n}{n_0} - \dfrac{n_0}{n}\right)$ is nearly proportional to x. Over this range, therefore, the curve of $\dfrac{I}{I_0}$, θ is symmetrical about $x = 0$; and to convert the

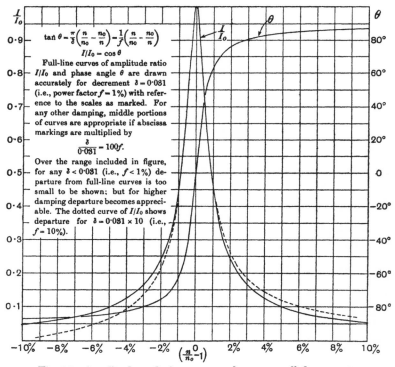

Fig. 36. Amplitude and phase curves for any small decrement.

resonance curve for decrement δ_1 into the curve for another decrement δ_2, it is only necessary to change the x-scale in the ratio $\delta_1 : \delta_2$. This is indicated in Fig. 36, which is drawn accurately and may be used for finding the amplitude and phase near resonance of the forced oscillation in a circuit of any practical decrement†.

* An instance will be found in XIV-7.

† Or power factor f. In Fig. 36, f and $\dfrac{\delta}{\pi}$ are to be taken as interchangeable. See Subsection 2 (f) below.

2. UNDAMPED IMPRESSED E.M.F.

Instead of varying n, the frequency of the impressed E.M.F. in Fig. 32, it is often convenient to keep n constant and to vary the resonant frequency n_0 of the circuit by adjustment of L or C, especially C. If the curves are plotted with values of n_0 as abscissa (instead of n), it is obvious from the foregoing remarks that the middle parts—say over the range $\frac{n}{n_0} = (1 \pm 2\,°/_\circ)$—are sensibly unchanged. It is, however, often more convenient to plot directly with values of C as abscissa. Now

$$C = \frac{1}{4\pi^2 n_0^2 L};$$

$$\therefore \frac{dC}{dn_0} = \frac{-2}{4\pi^2 L n_0^3},$$

$$\therefore \frac{dn_0}{n_0} = -\tfrac{1}{2} \frac{dC}{C}.$$

Hence in finding the decrement from the shape of the I, C resonance curve, the formula

$$\delta = \frac{\pi}{m} \cdot \tfrac{1}{2} \frac{C_2 - C_1}{C_0} *$$

is to be substituted for the previous formula

$$\delta = \frac{\pi}{m} \cdot \frac{n_1 - n_2}{n_0}.$$

2 (f). *Decrement and power factor.* The decrement in its fundamental sense must refer to an oscillatory system, a complete circuit. Also, the power factor of a complete circuit (at the resonance frequency) must always be unity. But on expressing the decrement δ as a function of the dimensions R, L, C of the circuit, viz.

$$\delta = \frac{R}{2nL}, \quad \text{or} \quad \delta = \pi \sqrt{\frac{C}{L}} R, \quad \text{or} \quad \delta = 2\pi^2 nCR,$$

it is seen that *the decrement of the circuit* (when not impractically heavily damped) is sensibly equal to π times *the power factor of a part of the circuit*, viz. of either the inductance L or the capacitance C if all the resistance R were associated therewith. For the power factor f of a coil L, R or a condenser C, R carrying alternating

---

* This is accurate to within $1\,°/_\circ$ if $\frac{C_2 - C}{C_0} < \tfrac{1}{5}$, while the formula for variation of n is accurate over any range. See E. B. Moullin, *Radio frequency measurements* (1926), p. 117.

current of frequency n is $\cos\theta$ (Fig. 37); and $\cos\theta$ differs from $\frac{1}{\pi}\delta = \cot\theta$ by less than 1 %, if $\delta < 0.5$.

It is therefore permissible to take f and $\frac{1}{\pi}\delta$ as interchangeable; and it is customary, ignoring the basic meanings of the terms, to speak indiscriminately of the decrement and the power factor of either a complete circuit or a part of a circuit such as a coil or a

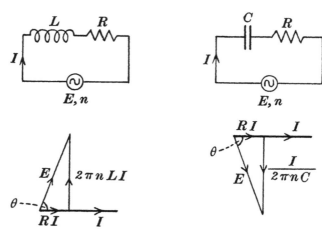

Fig. 37.

condenser. Accordingly, a coil of inductance L and resistance R_1 is said to have decrement $\delta_1 = \dfrac{R_1}{2nL}$ and power factor $f_1 = \dfrac{R_1}{2\pi nL}$; and a condenser of capacitance C and resistance R_2 to have decrement $\delta_2 = 2\pi^2 nCR_2$ and power factor $f_2 = 2\pi nCR_2$; and the circuit formed by connecting them in series to have decrement $\delta = \delta_1 + \delta_2$ and power factor $f = f_1 + f_2$.

In dealing with steady-state conditions, power factors are, on the whole, more convenient than decrements in that they lead to economy in the use of the symbol π.

3. The wavemeter

It is obvious that if the circuit of Fig. 32 is provided with some form of current measurer such as a hot-wire ammeter inserted in the circuit, and if the frequency n of the impressed E.M.F. is varied, or if the resonance frequency n_0 of the circuit is varied by alteration of L or C, the ammeter will indicate when $n = n_0$ by showing then its maximum deflection. The values of L and C being variable and known, the LC circuit then constitutes a frequency meter, which is the more sensitive as its decrement is smaller.

In Fig. 32 the E.M.F. was shown diagrammatically as impressed upon the circuit by a generator actually connected in series with it. A more common arrangement is that in which the E.M.F. is induced in the

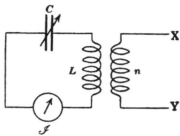

Fig. 38. Wavemeter.

inductance L by placing it in the magnetic field of a high-frequency current, as in Fig. 38. Provided that the reaction of the frequency meter on the circuit XY, the frequency n of whose current is to be measured, is small enough not appreciably to affect that current, n can be determined by varying the calibrated condenser C until the observed current $\mathscr{I}$ is a maximum. A calibrated circuit of this kind is commonly called a *wavemeter*, each graduation on the scale being generally the wavelength (in metres) which corresponds to the particular natural frequency of the circuit at that adjustment. It is in fact customary to speak of the wavelengths of high-frequency currents even when no waves are contemplated: what is signified is, of course, the wavelength of the radiation in space which would correspond with the particular frequency under consideration. This custom has arisen from the fact that we are seldom able directly to count the number of cycles per second, whereas we do often measure lengths of waves more or less directly, and perform gross operations such as cutting a length of wire for an antenna equal to some large fraction (e.g. $\frac{1}{4}$) of the wavelength. Wavelength, in this sense, and frequency are tacitly treated as interchangeable according to the relation

$$\lambda \times n = 3 \times 10^{10} \text{ cm per sec.}*$$

---

* Velocity of light = $2\cdot 997 \times 10^{10}$ cm/sec. There is a growing tendency to abandon wavelength in favour of frequency; partly, no doubt, because it is logically more accurate to do so; but largely because the closeness with which the modern telephone and C.W. telegraph transmitters may be packed together in the wireless spectrum without mutual interference depends more upon the absolute difference of their frequencies than upon the ratio of their wavelengths or frequencies. Whatever their wavelengths, if two transmitters differ in frequency

It is convenient to remember that 300 metres corresponds with a frequency of one million cycles per second.

A wavemeter provided with an instrument to *measure* the R.M.S. current, as in Fig. 38, can be used to obtain resonance curves. But any indicator capable of showing merely when the current is a maximum suffices for measuring the wavelength.

Other forms of wavemeter are based on the mechanical frequency of a tuning fork or a quartz crystal. (See IX–6(b) and 6(c).)

4. THE ANTENNA AS OSCILLATORY CIRCUIT

Strictly, no real circuit can have the ideal concentration of capacitance indicated in our diagrams; every centimetre of wire in the circuit has capacitance with respect to earth and with respect to every other centimetre of wire. But the "stray" capacitances γ_1, γ_2, γ_3 etc. in A, Fig. 39, are usually kept small—it is one of the

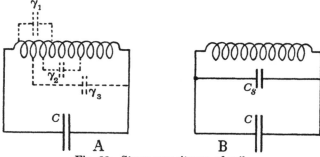

Fig. 39. Stray capacitance of coil.

controlling considerations in design—and can in practice either be ignored, or be sufficiently accurately represented by an imaginary single small capacitance C_s placed across the ends of the coil as in B, Fig. 39.

But in antennas, whose essential radiative property is conferred by their widespread configuration, there is a distribution and intermixture of inductance, capacitance and resistance which cannot be

---

by not less than 10 kilocycles per second (10 kc/s), their mutual heterodyne note (see VI–7) is almost inaudible. In Germany 1 c/s is called 1 *Hertz*. A fancy name hardly seems wanted; but it is preferable to the misleading term *cycle* or *kilocycle*, too common in America and England. The author advocates the use of kc/s, spoken *kay see ess*. This has the correct dimensions, T^{-1}, and is actually easier to say than is *kilocycle*.

4. THE ANTENNA AS OSCILLATORY CIRCUIT

disregarded. The most effective antenna, *qua* radiator, is an isolated straight wire oscillating at its natural frequency, with a potential node at its mid point (A, Fig. 40); or its equivalent for use on the ground, an earthed vertical wire oscillating with node at the bottom end (B, Fig. 40). The current amplitude in the wire then varies from a maximum at the middle or bottom to zero at the ends or top. The same is true of the antenna when loaded by a coil at the bottom (C, Fig. 40). Even with the addition of a flat top (D, Fig. 40), the current amplitude, although no longer zero at the top of the upload, must vary somewhat from point to point along it. In an antenna, therefore, owing to the non-localisation of the

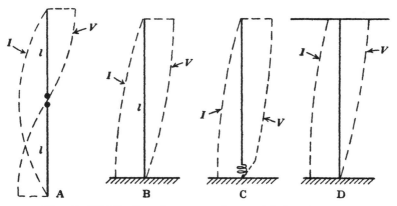

Fig. 40. Distribution of current and potential along antennas.

capacitance at one spot, there is no specifiable single value of the current at any instant. It is not one circuit carrying at any instant an assignable current i, but an infinitely numerous congeries of infinitesimal elements.

A further complication in the antenna is in respect to the location of the sink of power, indicated in our diagram by a resistance R (e.g. Figs. 32 and 49); for there are important losses of power by radiation and by currents in the ground, buildings, etc. near the antenna. Moreover the amount of these losses may vary rapidly with the frequency of the currents flowing.

An antenna, therefore, is not really *an* oscillatory system, but a collection of many oscillatory systems closely linked together. Nevertheless, when we are dealing with the steady-state condition, or something very like it such as a very lowly damped train of oscillation, a close approximation over a limited range of frequency is got by substituting an ideal *equivalent concentrated circuit* for the actual distributed circuit. When an antenna (Fig. 41) takes the place of a circuit with concentrated L, C and R, it may be treated as the equivalent of the concentrated circuit provided that L, C and R are chosen to satisfy the following conditions:

(i) $\dfrac{1}{2\pi\sqrt{CL}}$ equals the natural frequency of antenna circuit. This determines the product CL.

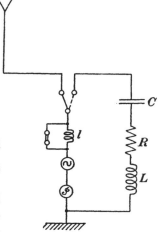

Fig. 41. Equivalent concentrated circuit.

(ii) The equivalent concentrated circuit and the actual antenna circuit are equally changed in natural frequency by the insertion of a small extra inductance l (or a large capacitance). With (i) this determines the values of C and L individually.

(iii) I^2R equals the power dissipated in the actual antenna, when the R.M.S. current at some specified point in it—invariably taken as the bottom—is also the R.M.S. current in the equivalent concentrated circuit. This determines R. But R, since it includes the radiation resistance, is apt to vary rapidly as the frequency is varied.

Whenever we speak without qualification of the capacitance, inductance, or resistance of an antenna circuit, we are tacitly referring to the equivalent concentrated circuit; just as when we speak of the length of a pendulum we really mean the length of the ideal simple equivalent pendulum. It should be remembered, moreover, that the equivalence is a limited one, and strictly refers only to the steady state within an indefinitely narrow range of frequencies about the particular frequency for which the above conditions (i), (ii) and (iii) are satisfied. The shock from an atmospheric will in general affect an antenna very differently from any concentrated circuit whatever.

In an antenna of the form shown at A in Fig. 42, consisting of a single wire (or uniform group of wires), the distribution of inductance and capacitance is very thorough; and it is nearly as thorough

4. THE ANTENNA AS OSCILLATORY CIRCUIT

in the form shown at B. When oscillating at its natural frequency (i.e. without loading inductance or capacitance inserted), the distribution is as indicated at B, Fig. 40. With decrease of frequency by the addition of loading inductance, the distribution changes to that of C, Fig. 40, and the capacitance increases, at first rapidly. At

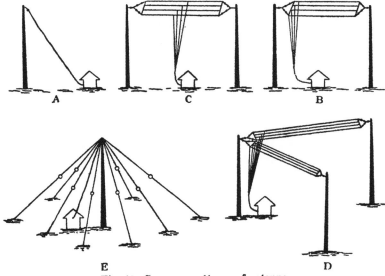

Fig. 42. Common patterns of antenna.

frequencies much lower than the unloaded frequency, the potential is sensibly uniform along the antenna, and the capacitance becomes indistinguishable from the electrostatic capacitance of the antenna to earth. At the other extreme, in an umbrella antenna (E, Fig. 42) the inductance is fairly well concentrated in the uplead, and the capacitance in the radial wires at the top; the equivalent concentrated capacitance is therefore nearly independent of the amount of inductance loading.

The change of equivalent concentrated capacitance with frequency is well illustrated in Fig. 43. The full-line curve gives the measured values of antenna reactance, X ohms, plotted against $p = 2\pi n$ sec.$^{-1}$ for an antenna of rather irregular form * but somewhat resembling B in Fig. 42.

* Two parallel wires, about 80 m long, stretching over part of the Engineering Laboratory, Cambridge, from near roof level at one end to a high chimney stack at the other, and with a rather wandering single-wire downlead.

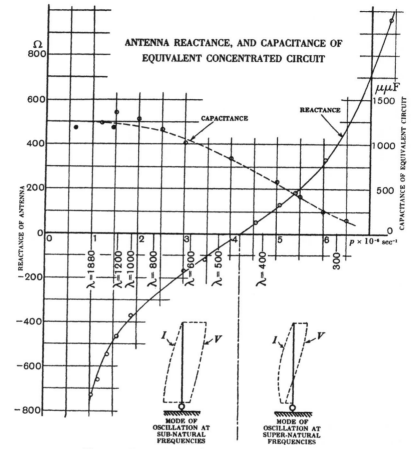

Fig. 43. Dependence of capacitance on frequency.

If L and C are the equivalent inductance and capacitance of the antenna at frequencies near $\dfrac{p}{2\pi}$,

$$X = pL - \frac{1}{pC}.$$

$$\therefore p\frac{dX}{dp} = pL + \frac{1}{pC},$$

$$\therefore \frac{2}{pC} = p\frac{dX}{dp} - X;$$

i.e. $C = \dfrac{2}{p\left(p\dfrac{dX}{dp} - X\right)}.$

4. THE ANTENNA AS OSCILLATORY CIRCUIT

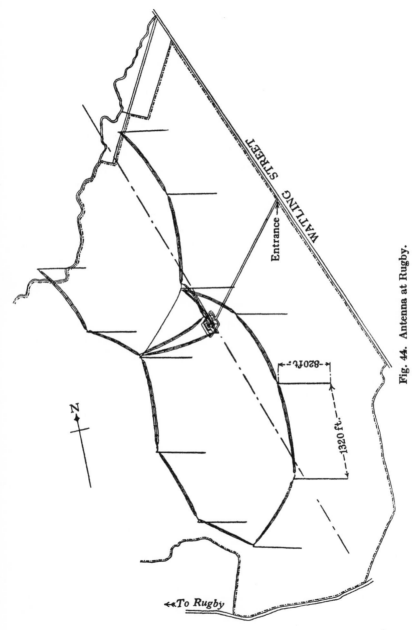

Fig. 44. Antenna at Rugby.

The broken-line curve shows the value of C, so calculated*. As the frequency is raised above about half the natural frequency (which is at $X = 0$, $p = 4\cdot 2 \times 10^6$ sec.$^{-1}$, $\lambda = 448$ m), the capacitance of the equivalent concentrated circuit falls rapidly; and at frequencies above the natural it assumes very low values—as would be expected when the mode of oscillation (indicated in the Figure) is considered.

The capacitance of different antennas for medium and long waves may range from (say) the 200 $\mu\mu$F of a small broadcasting receiving antenna; through the 1800 $\mu\mu$F of the high-power broadcast transmitting antenna of Plate I; to the 45000 $\mu\mu$F of the great antenna at Rugby (Fig. 44†), whose flat top consists of nearly 3 miles of a "sausage" of wires 12 feet in diameter supported on 12 masts 820 feet high.

The resistances of antennas also vary widely. Firstly, the radiation resistance (with inductance loading) may approach 40Ω‡; but may be very low, e.g. 0·055Ω at Rugby§. Secondly, the component of the

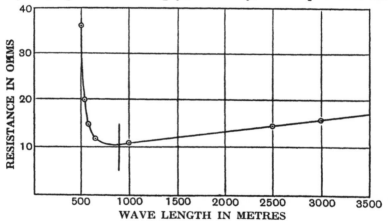

Fig. 45. Total resistance of antenna.

resistance due to losses in the ground under the antenna, in the masts and stays, neighbouring trees, walls, etc. may be large or small.

* Except that the solid dots indicate a more accurate determination by another method suitable for long wavelengths.

† From E. H. Shaughnessy, "Rugby radio station," *Journal I.E.E.* Vol. 64 (1926), p. 683, by permission of the Institution.

‡ For a vertical earthed wire oscillating at its natural frequency; see p. 33. When the antenna is loaded with series capacitance so as to oscillate at less than the natural wavelength, higher resistances are obtained; see S. Ballantine, "On the radiation resistance of a simple vertical antenna at wavelengths below the fundamental," *Proc. Inst. Radio Engrs.* Vol. 12 (1924), p. 823.

§ See p. 34.

5. DAMPED IMPRESSED E.M.F.

Fig. 45* gives the measured resistance of the same antenna as that to which Fig. 43 refers; and Fig 46† gives the measured resistance of the Rugby antenna. The rise of resistance with increasing wavelength seen in Fig. 45 is, of course, attributable to the losses, since the component due to radiation is decreasing.

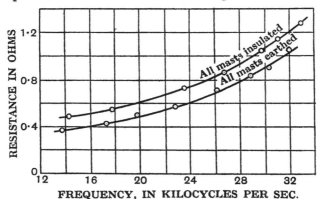

Fig. 46. Total resistance of Rugby antenna.

5. Damped impressed E.M.F.

5 (a). *Free oscillation.* If a charged condenser is discharged through a circuit possessing inductance (and not too much resistance), a damped oscillation takes place. Thus for the $L_1 C_1 R_1$ circuit of Fig. 47, when the switch S is closed the equation of E.M.Fs. is

$$L_1 \frac{dq_1}{dt^2} + R_1 \frac{dq_1}{dt} + \frac{1}{C_1} q_1 = 0,$$

whose solution (viz. the complementary function of p. 65) is

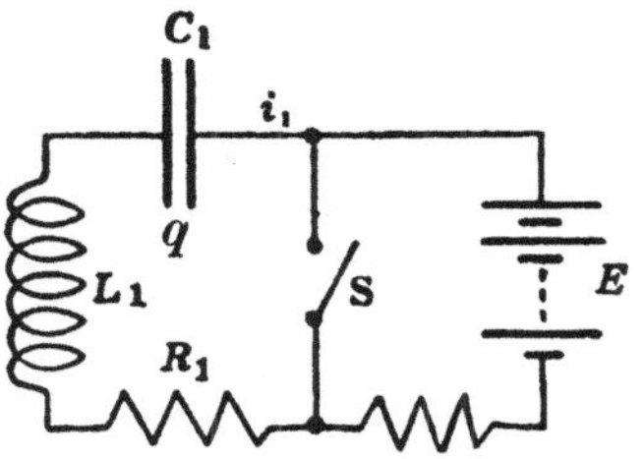

Fig. 47. Free oscillation.

$$q_1 = Q_1 \epsilon^{-b_1 t} \cos(p_1 t + \phi),$$
$$i_1 = -p_0 Q_1 \epsilon^{-b_1 t} \sin\left(p_1 t + \phi + \sin^{-1}\frac{b_1}{p_0}\right),$$

---

* From E. B. Moullin, *loc. cit.* p. 78, Fig. 106, by permission of the publishers.
† From Hansford and Faulkner, *loc. cit.* p. 64, by permission of the Institution. Measurements in U.S.A. have shown that with the same current the Rugby antenna radiates a slightly stronger field (about 8 °/₀ more) when the masts are insulated than when they are earthed. This is what would be expected. But the rise of total resistance on insulating the masts is much more than the additional radiation accounts for, and seems to show that the insulation is imperfect.

where $b_1 \equiv \dfrac{R_1}{2L_1}, \quad p_1 \equiv \sqrt{\dfrac{1}{C_1L_1} - \dfrac{R_1^2}{4L_1^2}}, \quad p_0^2 \equiv p_1^2 + b_1^2$

and Q_1 and ϕ depend on the initial conditions.

In the circumstances of Fig. 47, if t is measured from the instant of closing the switch, the initial conditions are

$$t = 0, \quad q_1 = EC_1, \quad i_1 = 0,$$

giving $Q_1 = \dfrac{EC_1\sqrt{p_1^2 + b_1^2}}{p_1}$ and $\tan \phi = -\dfrac{b_1}{p_1}$.

Hence
$$i_1 = -\dfrac{p_0^2}{p_1} EC_1 e^{-b_1 t} \sin p_1 t,$$
$$\doteqdot -p_1 EC_1 e^{-b_1 t} \sin p_1 t$$

if the circuit has not an excessive decrement, viz. if $\delta^2 \ll 8\pi^{2*}$. Fig. 48 shows this damped oscillation (reversed in sign), drawn to scale for the case
$$\dfrac{b_1}{p_1} = \dfrac{0\cdot 06}{2\pi}, \quad \text{i.e. } \delta_1 = 0\cdot 06.$$

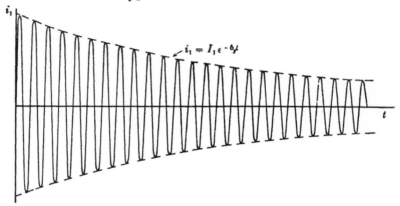

Fig. 48. Damped oscillation.

The oscillation dies away asymptotically to zero as the initial energy of the charged condenser is dissipated in heating the resistance R_1.

* See p. 67.

5. DAMPED IMPRESSED E.M.F.

5 (b). *Loosely coupled circuit.* Let a secondary circuit $L_2 C_2 R_2$ now be *coupled* to the oscillating circuit of Fig. 47 by the existence of a small mutual inductance M between L_1 and L_2, as in Fig. 49. If the coupling is sufficiently loose to allow the reaction E.M.F. $M\frac{di_2}{dt}$ in the primary circuit to be ignored, the primary oscillation remains

$$i_1 = -p_1 E C_1 \epsilon^{-b_1 t} \sin p_1 t;$$

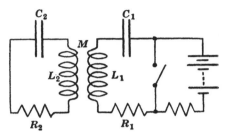

Fig. 49. Coupled circuits.

and there is impressed in the secondary circuit the damped E.M.F. $M\frac{di_1}{dt}$, which, with the previous approximation as to moderate damping, is

$$-p_1^2 M E C_1 \epsilon^{-b_1 t} \cos\left(p_1 t + \tan^{-1}\frac{b_1}{p_1}\right).$$

The equation to the oscillation in the secondary circuit is accordingly

$$L_2 \frac{d^2 q_2}{dt^2} + R_2 \frac{dq_2}{dt} + \frac{1}{C_2} q_2 = p_1^2 M E C_1 \epsilon^{-b_1 t} \cos\left(p_1 t + \tan^{-1}\frac{b_1}{p_1}\right).$$

The complete solution of this equation is of the form

$$q_2 = A \epsilon^{-b_1 t} \cos(p_1 t + \alpha) + B \epsilon^{-b_2 t} \cos(p_2 t + \beta);$$

that is, the secondary circuit is the seat of two damped oscillations, one with the frequency and damping proper to the primary alone, and the other with the frequency and damping proper to the secondary alone.

It can be shown that in the special case of synchronism between the two circuits, viz. if $p_1 = p_2 = p$, the secondary current becomes

$$i_2 = \frac{pMI_1}{2L_2(b_2 - b_1)}(\epsilon^{-b_1 t} - \epsilon^{-b_2 t}) \sin(pt + \gamma).$$

This expression may be rewritten in a form showing the process of building up and decay more clearly in terms of the number of periods N (instead of time t) from the beginning, by making the substitutions
$$p = 2\pi n,$$
$$pt = 2\pi N,$$
$$b = n \cdot \delta,$$
$$bt = N \cdot \delta.$$

Then
$$i_2 = \frac{\pi M I_1}{L_2(\delta_2 - \delta_1)}(\epsilon^{-\delta_1 N} - \epsilon^{-\delta_2 N}) \sin(2\pi N + \gamma).$$

Fig. 50 shows the course of an oscillation of this character, together with the primary oscillation producing it, for the decrements
$$\delta_1 = 0\cdot06,$$
$$\delta_2 = 0\cdot20.$$

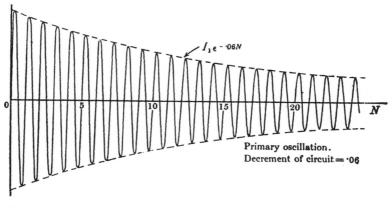

Primary oscillation.
Decrement of circuit = ·06

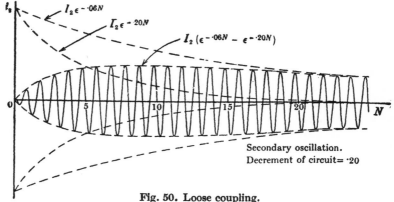

Secondary oscillation.
Decrement of circuit = ·20

Fig. 50. Loose coupling.

5 (c). *Resonance and damped E.M.F.* If the primary discharge is repeated at intervals of time T long enough to allow the oscillations practically to have died away between one discharge and the next, but short enough to allow an ammeter in the secondary circuit to take up a steady reading, this reading will be $\mathscr{I}_2$ where

$$\mathscr{I}_2{}^2 = \frac{1}{T}\int_0^T i_2{}^2 \, . \, dt.$$

By noting the ammeter reading with various slight amounts of distuning between p_1 and p_2 from the adjustment giving $\mathscr{I}_2 =$ max., resonance curves for this R.M.S. current $\mathscr{I}_2$ can be plotted, as in 2 (e) of this Chapter. It is fairly obvious on physical grounds that an increase of either of the decrements δ_1, δ_2 of the two circuits alone must reduce the sharpness of the resonance curve; and indeed it can be proved that the shape of the resonance curve for $\mathscr{I}_2$ in the circuit of decrement δ_2 provoked by the impressed damped E.M.F. of decrement δ_1 depends on the sum of the decrements $(\delta_1 + \delta_2)$ in the same way as the shape in Figs. 35 and 36 depended on the single decrement δ^*. If, therefore, the switch in Fig. 49 is closed and opened at regular intervals, and a resonance curve is plotted by observing $\mathscr{I}_2$ as C_2 is varied, we have

$$\delta_1 + \delta_2 = \frac{\pi}{2m} \frac{(C_2)_2 - (C_2)_1}{(C_2)_0},$$

as on p. 73.

5 (d). *Practical equivalence of methods of distuning.* With low values of $(\delta_1 + \delta_2)$ a practically indistinguishable result would be given by varying C_1 instead of C_2; for although the initial charge EC_1, which ought to be held constant during the distuning operations, would be varied slightly, the change of ordinate due to this change of initial charge would be very small compared with the change of ordinate due to the distuning *per se*. Thus if $(\delta_1 + \delta_2)$ were 0·2, the change of C_1 from resonance would be $\pm 6\tfrac{1}{2}\,\%$ for $m = 1$. The resonance curve observed would be changed from the full-line (correct) curve of Fig. 51 (not to scale) to the dotted curve, where

$$\mathrm{AD} \fallingdotseq \mathrm{BE} \fallingdotseq 6\tfrac{1}{2}\,\%\ \text{of OG}.$$

---

* For a range of $\left(\dfrac{n_1}{n_2} - 1\right)$ up to about $\pm\, 5\,\%$ or $10\,\%$, and with the familiar approximation $\left(\dfrac{\delta}{\pi}\right)^2 \ll p^2$. See G. W. Pierce, *loc. cit.* p. 11, Book I, Chaps. IX and X.

In using A'B' instead of AB, the decrement would be calculated as $\pi \dfrac{A'B'}{FC'}$ instead of $\pi \dfrac{AB}{FC}$; and the error would be about

$$\frac{AA'}{FC} = 6\tfrac{1}{2}\% \times \frac{AA'}{AC} \fallingdotseq 1\%.$$

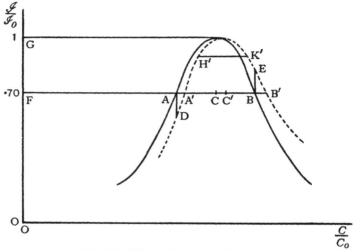

Fig. 51. Effect of manner of tuning.

As regards wavelength measurements, it is only the summit of the resonance curve which is sought, and this is not misplaced by adjusting C_1 instead of C_2. But the experimental method of locating the summit is, of course, to assume it is midway between two observed points H', K' perceptibly equally far below it. The error is therefore reduced below the error in the decrement measurement the more H'K' is brought above A'B' towards the summit. With a sensitive indicator (not rectifier, telephone and ear) H'K' could be brought very near the summit, and the wavelength error would be very small.

We have now examined the effect on the current magnitude in an oscillatory circuit such as $L_2 C_2 R_2$ in Fig. 49 of small distunings around the resonance condition

(i) by variation of frequency n of a sustained undamped impressed E.M.F. (Subsection 2(e));

(ii) by variation of C_2, when the impressed E.M.F. is in the form of repeated equal damped trains of decrement δ_1 (Subsection 5(c)); and

(iii) by variation of C_1, thereby changing not only the frequency of the damped E.M.F. $\left(\propto \dfrac{1}{\sqrt{C_1}}\right)$ but also, as an undesired concomitant, its magnitude ($\propto C_1$).

We have seen that, provided decrements are moderate, the resonance curves are always—even in the last case—sensibly the same, depending only on the total decrement $(\delta_1 + \delta_2)$. Generalising from these investigations, the following practical rules may be formulated. With low decrements, it matters not where the resistance is located in the circuit; and sensibly the same change from the resonance current is produced by the same small change in frequency, whether it be made by change of n or C_1 or C_2 or L_1 or L_2. The frequency may be regarded as the resonance frequency $\dfrac{1}{2\pi}\sqrt{\dfrac{1}{CL}}$ or the natural frequency $\dfrac{1}{2\pi}\sqrt{\dfrac{1}{CL} - \dfrac{R^2}{4L^2}}$, without practical distinction*.

6. Tight coupling

If the two circuits of Fig. 49 are more closely coupled together by increasing their mutual inductance M, the reaction E.M.F. $M\dfrac{di_2}{dt}$ becomes great enough to affect appreciably the primary current. The free oscillations are given by the simultaneous equations

$$\left(L_1 D^2 + R_1 D + \frac{1}{C_1}\right)q_1 = -M D^2 q_2,$$

$$\left(L_2 D^2 + R_2 D + \frac{1}{C_2}\right)q_2 = -M D^2 q_1,$$

where D stands for the operator $\dfrac{d}{dt}$. We have seen† how slight is the effect of the resistance on the frequency of a single lowly damped circuit, and we may readily suppose that the resistance terms in

---

* On the other hand, the cumbersomeness of the algebraic treatment of oscillatory circuits may be considerably affected by the location of the resistance in the circuit, and by the choice between these two special frequencies. For example, the solution with given initial conditions of the differential equation on p. 85 for the circuit of Fig. 49, while troublesome for the case $p_1 = p_2$, is more troublesome for the case $\dfrac{1}{C_1 L_1} = \dfrac{1}{C_2 L_2}$.

† P. 67.

these equations likewise have negligible influence on the frequencies of the oscillations which occur. Neglecting resistances, these may be found as follows:

$$\begin{cases} \left(L_1 D^2 + \dfrac{1}{C_1}\right) q_1 = -M D^2 q_2 & \quad\ldots\ldots\ldots\ldots(1), \\ \left(L_2 D^2 + \dfrac{1}{C_2}\right) q_2 = -M D^2 q_1 & \quad\ldots\ldots\ldots\ldots(2). \end{cases}$$

Differentiating twice,

$$\begin{cases} \left(L_1 D^4 + \dfrac{1}{C_1} D^2\right) q_1 = -M D^4 q_2 & \quad\ldots\ldots\ldots\ldots(3), \\ \left(L_2 D^4 + \dfrac{1}{C_2} D^2\right) q_2 = -M D^4 q_1 & \quad\ldots\ldots\ldots\ldots(4). \end{cases}$$

Eliminating q_2 from (3) by the use of (1), (2) and (4), we obtain

$$\left[\left(L_1 - \frac{M^2}{L_2}\right) D^4 + \left(\frac{1}{C_1} + \frac{L_1}{L_2 C_2}\right) D^2 + \frac{1}{L_2 C_1 C_2}\right] q_1 = 0.$$

Writing $\quad p_1^2 \equiv \dfrac{1}{L_1 C_1}, \quad p_2^2 \equiv \dfrac{1}{L_2 C_2}, \quad k^2 \equiv \dfrac{M^2}{L_1 L_2},$

this becomes

$$[(1 - k^2) D^4 + (p_1^2 + p_2^2) D^2 + p_1^2 p_2^2] q_1 = 0.$$

Hence $\quad q_1 = A \sin(p_a t + \alpha) + B \sin(p_b t + \beta),$

where A, α, B, β are any constants determined by initial conditions, and p_a^2, p_b^2 are the roots of

$$(1 - k^2) p^4 - (p_1^2 + p_2^2) p^2 + p_1^2 p_2^2 = 0,$$

i.e. $\quad p_a^2, p_b^2 = \dfrac{(p_1^2 + p_2^2) \mp \sqrt{(p_1^2 + p_2^2)^2 - 4(1 - k^2) p_1^2 p_2^2}}{2(1 - k^2)},$

i.e. $\quad n_a, n_b = \sqrt{\dfrac{(n_1^2 + n_2^2) \mp \sqrt{(n_1^2 + n_2^2)^2 - 4(1 - k^2) n_1^2 n_2^2}}{2(1 - k^2)}},$

where n_a, n_b are the frequencies of the two terms of the primary oscillation. On solving for q_2 instead of q_1, the same two frequencies are found in the secondary oscillation.

An examination of this expression shows that if n_1 is the smaller of the two natural frequencies n_1 and n_2, then $n_a < n_1$ and $n_b > n_2$. If the two circuits are separately synchronised, i.e. if $n_1 = n_2 = n$, the expression simplifies to

$$n_a = \frac{n}{\sqrt{1 + k}},$$

$$n_b = \frac{n}{\sqrt{1 - k}}.$$

6. TIGHT COUPLING

The resistances we have neglected affect the frequencies n_a and n_b to a negligible extent provided that $\delta_1^2 \ll \dfrac{p_1^2}{4\pi^2}$ and $\delta_2^2 \ll \dfrac{p_2^2}{4\pi^2}$; they do determine the dampings of the two oscillations in each of the circuits.

The oscillation in either circuit may be examined experimentally by taking a resonance curve with a loosely coupled wavemeter, as already explained. The two frequencies are then revealed by the appearance of two peaks in the resonance curve. Fig. 52 shows such

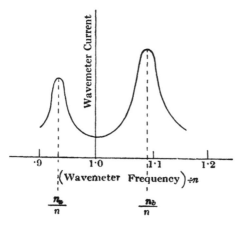

Fig. 52. Resonance curve with close coupling.

a curve. As the mutual inductance M, and therefore the *coefficient of coupling* $k \equiv \dfrac{M}{\sqrt{L_1 L_2}}$, between the two circuits is gradually reduced, the two peaks of the resonance curve draw gradually together, until when the coupling is very *light* or *loose* they coalesce.

The coexistence in each circuit of two oscillations of unequal frequencies implies the occurrence of *beats*. The energy in each circuit therefore waxes and wanes with the beat frequency $(n_b - n_a)$; and since the total energy is constant (in the absence of resistances) or gradually declining (in the presence of resistances), the beat maxima in one circuit coincide with the beat minima in the other. If the two circuits are tuned, this interchange of activity is complete, and with small dampings may occur repeatedly before the total energy has fallen to a negligible value. Fig. 53 shows the alternate waxing and waning of the primary and secondary amplitudes in a pair of tuned coupled circuits oscillating in this manner.

The precise shapes of the curves of current or P.D. (P.D. in Fig. 53) need not concern us here*. The dotted envelopes, showing the changing amplitudes of successive cycles, are the important feature, and indicate how the energy of the oscillation repeatedly changes its seat from one circuit to the other.

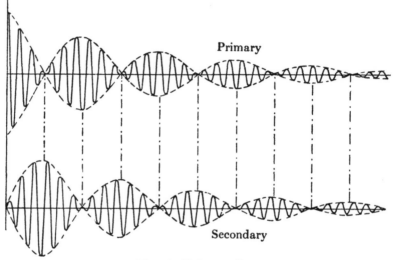

Fig. 53. Tight coupling.

We have examined the coupled circuits from the standpoint of using them to generate oscillations by operation of the switch in Fig. 49, and have found the frequencies of the two oscillations which occur. These are the natural frequencies of the coupled system; and they are therefore also the resonant frequencies of the system (to a very close approximation with practical decrement values). If a steady alternating E.M.F. is impressed in one of the circuits, the response rises to a maximum when the frequency of the impressed

---

* Taking account of damping the mathematical analysis is very complicated, even in special cases, and probably most engineers will find a few experiments with a simple mechanical model more illuminating. (See Section 9 of this Chapter.) A full mathematical treatment of the free oscillations in two coupled circuits is given by Pierce, *loc. cit.* p. 11, Book I, Chaps. vii–x.

E.M.F. equals *either* n_a or n_b. The resonance curve possesses two peaks, and their summits (at $n = n_a$ and $n = n_b$) may be separated more or less by increasing more or less the coefficient of coupling k. k, of course, may have any value between 0 and 1. If $n_1 = n = n_2$, and $k = 0.9$, then $n_a = 0.72\,n$ and $n_b = 3.2\,n$.

7. Methods of coupling circuits

In the last Section the particular form of coupling which is provided by a mutual inductance between the two circuits (Fig. 49) has been examined. This form of coupling is of the commonest occurrence, and possesses the two advantages that there is no conductive connection between the two circuits*, and that the closeness of the coupling is easily variable without incidental alteration of the natural frequencies of the individual circuits. Thus in Fig. 49 the inductances L_1, L_2 may be those of two mechanically separate coils of wire wound on separate frames (e.g. ebonite or cardboard cylinders, one sliding into the other), and their mutual inductance M may be varied by

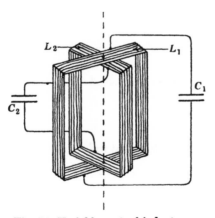

Fig. 54. Variable mutual inductance.

merely varying their relative position. Fig. 54 shows one form of construction, in which one coil can be rotated within the other, so

---

* This is often a great practical convenience, particularly where alternating and direct currents and E.M.Fs. are superposed in the same circuit, as in triode dispositions.

that M varies from zero when their planes are perpendicular to a maximum when their planes are coincident.

There are obviously other ways of coupling circuits than by mutual inductance, and some are of considerable practical importance. In examining them, it will be convenient to use the coupling by mutual inductance as a standard for comparison; and in the rest of this Section we shall refer to this type as "mutual coupling," and to the formula for the natural frequencies as the "mutual formula" (Fig. 55).

Fig. 55. Mutual inductance coupling.

$$p^4[L_1L_2C_1C_2 - M^2C_1C_2] - p^2[L_1C_1 + L_2C_2] + 1 = 0.$$

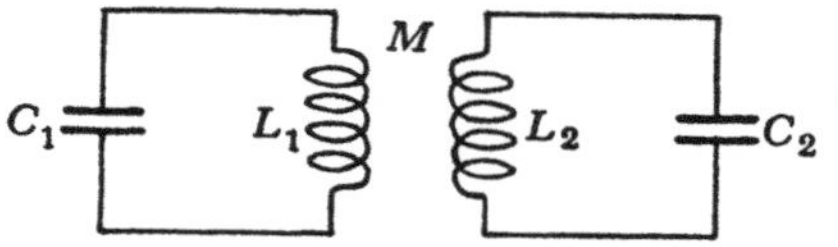

If $L_1C_1 = L_2C_2 = LC$, then

$$p_a^2 = \frac{1}{LC(1+k)}, \quad p_b^2 = \frac{1}{LC(1-k)}.$$

Let
$$\omega_1^2 \equiv \frac{1}{L_1C_1}, \quad \omega_2^2 \equiv \frac{1}{L_2C_2}, \quad k^2 \equiv \frac{M^2}{L_1L_2};$$

then
$$p_a^2, p_b^2 = \frac{(\omega_1^2 + \omega_2^2) \pm \sqrt{(\omega_1^2 + \omega_2^2)^2 - 4(1-k^2)\omega_1^2\omega_2^2}}{2(1-k^2)}$$

$$\equiv \text{"Mutual formula."}$$

Fig. 56 A. Internal inductance coupling.

$$p^4[(L_1+L_0)(L_2+L_0)C_1C_2 - L_0^2C_1C_2] - p^2[(L_1+L_0)C_1 + (L_2+L_0)C_2] + 1 = 0.$$

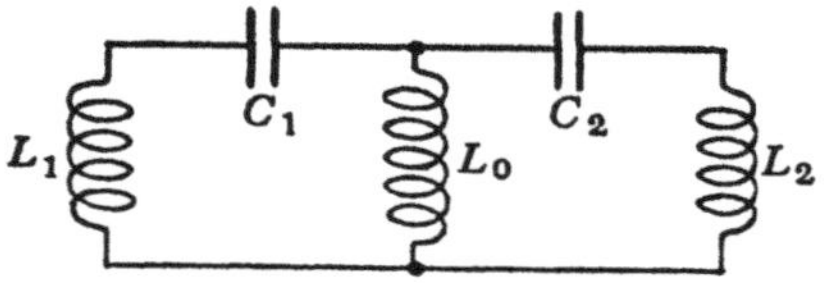

If $L_1C_1 = L_2C_2 = LC$, then

$$p_a^2 = \frac{1}{LC}, \quad p_b^2 = \frac{1}{LC + L_0(C_1+C_2)}.$$

Let
$$\omega_1^2 \equiv \frac{1}{(L_1+L_0)C_1}, \quad \omega_2^2 \equiv \frac{1}{(L_2+L_0)C_2}, \quad k^2 \equiv \frac{L_0^2}{(L_1+L_0)(L_2+L_0)};$$

then
$$p_a^2, p_b^2 = (\text{Mutual formula}).$$

7. METHODS OF COUPLING CIRCUITS

Fig. 56 B. Internal capacitance coupling.

$$p^4 [L_1 L_2 C_1 C_2 C_0^2] - p^2 [L_1 C_1 C_0 (C_2 + C_0) + L_2 C_2 C_0 (C_1 + C_0)] + [(C_1 + C_0)(C_2 + C_0) - C_1 C_2] = 0.$$

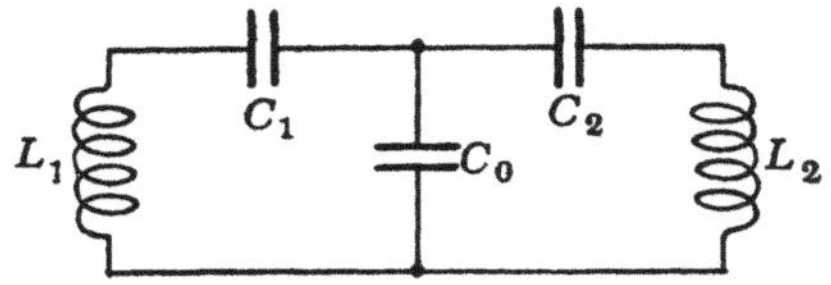

If $L_1 C_1 = L_2 C_2 = LC$, then

$$p_a^2 = \frac{1}{LC}, \quad p_b^2 = \frac{1}{LC} \cdot \frac{C_1 + C_2 + C_0}{C_0}.$$

Let $\omega_1^2 \equiv \dfrac{1}{L_1 \dfrac{C_1 C_0}{C_1 + C_0}}, \quad \omega_2^2 \equiv \dfrac{1}{L_2 \dfrac{C_2 C_0}{C_2 + C_0}}, \quad k^2 \equiv \dfrac{C_1 C_2}{(C_1 + C_0)(C_2 + C_0)};$

then $p_a^2, p_b^2 = $ (Mutual formula) $(1 - k^2)$.

Fig. 57 A. External capacitance coupling.

$$p^4 [L_1 L_2 (C_1 + C_0)(C_2 + C_0) - L_1 L_2 C_0^2] - p^2 [L_1 (C_1 + C_0) + L_2 (C_2 + C_0)] + 1 = 0.$$

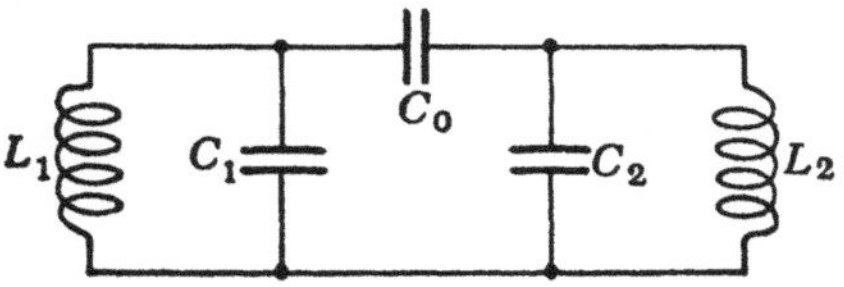

If $L_1 C_1 = L_2 C_2 = LC$, then

$$p_a^2 = \frac{1}{LC}, \quad p_b^2 = \frac{1}{LC + (L_1 + L_2) C_0}.$$

Let $\omega_1^2 \equiv \dfrac{1}{L_1 (C_1 + C_0)}, \quad \omega_2^2 = \dfrac{1}{L_2 (C_2 + C_0)}, \quad k^2 \equiv \dfrac{C_0^2}{(C_1 + C_0)(C_2 + C_0)};$

then $p_a^2, p_b^2 = $ (Mutual formula).

Fig. 57 B. External inductance coupling.

$$p^4 [L_1 L_2 L_0^2 C_1 C_2] - p^2 [L_1 L_0 C_1 (L_2 + L_0) + L_2 L_0 C_2 (L_1 + L_0)] + [(L_1 + L_0)(L_2 + L_0) - L_1 L_2] = 0.$$

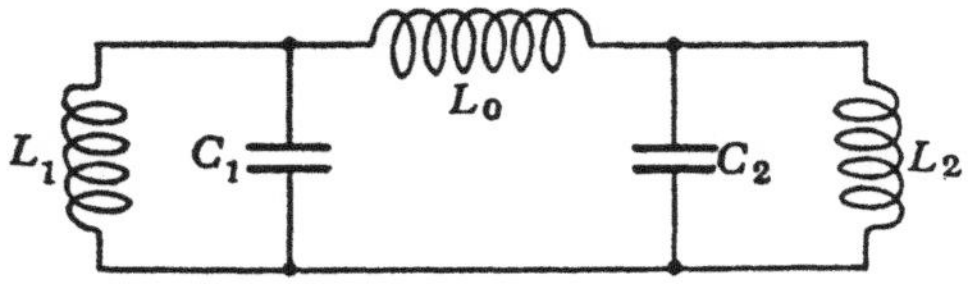

If $L_1 C_1 = L_2 C_2 = LC$, then

$$p_a^2 = \frac{1}{LC}, \quad p_b^2 = \frac{1}{LC} \cdot \frac{L_1 + L_2 + L_0}{L_0}.$$

Let $\omega_1^2 \equiv \dfrac{L_1 + L_0}{L_0 C_1}, \quad \omega_2^2 = \dfrac{L_2 + L_0}{L_2 L_0 C_2}, \quad k^2 \equiv \dfrac{L_1 L_2}{(L_1 + L_0)(L_2 + L_0)};$

then $p_a^2, p_b^2 = $ (Mutual formula) $(1 - k^2)$.

96   IV. OSCILLATORY CIRCUITS

Figs. 56 A and 56 B show two methods of coupling by means of a common internal path, which might be of mixed character but is here the pure inductance L_0 or the pure capacitance C_0; and Figs. 57 A and 57 B show the two corresponding methods employing a common external* path.

In Fig. 56 A the interaction between the two circuits is of precisely the same kind as in mutual coupling. If the simultaneous differential equations of Fig. 56 A are written down, it will be seen at once that they differ from the equations for Fig. 55, viz. the equations on p. 89, only in that $(L_1 + L_0)$, $(L_2 + L_0)$ and L_0 are substituted for L_1, L_2 and M respectively. With the corresponding meanings of p_1^2, p_2^2, and k^2, therefore, the natural frequencies are given by the mutual formula unchanged.

In the other couplings the equations for p^2 are obtained by the same general method already followed in mutual coupling. The Figures show the results of the analysis in each case. For external capacitance coupling (Fig. 57 A) the steps by which the result is obtained are reproduced below as an example of the calculation of the natural frequencies in any compound circuit.

Equating the total E.M.F. round a closed path to zero, and writing $\mathbf{D} \equiv \dfrac{d}{dt}$, with reference to Fig. 58 we have:

for ABCD, $\qquad L_1 \mathbf{D}^2 (q_1 - q_0) + \dfrac{1}{C_1} q_1 = 0 \quad \ldots\ldots\ldots\ldots\ldots(1);$

for EFGH, $\qquad L_2 \mathbf{D}^2 (q_2 + q_0) + \dfrac{1}{C_2} q_2 = 0 \quad \ldots\ldots\ldots\ldots(2);$

for CDHG, $\qquad \dfrac{1}{C_1} q_1 + \dfrac{1}{C_0} q_0 - \dfrac{1}{C_2} q_2 = 0 \ldots\ldots\ldots\ldots\ldots(3).$

Eliminating q_0 from (1) and (2) by use of (3):

$$\begin{cases} \left[\left(L_1 + \dfrac{L_1 C_0}{C_1}\right) \mathbf{D}^2 + \dfrac{1}{C_1}\right] q_1 = \dfrac{L_1 C_0}{C_2} \mathbf{D}^2 q_2 \quad \ldots\ldots\ldots(4), \\ \left[\left(L_2 + \dfrac{L_2 C_0}{C_2}\right) \mathbf{D}^2 + \dfrac{1}{C_2}\right] q_2 = \dfrac{L_2 C_0}{C_1} \mathbf{D}^2 q_1 \quad \ldots\ldots\ldots(5). \end{cases}$$

---

* It is convenient to have some epithets to distinguish the class of coupling of Figs. 56 from that of Figs. 57. *Internal* and *external* will serve, although these descriptions are not logically explicit.

7. METHODS OF COUPLING CIRCUITS

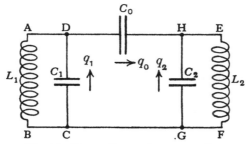

Fig. 58. External capacitance coupling.

Differentiating twice:

$$\begin{cases} \left[\left(L_1+\dfrac{L_1 C_0}{C_1}\right) \mathbf{D}^4 + \dfrac{1}{C_1}\mathbf{D}^2\right] q_1 = \dfrac{L_1 C_0}{C_2}\mathbf{D}^4 q_2 \quad \ldots\ldots\ldots(6), \\ \left[\left(L_2+\dfrac{L_2 C_0}{C_2}\right) \mathbf{D}^4 + \dfrac{1}{C_2}\mathbf{D}^2\right] q_2 = \dfrac{L_2 C_0}{C_1}\mathbf{D}^4 q_1 \quad \ldots\ldots\ldots(7). \end{cases}$$

Eliminating q_2 from (7) by use of (6) and (4):

$$[\{L_1 L_2 (C_1+C_0)(C_2+C_0) - L_1 L_2 C_0^2\}\mathbf{D}^4 + \{L_1(C_1+C_0) \\ + L_2(C_2+C_0)\}\mathbf{D}^2 + 1] q_1 = 0 \quad \ldots\ldots(8).$$

The frequencies are then found as those values of p which will make $q_1 = A(\sin pt + \alpha)$ satisfy this equation; and since

$$\mathbf{D}^2[A\sin(pt+\alpha)] = -p^2[A\sin(pt+\alpha)],$$

we have only to substitute $-p^2$ for $\mathbf{D}^2$ and p^4 for $\mathbf{D}^4$, and to solve as a quadratic in p^2. The coefficients are, however, at present unwieldy, and we must endeavour to compress them*.

---

* From the physical standpoint it is obvious that we should like to find p_a^2 and p_b^2 in terms only of the frequencies $\dfrac{\omega_1}{2\pi}$ and $\dfrac{\omega_2}{2\pi}$ of two single circuits, and some number k^2 expressing the closeness of their connection. Doubtless the most convenient and interesting single circuits to select from the meshwork ABCGFEHD would be ABCD and EFGH, for these offer the advantage of being real circuits existing before the coupling condenser C_0 was introduced ($C_0=0$) and having frequencies independent of C_0. However, in the general case where ABCD and EFGH are *any* circuits of low damping, it seems impossible to do this; and we must adopt whatever expressions for ω_1^2 and ω_2^2 will produce the desirable algebraic simplification. In an interesting chapter on "Special Circuits" in L. S. Palmer's *Wireless Principles and Practice* (1928), a number of formulas for frequencies of coupled circuits are collected and discussed; but they refer only to special cases, such as $L_1 C_1 = L_2 C_2$.

Writing

$$\omega_1^2 \equiv \frac{1}{L_1(C_1+C_0)}, \quad \omega_2^2 \equiv \frac{1}{L_2(C_2+C_0)}, \quad k^2 \equiv \frac{C_0^2}{(C_1+C_0)(C_2+C_0)},$$

on dividing throughout by $L_1L_2(C_1+C_0)(C_2+C_0)$ equation (8) simplifies at once to

$$(1-k^2)p^4 - (\omega_1^2 + \omega_2^2)p^2 + \omega_1^2\omega_2^2 = 0;$$

$$\therefore p_a^2, p_b^2 = \frac{(\omega_1^2+\omega_2^2) \pm \sqrt{(\omega_1^2+\omega_2^2)^2 - 4(1-k^2)\omega_1^2\omega_2^2}}{2(1-k^2)}.$$

It is to be remembered that this formula, and the others like it given in Figs. 55, 56 and 57, are true for circuits constructed with

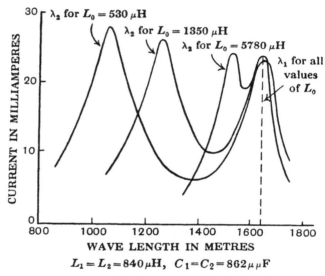

Fig. 59. Resonance curves with external inductance coupling.

any inductance and capacitance values; the only approximation that has been made is that the resistances actually present are small enough to have no appreciable effect on the natural frequencies. The general results may, of course, be applied to special cases. An interesting special case is when $L_1C_1 = L_2C_2 = LC$ (say). The values of p^2 are then easily found from the quadratic equation for p^2. These also are given in the Figures. It is to be noticed that in all four couplings of Figs. 56 and 57 (but not in that of Fig. 55) one of the natural frequencies of the coupled circuits is then $\dfrac{1}{2\pi\sqrt{LC}}$, i.e. is independent of the value of the coupling inductance or capacitance. This is illustrated in Fig. 59*, which shows some resonance curves found experimentally for the case of external inductance coupling (Fig. 57 B).

8. The steady state in compound circuits

8 (a). *Method of solution.* In calculating the final state in circuits subjected to a continued sinoidal E.M.F.†, only the particular integrals of the differential equations are required; consequently the differential equations may be dispensed with altogether. Thus if the sinoidal current of instantaneous value i flows between A and B in Fig. 60, the P.D. between A and B is

$$e = L\frac{di}{dt} + Ri + \frac{1}{C}\int i\,dt.$$

Fig. 60.

But since i is of the form $i = I \sin pt$, the operator $\dfrac{d}{dt}$ (or **D**) has the effect of multiplying by p and advancing in phase by $90°$; and conversely $\left(\dfrac{1}{p}, -90°\right)$ with the operator $\int (\ )\,dt$ (or $\mathbf{D}^{-1}$). This is expressed by the familiar rotating vector, or clock, diagram included in Fig. 60; or by the algebraic equivalent

$$E = \left(jpL + R + \frac{1}{jpC}\right)I,$$

where E and I here stand for the vector quantities (and not for the mere amplitudes), and the operator j effects an advance in phase of $90°$ and may be treated algebraically as $j \equiv \sqrt{-1}$. The use of the $j \equiv \sqrt{-1}$ reduces the calculations to simple algebra.

* From L. S. Palmer, *loc. cit.* p. 97; by permission of the publishers.

† Or to any periodic E.M.F., since this may be analysed into a Fourier series of continued sinoidal terms, the effects of which may be calculated separately and summed.

The investigation of the steady state by this algebraic method is always straightforward, and in numerical calculations the working need never be cumbersome; but when working in algebraic symbols the expressions often become very unwieldy*. Fortunately the approximation $R^2 \ll p^2 L^2$, i.e. $\delta^2 \ll 4\pi^2$, which is permissible in most wireless circuits, often introduces great simplifications. The method is illustrated below by the derivation of formulas for the steady-state condition in two circuits of practical interest.

8 (b). *Mutual coupling for maximum secondary current.* In the arrangement of Fig. 61, taking C_1, C_2 and M as variable, let us find the conditions for obtaining the largest secondary current I_2 with a given E.M.F. E, $\dfrac{p}{2\pi}$ impressed in the primary circuit.

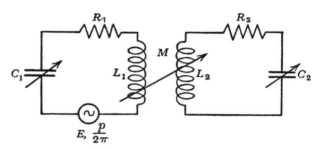

Fig. 61.

$$\begin{cases} \left(R_1 + jpL_1 + \dfrac{1}{jpC_1}\right) I_1 + jpMI_2 = E, \\ \left(R_2 + jpL_2 + \dfrac{1}{jpC_2}\right) I_2 + jpMI_1 = 0. \end{cases}$$

$$\therefore \; -\left(R_1 + jpL_1 + \dfrac{1}{jpC_1}\right) \dfrac{R_2 + jpL_2 + \dfrac{1}{jpC_2}}{jpM} I_2 + jpMI_2 = E,$$

i.e. $\left[\dfrac{\delta_1}{\pi} + j\left(1 - \dfrac{1}{p^2 L_1 C_1}\right)\right] \left[\dfrac{\delta_2}{\pi} + j\left(1 - \dfrac{1}{p^2 L_2 C_2}\right)\right] + \dfrac{M^2}{L_1 L_2}$

$$= -\dfrac{jM}{pL_1 L_2} \cdot \dfrac{E}{I_2}.$$

* Thus $\dfrac{(1+j2)(3+j4)}{(5+j6)} = (0{\cdot}57 + j\, 1{\cdot}31)$; but

$\dfrac{(a+jb)(c+jd)}{e+jf} = \dfrac{1}{e^2+f^2}[(ace - bde + adf + bcf) + j(ade + bce - acf + bdf)].$

8. THE STEADY STATE IN COMPOUND CIRCUITS

For choice of C_1 and C_2, $\dfrac{E}{I_2}$ has minimum magnitude when $p^2 L_1 C_1 = 1 = p^2 L_2 C_2$. Then

$$\frac{E}{I_2} = j p L_1 L_2 \frac{\dfrac{\delta_1 \delta_2}{\pi^2} + \dfrac{M^2}{L_1 L_2}}{M}.$$

For choice of M this is a minimum when

$$\frac{M^2}{L_1 L_2} = \frac{\delta_1 \delta_2}{\pi^2},$$

i.e. when the coefficient of coupling $k \equiv \sqrt{\dfrac{M^2}{L_1 L_2}} = \dfrac{\sqrt{\delta_1 \delta_2}}{\pi}$.

In ordinary wireless circuits δ_1 and δ_2 may commonly have values such as $0\cdot031$; with these values, for maximum secondary current $k = \tfrac{1}{100}$. That is to say, the greatest power is transferred to the secondary circuit when the mutual inductance between primary and secondary is only 1 °/₀ of the greatest value it can have (viz. $\sqrt{L_1 L_2}$)—a value which sensibly does obtain with the iron-cored transformers of ordinary power circuits. Although we are here dealing with the same steady-state sinoidal conditions with which the low-frequency power engineer is for the most part concerned, the fact that our circuits are oscillatory circuits of low damping has changed the desirable coefficient of coupling from unity to 1 °/₀. Expressed otherwise, transformer magnetic leakage, ignored as a first approximation in ordinary transformer calculations, is here a controlling feature.

8 (c). *Impedance of a rejector circuit.* A rejector circuit* AB of rather general form is shown in Fig. 62. Let E be the P.D. of frequency $\dfrac{p}{2\pi}$ maintained between A and B; and let I_1, I_2 be the currents in the C and D paths respectively. It is required to find the impedance Z between A and B.

For circuit ACB,

$$\left(\frac{1}{jpC} + R + jpL'' \right) I_2 - jpMI_1 = E \quad \ldots\ldots\ldots\ldots(1).$$

For circuit ADB, $\qquad jpLI_1 - jpMI_2 = E \quad \ldots\ldots\ldots\ldots(2).$

---

* So called because, as will be seen, it offers a very large impedance between the two points AB to a P.D. of special frequency maintained across these points.

From (1) and (2):
$$\left[R + j\left(pL'' + pM - \frac{1}{pC}\right)\right]I_2 = jp(L' + M)I_1 \quad \ldots\ldots(3).$$

From (2) and (3):
$$E = jpI_2\left[-M + L'\frac{R + j\left(pL'' + pM - \frac{1}{pC}\right)}{jp(L' + M)}\right] \quad \ldots\ldots(4).$$

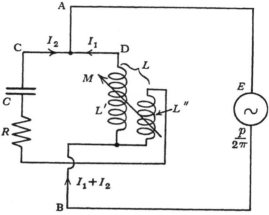

Fig. 62. A rejector circuit.

From (3) and (4):
$$Z \equiv \frac{E}{I_1 + I_2} = \frac{-p^2(L'L'' - M^2) + \dfrac{L'}{C} + jpL'R}{jp(L' + L'' + 2M) - \dfrac{j}{pC} + R},$$

$$= \frac{\left[\dfrac{L'}{C} - p^2(L'L'' - M^2)\right] + j[pL'R]}{[R] + j\left[p(L' + L'' + 2M) - \dfrac{1}{pC}\right]},$$

$$= pL'\frac{\left[\dfrac{1}{pC} - p\left(L'' - \dfrac{M^2}{L'}\right)\right] + j[R]}{[R] + j\left[pL - \dfrac{1}{pC}\right]} \quad \ldots\ldots\ldots\ldots(5),$$

where $L \equiv L' + L'' + 2M$. So far this analysis is accurate.

8. THE STEADY STATE IN COMPOUND CIRCUITS

Let $p_0^2 \equiv \dfrac{1}{LC}$. If $R^2 \ll p_0^2 L^2$, i.e. if $\delta^2 \ll 4\pi^2$, at $p \eqsim p_0$ the magnitude of the denominator changes rapidly with p or L or C, passing through its minimum value, R, when $pL = \dfrac{1}{pC}$; whereas the magnitude of the numerator does not change rapidly, since $\dfrac{M^2}{L'}$ cannot exceed L''. Hence the magnitude of the impedance is nearly a maximum when $p = p_0$. Then

$$Z_0 = p_0 L' \frac{\left[\dfrac{1}{p_0 C} - p_0 \left(L'' - \dfrac{M^2}{L'}\right)\right] + j[R]}{R},$$

$$= p_0 L' \frac{\left[1 - \dfrac{L'L'' - M^2}{L'L}\right] + j\left[\dfrac{\delta}{\pi}\right]}{\dfrac{\delta}{\pi}},$$

$$\eqsim p_0 L' \frac{\pi}{\delta}\left(1 - \frac{L'L'' - M^2}{L'L}\right) \quad \ldots\ldots\ldots\ldots\ldots\ldots(6),$$

i.e. a pure resistance many times as great as the reactance of the coil L' bridging A and B.

Near resonance, when $p = p_0(1 - x)$ (say), where $x \ll 1$ (in magnitude),

$$Z_x \eqsim Z_0 \frac{R}{R + j\left(pL - \dfrac{1}{pC}\right)},$$

$$= Z_0 \frac{R}{R - j 2xp_0 L},$$

$$= \frac{Z_0}{1 - j\dfrac{2\pi x}{\delta}} \quad \text{where} \quad \delta \equiv \frac{R}{2n_0 L} \quad \ldots\ldots\ldots\ldots(7).$$

It is often convenient to express the impedance Z_x as a resistance $R_{\text{ser.}}$ and a reactance $X_{\text{ser.}}$ in series, i.e. to write

$$Z_x = R_{\text{ser.}} + jX_{\text{ser.}}.$$

Equation (7) shows that

$$R_{\text{ser.}} = aZ_0 \text{ and } X_{\text{ser.}} = bZ_0,$$

where $\quad a = \dfrac{1}{1 + 4\pi^2 \left(\dfrac{x}{\delta}\right)^2} \quad$ and $\quad b = \dfrac{2\pi \dfrac{x}{\delta}}{1 + 4\pi^2 \left(\dfrac{x}{\delta}\right)^2}.$

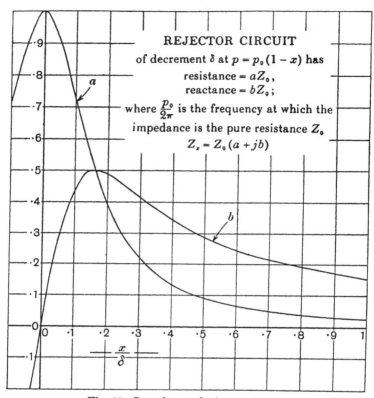

Fig. 63. Impedance of rejector circuit.

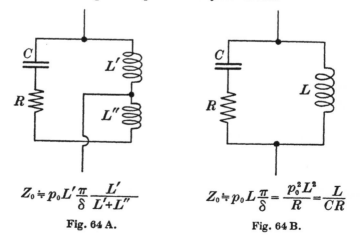

Fig. 64 A.   Fig. 64 B.

$$Z_0 \doteqdot p_0 L' \frac{\pi}{\delta} \frac{L'}{L'+L''} \qquad Z_0 \doteqdot p_0 L \frac{\pi}{\delta} = \frac{p_0^2 L^2}{R} = \frac{L}{CR}$$

REJECTOR CIRCUIT of decrement δ at $p = p_0(1-x)$ has
resistance $= aZ_0$,
reactance $= bZ_0$;
where $\frac{p_0}{2\pi}$ is the frequency at which the impedance is the pure resistance Z_0
$$Z_z = Z_0(a+jb)$$

The coefficient b is a maximum, viz. $\frac{1}{2}$, when $2\pi\frac{x}{\delta}=1$; that is, the reactance of the rejector circuit is a maximum (+ or −), viz. $\frac{1}{2}Z_0$, when the distuning fraction x (+ or −) is equal to $\frac{1}{2\pi}$ times the decrement of the circuit. Values of a and b are plotted in Fig. 63.

Two special cases of the general circuit of Fig. 62 are very common. When $M=0$, we have Fig. 64A; and when also $L''=0$, we have Fig. 64B.

With the approximation already used, viz. $\delta^2 \ll 4\pi^2$, it is of no significance where the resistance R is located in the circuit, or whether the tuning (near resonance) is effected by adjustment of p or L or C. If $\delta = 0.031$, $\frac{\pi}{\delta} = 100$, and at the resonant frequency the impedance of the combination in Fig. 64B is a resistance 100 times the reactance of the coil or the condenser alone.

9. MECHANICAL MODEL OF COUPLED OSCILLATORY CIRCUITS

The mathematical analysis of the behaviour of coupled circuits is either so complicated or so incomplete that a simple mechanical model with which one can play may be of value in throwing light on the phenomena, and even in obtaining solutions to numerical problems.

Fig. 65 shows a pendulum model, very easily constructed, with which a great variety of experiments can be made. If the length of a pendulum is l, and the bob has mass m, weight mg, and horizontal displacement s, it represents an electric circuit LCR thus:

Electrical	Mechanical
q	s
i	$\frac{ds}{dt}$
L	m
C	$\frac{l}{mg}$
R	Size of damping vane

The pendulums may be about a metre long, and when they represent circuits accurately tuned together, each can conveniently be adjusted to have a period of exactly two seconds. Since the ratio $\frac{L}{C}$ is proportional in the model to m^2, a change of $\dfrac{L}{C}$ without change of LC is readily represented. More or less damping can be added by means of paper wind-vanes fixed to the strings. The coupling between the two oscillators is increased by lowering the point where the two strings are tied together, being zero when they are not so tied provided that the supporting beam is sensibly rigid. The type of coupling represented is internal capacitance coupling (Fig. 56 B); but this is probably not of much significance. If the bob of one pendulum is made much more massive than the other, and is lightly damped, a very close approximation to a steady (undamped) primary oscillation is obtained.

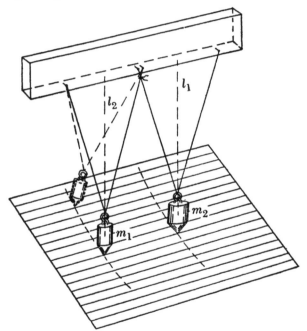

Fig. 65. Coupled pendulums.

CHAPTER V

THE PRODUCTION OF HIGH-FREQUENCY CURRENTS

1. Spark Transmitters

1 (a). *General.* Our study in the last Chapter of the properties of a circuit comprising capacitance and inductance—and not too much resistance—has shown that it is only necessary to disturb such a circuit in any way, and then to leave it alone, to obtain an oscillation of any frequency desired. The current so obtained is not, it is true, a steady alternating current, but a damped oscillation as portrayed in Fig. 48. However, if the damping is slight, a succession of damped oscillations is for many purposes substantially equivalent to a steady oscillation of the same frequency.

The first wireless waves ever observed—by D. E. Hughes in 1879—were produced by this method; it was the only method for many years after wireless telegraphy had reached a commercial stage; and it remains in use in a large fraction—reckoned by number, though not by power—of wireless stations to-day*. When only very small power is involved, the requisite disturbance of the oscillatory circuit can be effected by some repeated switching operation, such as occurs at the vibrating contacts of a buzzer (Fig. 71); but for the powers involved in producing the high-frequency currents required in transmitting stations a spark gap, repeatedly ruptured by a spark and restored, is always resorted to.

Under ordinary conditions, air and other gases are almost perfect insulators; but when subjected to sufficiently intense electric stress a gas becomes ionised and allows a current to flow. High temperature of the gas, and the projection into it of electrons emitted from a hot solid body or otherwise, tend to produce or maintain the ionisation. If therefore the P.D. between two electrodes spaced a short distance apart in air is gradually raised from a low value, at first no current flows between them; but when the P.D. becomes so great that the electric field is strong enough to ionise the gas, a current begins to flow, and the effect of this current is itself to increase the conductance of the air gap. If the ratio between the P.D. and the current across the gap is called the resistance of the gap, the resistance is a

* But the spark method is obsolescent. An international conference at Washington in 1927 recommended the abolition of all spark transmitters by the end of 1939.

108 V. THE PRODUCTION OF HIGH-FREQUENCY CURRENTS

function of the current, decreasing as the current increases*. As long as the current continues, the state of ionisation is maintained despite the cessation of the intense electric field required to initiate the ionisation. Thus in order to start a discharge across a $\frac{1}{4}''$ gap between copper electrodes, a P.D. of nearly 20,000 volts might be needed; but once started, a current of 10 amperes would be maintained by a P.D. of only about 50 volts. If the current ceases, even for a very brief period, the ionisation disappears, and the insulating properties of the gap are recovered.

A spark gap may therefore be used as a form of high-tension switch which, without moving parts, closes automatically when the P.D. across it rises to the disruptive value, and opens again when the current through the spark becomes too small to maintain the ionisation on which the conductivity of the gap depends. Thus a spark gap may replace the mechanical switch S in Fig. 47. Provided that the P.D. E of the battery or other source charging the condenser C_1 is high enough to jump the spark gap, as C_1 charges up the P.D. across the gap gradually rises towards the value E. When it reaches the disruptive value† the gap breaks down, and its resistance falls from a sensibly infinite value to some low value of the order of an ohm. The condenser C_1 thereupon discharges through $L_1 R_1$ and the gap with a damped oscillation very much as in Fig. 48‡. When the oscillation has died down to a very small amplitude, there is nothing to maintain ionisation, and the insulating property of the gap is restored. When the condenser is again sufficiently charged up, a new spark occurs, and the whole process repeats itself indefinitely.

If the battery and series resistance of Fig. 47 are replaced by the more convenient spark coil, or alternator with step-up transformer, the arrangement becomes a spark generator, which until a few years ago was by far the commonest of all forms of

* Hence the familiar "negative resistance" of the ordinary electric arc. See Subsection 4 (a) in this Chapter.

† The sparking P.D. depends somewhat on the shape of the electrodes, and is roughly proportional to the length of the gap. For air under ordinary atmospheric conditions, the P.D. necessary to jump a $\frac{1}{4}''$ gap varies from about 8000 volts between sharp needle points to about 20,000 volts between large smooth spheres.

‡ Any difference is due to the fact that the resistance of the gap is not constant, but depends upon the current flowing, and on other factors determining the state of ionisation. The amplitude therefore does not fall off strictly according to the exponential law of Fig. 48; but this forms a sufficiently close approximation for most purposes.

1. SPARK TRANSMITTERS

generator of high-frequency current*. The circuits of a typical simple spark transmitter are shown in Fig. 66. A is a low-frequency alternator (of say 50–500 periods per second) connected at will through the key K to the step-up transformer T which provides current at high tension to charge the condenser C_1. When a spark occurs across the gap G, the high-frequency oscillation in the spark circuit $L_1 C_1 G$ induces a powerful oscillation in the tuned antenna (secondary) circuit by virtue of the mutual inductance M between the two circuits. Energy is radiated from the antenna in all directions over the surface of the Earth at a rate which is proportional to the square of the current $\mathscr{I}_2$ indicated by a hot-wire ammeter inserted at the base of the antenna circuit.

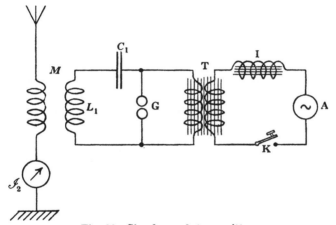

Fig. 66. Simple spark transmitter.

According as the coefficient of coupling† between the spark circuit and the antenna circuit is small or large, the oscillations in the two circuits are as in Fig. 50 or Fig. 53 respectively. The condition of Fig. 53 is objectionable, and in fact forbidden by International Convention, because the radiation contains waves of two lengths, only one of which can be fully utilised at the receiving station, while the other may be an unnecessary nuisance to other receiving stations within range. On the other hand, if the coupling

* Witness the German *Funkentelegraphie* (*Funk* = spark), which was almost synonymous with *wireless telegraphy*. German engineers seem unable to abandon this word, and *Rundfunk* is the absurd name commonly used for the broadcast telephony service.

† See p. 91.

is very loose, the transference of energy from spark circuit to antenna is very slow, and poor efficiency results; for as long as oscillation continues in the spark circuit power is being wasted in the spark gap, whose resistance is far from negligible. Thus to get rid of the energy quickly tight coupling is desired; but to keep the singleness of wavelength loose coupling is necessary. The compromise adopted may be a coefficient of coupling $\dfrac{M}{\sqrt{L_1 L_2}}$ in the region of 6 per cent.; in which case, if two wavelengths were detectable at all, they would be about 97 per cent. and 103 per cent. of the common wavelength of the two circuits when separated. Actually, however, with the coefficient of coupling used, the antenna circuit generally has a decrement high enough to prevent two peaks from showing in the resonance curve.

As soon as the oscillation has died away in the spark circuit— e.g. in $\dfrac{1}{10,000}$ second or so—the gap de-ionises and is again an insulator; the condenser C_1 (Fig. 66) receives a fresh charge from the alternator; and the process repeats itself as long as the key is held depressed. The Morse dots and dashes thus consist of short or long series of sparks, each spark setting up in the antenna a brief and violent oscillation followed by a relatively lengthy period of quiescence before the next spark occurs. The rate at which the sparks follow each other depends on the time taken to recharge the condenser to the sparking voltage, and this depends on the alternator frequency and on the length of the gap. The older spark transmitters made use of alternators of ordinary commercial frequency such as 50 c/s; the spark gap is then set so that it breaks down at a voltage much less than the maximum which would be produced during a non-sparking cycle, and several sparks occur at irregular intervals during the half-cycle.

A trouble which is liable to occur is the formation of an arc across the spark gap; that is, after the gap has been rendered conducting by the disruptive spark and high-frequency oscillation, an arc may be established, fed directly from the transformer and alternator, thus wasting power and preventing the recovery of the insulating property of the gap. The iron-cored inductance I shown in Fig. 66 inserted between alternator and transformer tends to inhibit such arcing by retarding the re-charge of the condenser during and immediately after the oscillatory discharge through the

1. SPARK TRANSMITTERS

spark gap. An additional measure against arcing which has sometimes been adopted is the use of a rotary spark gap, in which the electrodes are made to approach and separate at high speed several times per half period of the alternator. A rotary gap tears out any incipient arc, and the incidental ventilation and cooling tend to prevent any arc from forming. Such a gap, moreover, tends to equalise the intervals between successive sparks.

Some notion of practical dimensions in a simple spark transmitter may be got from the following figures referring to a typical example

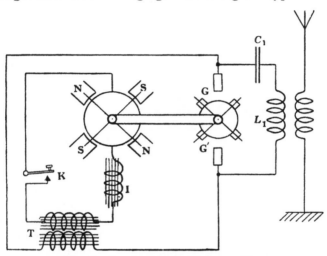

Fig. 67. Synchronous spark transmitter.

of a kind which has been much used in small ship or shore stations for commercial work.

Wavelength = 600 m.   Frequency = 5×10^5 c/s.
Alternator output = $1\frac{1}{2}$ kilowatts at 50 c/s.
Average spark rate = 500 per sec.
Sparking P.D. = 10,000 V. $C_1 = 0.05$ μF.
Antenna current = 6 A (R.M.S.).

Plate V shows a form of fixed spark gap used in such a transmitter. The high-tension condenser C_1 might be constructed of alternate glass and zinc plates in a galvanised iron tank filled with oil. In transmitters of low power, mica condensers sealed in wax are often used.

1 (b). *Synchronous sparking.* In a more modern form of spark transmitter, the spark takes place at a rotary gap mounted on the

shaft of the alternator, as illustrated diagrammatically in Fig. 67. By making the number of moving electrodes equal to the number of poles, one spark per half cycle is obtained. The alternator frequency is commonly 200–500 c/s*.

The working condition to be aimed at is that in which the spark occurs at an instant of maximum condenser P.D., when the charging current from the transformer is zero. It may be seen from IV-2 (d) that this condition is realised if the alternator circuit (i.e. the compound circuit $AITC_1$ of Fig. 66) resonates with the alternator frequency, and, by adjusting the position of the stationary electrodes, a spark is made to occur at the instants $0, \dfrac{1}{2n}, \dfrac{1}{n}$ ... etc., where n is the frequency of the alternator. Referring to Fig. 34, a spark occurs, and a fresh start is made with zero charge and zero charging current in the condenser, when the curve first meets the axis. Alternatively, it may be arranged that the spark shall occur only every several half periods, at the second, third, fourth... etc. passage through zero in Fig. 34†. The required resonance condition is given by $4\pi^2 n^2 s^2 LC_1 = 1$, where s is the step-up turns ratio of the transformer, and L is the inductance of I plus the synchronous inductance of the alternator and the primary equivalent leakage inductance of the transformer (L_1 being, of course, negligible in comparison). By these means, precisely regular sparking is obtained, quite free from arcing, and with a high power-factor at the alternator terminals. Fig. 68‡ depicts a low-frequency half cycle under these conditions; the spark is there supposed to occur exactly at $pt = 180°$, whereupon the whole process repeats itself in the reverse direction.

This system is known as the synchronous spark system; it has been very widely used by the Marconi Co., for powers between ½ kW and upwards of 100 kW. In high-power spark transmitters, the practice has been to insert a signalling switch in series with the high-tension winding of the transformer. The switch is controlled through relays by the Morse key, and is provided with powerful air blasts to minimise arcing at the breaks.

The following figures refer to an installation for transmitting to

* The alternator is often of the inductor type. See Subsection 5 (b) of this Chapter.

† This was the practice in the old high-power, slow-spark, time-signal transmissions at Eiffel Tower.

‡ From L. B. Turner, "The low-frequency circuit in spark telegraphy," *Electrician*, Vol. LXIX (1912), p. 694.

ships on a wavelength of 600 m. Alternator, 300 c/s; 10 kW; 500 V, reducible by field rheostat for reducing power radiated. Step-up transformer of ratio 1 : 21; inductance in primary circuit, 0·0104 H. Spark-circuit condenser, 0·06 μF; inductance, 1·7 μH. Sparking P.D., 23·5 kV; 600 sparks per sec. Power factor across transformer, 0·78.

While the oscillation is taking place, enormous currents flow back and forth across the spark gap, with consequent large wastage

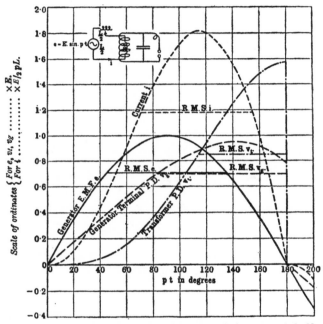

The v_g curve is for case when inductance of alternator is half the total inductance.

Fig. 68. Low-frequency resonance; one spark per half cycle.

of power there. For example, in the typical transmitter specified on p. 111, the amplitude of the spark current at the beginning of the oscillation is 1000 A. The mean resistance of the spark gap during the oscillation might be about half an ohm. The sooner the energy can be got out of the spark circuit, wherein no useful work is done by the current, into the antenna circuit, wherein the ohmic losses can be kept smaller and whence useful radiation occurs, the more efficient the transmitter. We have seen that tightening the coupling hastens the transference of energy from primary to secondary; but that retransference from secondary to primary may then occur (Fig. 53). If a tight coupling could be used, and if the spark circuit could be removed or open-circuited immediately it had passed all its energy into the antenna circuit for the first time, we should have the advantage of tight coupling without its defect.

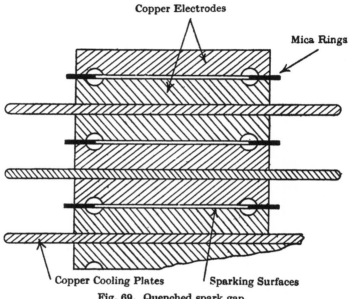

Fig. 69. Quenched spark gap.

1 (c). *Quenched spark.* This is precisely what is effected by the use of a special form of spark gap called a "quenched gap." This form of gap appears to have been introduced by E. von Lepel, but its action was investigated and explained first by M. Wien. The ordinary spark gap, between two small metal electrodes separated ¼ inch or so in open air, does not become de-ionised quickly enough to prevent re-ignition after the first beat node of current across it. The essential feature of the quenched gap is that the rate of de-ionisation is very much raised. In the form of quenched gap developed by the Telefunken Company, the chief exploiters of quenched spark working, this is achieved by subdividing the gap into a number of very short paths between large well-cooled copper or silver surfaces spaced about 0·2 mm apart by mica insulating separators, as shown in Fig. 69. Owing to the shape and extent of

the sparking surfaces, and their proximity to every portion of the ionised gas, the cooling and therefore the de-ionising of the gas are extremely rapid. Moreover, enclosing the spark gap from the open atmosphere with only very small volume of air within is found to add to the quenching effect, presumably owing to the sudden rise of gas pressure when the spark occurs. About 1200 V per gap is required to produce a spark, and as many gaps are used in series as will bring the total sparking P.D. nearly up to the maximum P.D. available. A complete spark gap of this type is shown in Plate VI.

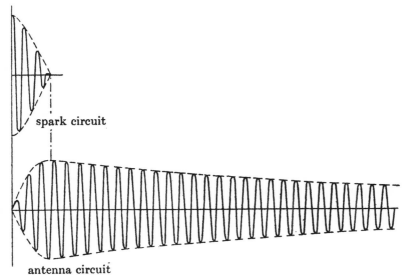

Fig. 70. Oscillation with quenched spark.

Any desired number of the component gaps can be short-circuited by spring clips inserted between the cooling plates; three of the seven gaps are so short-circuited in the photograph.

With such a quenched gap, it is found practicable to use a coupling coefficient of about 20 per cent. and yet avoid re-ignition after the first beat node. The spark circuit is then extinguished after about $2\frac{1}{2}$ periods, leaving a large proportion of the initial energy in the antenna circuit, which then oscillates with its own independent wavelength and decrement. Fig. 70 shows approximately the trains of oscillation in the two circuits for the case of antenna circuit decrement 0·06, a practical value. The current scales

116 V. THE PRODUCTION OF HIGH-FREQUENCY CURRENTS

in the Figure bear no relation to each other; the relative values of the currents must depend, of course, on the relative values of the capacitances of the primary and secondary circuits.

Resonance of the alternator circuit is even more important in a transmitter of this type than with the rotary synchronous unquenched spark gap previously described. For with the fixed gap there is no mechanical determination of the instant of sparking, as there is with the rotary gap; moreover, any trace of arcing would obviously prevent the rapid de-ionisation or quenching which is required.

Various forms of quenched spark gap have been constructed beside the one illustrated in Fig. 69 and Plate VI. The sparking surfaces have been made to move at high speed relative to each other; and the gap has been flooded with oil or other liquid, or hydrogen. In all, the object aimed at is to get the gap de-ionised within a very few periods. Anti-arcing devices are sometimes confused with quenching devices proper, but they should be clearly distinguished. Calculation shows at once that, except with the small attenuation factors $\epsilon^{-\frac{R}{2L}t}$ * associated with very great wavelengths, it is not practicable to quench early in the oscillation by the mechanical separation of the electrodes in rotary gaps, or by the violent air blasts sometimes employed to ventilate spark gaps. Thus if the wavelength is 1000 m (3×10^5 c/s), and the primary oscillation is to last for about three cycles as in Fig. 70, the spark must be quenched $\frac{1}{100,000}$ second after it is ignited. The prevention of arcing is another matter. Here the concern is, as it were, to sweep out the gap after one oscillation has finished and before another begins; and except with large wavelengths a large proportion of the spark period is available for so doing.

A non-quenched synchronous-spark ship transmitter of $1\frac{1}{2}$ kilowatts is shown in Plate VII. A rotary converter (D.C. to A.C., 500 c/s), with spark gap mounted on the shaft, is in the bottom compartment. On the panel front are mounted the remote-control starter, two wheel-handles for adjustment of wavelength and power, and the antenna ammeter. The high-frequency transformer (below) and antenna loading inductance (above) are seen within the cabinet.

* Or $\epsilon^{-\delta \cdot N}$, as on p. 86.

2. Excitation by buzzer

In making tests of apparatus, and in laboratory measurements, it is often desired to stimulate an oscillatory circuit into feeble damped trains of oscillation. This is conveniently done by means of a *buzzer* (small electro-magnetic vibrator). Portable wavemeters are often fitted with such a device.

In the circuit of Fig. 47, when (on p. 108) E was of the order of thousands of volts, we caused S repeatedly to open and close by constructing it as a spark gap; and if E is the 10,000 V and C_1 the 0·05 μF of our previous numerical example (p. 111), the energy $\frac{1}{2}C_1E^2$ released per spark is $2\frac{1}{2}$ W sec. But if E is (say) only 10 V, S may oe a mechanically vibrating pair of platinum-tipped contact points. The oscillatory energy per make is now $2\frac{1}{2}$ μW sec. If this energy could actually be released as an oscillation in $L_1C_1R_1$ it would be more than adequate for many measurements; but there would in practice be grave difficulty in effecting the transition from no contact to good contact quickly enough to avoid losing most of the oscillatory energy at the contact in the first few millionths of a second. It is easier to break a circuit very suddenly than to make it very suddenly. Accordingly in using a buzzer as the vibrating contact exciting oscillations of high frequency, it is usually preferable to cause it to break suddenly the path of a steady current I flowing in the inductance L, leaving the latter with the energy $\frac{1}{2}LI^2$ to be dissipated as an oscillation in the LC circuit. A suitable disposition is shown in Fig. 71*.

The exact course of the oscillation consequent on the sudden opening of the contact is given by the solution of the differential equation

$$L\frac{d^2q}{dt^2} + R\frac{dq}{dt} + \frac{1}{C}q = 0,$$

as on p. 65, but now with the initial conditions

$$t = 0, \quad q = 0, \quad i = I.$$

---

* When, as in the conveniently simple arrangement shown in Fig. 71, a single contact breaks both the current through L and the current through the buzzer magnet, it is essential to shunt the magnet by a resistance r to prevent the formation of a large (arcing) P.D. across the opening contact. An attempt to make the current in the highly inductive magnet fall rapidly to zero would necessarily fail. But even when such arcing is suppressed it seems that serious damping of the oscillatory circuit, owing to transitional poor conductance at the buzzer contact, is very liable to occur with frequencies above about 200 kc/s. See L. B. Turner, "Wavemeters," *Wireless World and Radio Review*, Vol. xv (1924), p. 381.

With the same approximation of not excessive damping, this yields

$$i \fallingdotseq I\epsilon^{-\frac{R}{2L}t} \cos pt.$$

That the oscillation now starts as a cosine, instead of a sine, function of time has no practical significance for most purposes. In the circuit of our previous example (p. 111), for the frequency 5×10^5 c/s ($\lambda = 600$ m), $L \fallingdotseq 2\mu$H. If the current I flowing across the buzzer contact at the instant of break is 1A, the oscillatory energy per break is now 1μW sec.

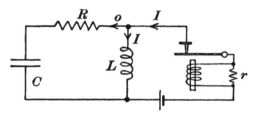

Fig. 71. Excitation by buzzer.

3. Continuous wave methods

We have seen that in spark transmitters a train of waves is radiated each time a spark occurs, and that betweenwhiles the power falls appreciably to zero for relatively long epochs. Since the amplitude of antenna current decreases according to an exponential function of the form ϵ^{-bt}, it never quite reaches zero; and if the damping is made sufficiently small and the sparks recur sufficiently rapidly, a new oscillation may be set up by a new spark before the oscillation from the preceding spark has died away to a negligible amplitude. If the new spark were made to occur at the right instant, the old and the new oscillations would be in phase, and a continuous oscillation more or less fluctuating in amplitude would be produced*. The antenna current would then approximate to a steady alternating current, the limiting case of a damped oscillation of vanishingly small decrement, and

---

* The practical difficulties in realising this process are very great, and could only be overcome with very long waves. The Marconi Co. did, however, build several high-power, long-wave stations of this type, e.g. at Stavanger (Norway) and Carnarvon. The system was a development from the synchronous rotary spark gap. It was a remarkable engineering achievement, but much better methods are now available.

3. CONTINUOUS WAVE METHODS

the advantages to be got from sharpness of tuning, discussed in IV-2 (e) and IV-5(c), would be developed to the maximum. The radiation in space corresponding to a dot or a dash would no longer consist of separated trains of waves, but would be of sensibly constant amplitude throughout the dot or dash except during the short periods of growth and dying away immediately following the depression and release of the signalling key. Such radiation is called *continuous wave* radiation, commonly abbreviated to *C.W.*; and the systems of wireless telegraphy in which the oscillation is continuous during the Morse dot or dash are called undamped or C.W. systems, in contrast to damped or spark systems.

Fig. 72 shows the condition during the signalling of a Morse dot, at A by a spark transmitter, and at B by a C.W. transmitter. It is not true to scale in that the true intervals between the separate wave trains in A would usually be much longer than the effective length of the train itself; and that therefore for comparable transmitter powers the amplitude in B would be much smaller than the maximum amplitude in A.

Owing to this very bad load factor in a spark transmitter, in both spark circuit and aerial; to the ever growing difficulties in dealing with higher powers (say above 100 kW) in a spark gap, and in switching such power on and off under the control of a high-speed signalling key; to the superiority of continuous over damped waves as

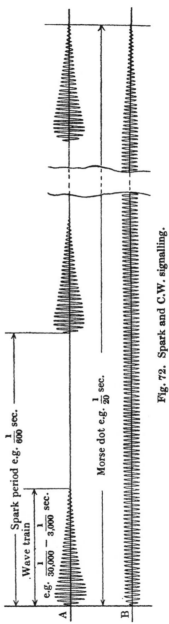

Fig. 72. Spark and C.W. signalling.

regards selectivity and manner of detection at the receiver; and—most recent and potent ground of all—to the need of C.W. to permit of telephonic modulation (as contrasted with telegraphic keying); for such reasons as these, the spark type of high-frequency generator has been replaced in modern practice by C.W. generators of one form or another.

The forms of C.W. generator available fall into three great categories—arc, alternator and thermionic tube (so-called valve). The last category is undoubtedly of preponderating interest and practical importance, on account of the great diversity, technical elegance and adaptability of triode methods. Its treatment is reserved until Chapter IX, following a consideration of the fundamental properties of the triode. Here we give a brief account of arc and alternator generators, both of which are used in many high-power, long-wave telegraphic (Morse) transmitters. They cannot be employed for the production of short waves, for which the triode holds the field; but for long-wave service they may well continue as serious competitors.

4. ARC GENERATORS

4 (a). *The arc as a conductor.* Historically, the first continuous wave generator in a practical form was a development of the "singing arc" of W. Duddell (1900). Duddell never succeeded in raising the frequency beyond the audible range, or in obtaining any but feeble oscillations. V. Poulsen in 1903 modified the arc and succeeded in producing oscillations not only of the high frequencies required for wireless telegraphy but also of relatively high power. The complete theory of the operation of the Poulsen arc is very complex[*]. In the physical conditions of the arc there are many independent variables competent to modify profoundly the cycle of operations, and various distinct modes of oscillation can occur.

The arc is a conductor for which Ohm's law has no significance except to define the term *resistance* as the ratio between P.D. and current—a term worth coining only because in most conductors this ratio is sensibly constant. In a metallic conductor an interchange of electrons between the atoms is always proceeding, and when a P.D. is maintained across the conductor, the current which flows

---

[*] The Poulsen arc has been deeply studied by P. O. Pedersen, who has published a number of papers on the subject in the *Proc. I.R.E.* 1917–1926.

4. ARC GENERATORS

is the drift of these temporarily free electrons under the electric field thereby set up. Whether the P.D. is applied or not, the electrons are there, ready to have their random motion converted into a directed drift through the conductor; and because their number is unaffected by the P.D., the rate of drift, the current produced, is proportional to the P.D. The characteristic curve between P.D. v and current i for such a conductor is a straight line through the origin, such as OA in Fig. 73.

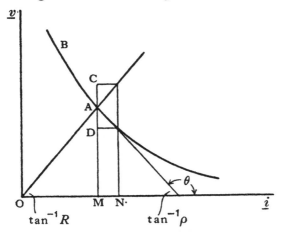

Fig. 73. Static characteristics of metallic conductor and arc.

Conduction through an arc is very different. The arc is a gaseous conductor; or, more precisely, is a body of non-conducting gas in which the carriers of electricity, the ions or wanderers, depend for their existence upon the current they convey. The bigger the current, the more than proportionately better is the conductor, and the curve for the arc is accordingly a curve of negative slope, such as BA in Fig. 72. An increment of current $\delta i = \mathrm{MN}$ in the metallic conductor is accompanied by a rise of P.D. $\delta v = \mathrm{AC} = R \cdot \delta i$, where R is the resistance of the conductor; whereas in the arc, the increment δi is accompanied by a fall of P.D. $\delta v = \mathrm{AD} = -\rho \cdot \delta i$, where $-\rho = \tan\theta = \dfrac{dv}{di}$ is the slope of the characteristic at A. The *slope resistance* or *differential resistance* is negative and of magnitude ρ*.

---

* In Fig. 79, the carbon-carbon arc in air (ordinary arc lamp) is seen to have a slope at 5A of about -2Ω. It is on account of the negative slope resistance that arc lamps are always run with a ballast resistance in series.

122 V. THE PRODUCTION OF HIGH-FREQUENCY CURRENTS

Consider an arc shunted by a resistance R burning steadily, with a total feed current I maintained constant. The constancy could ideally be ensured by feeding from a battery of very high voltage through a very large resistance, as in Fig. 74. Any change δi in arc current must be accompanied by the equal negative change $-\delta i$ in the shunt current; and these changes of current imply changes of P.D.

across AB, of $(-\rho)\,\delta i$, and
across DC, of $R\,(-\delta i)$.

Hence the imagined circulation δi round ABCD has evoked an E.M.F. in the opposite direction of magnitude

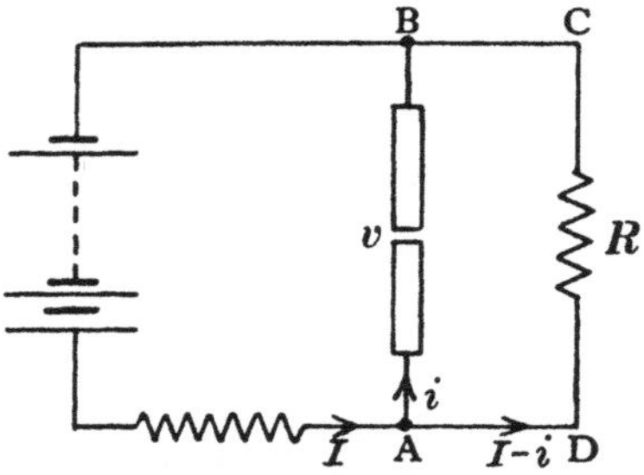

Fig. 74. Arc shunted by resistance.

$$R\,.\,\delta i - \rho\,.\,\delta i.$$

If this is positive there is stability; but if $\rho > R$, the slightest circulation of current between arc and shunt augments itself.

4 (b). *Duddell's singing arc.* In the Duddell singing* arc, the shunt connected across the arc consists of an inductance L and capacitance C in series (Fig. 75), with only so much resistance R as cannot be avoided. The substantial constancy of total feed current is (further) ensured by a large choke in the feed circuit. The characteristic curve of the arc can be represented fairly closely by a hyperbola

$$v = P + \frac{Q}{i_a},$$

where P and Q are constants suitably chosen for the particular arc. The magnitude ρ of the negative slope of the characteristic is then

$$\rho = \frac{Q}{i_a^{\,2}}.$$

Equating the sums of E.M.Fs. round $CARL$ to zero, and differentiating once, we get

$$L\frac{d^2 i_c}{dt^2} + R\frac{di_c}{dt} + \frac{1}{C}i_c - \frac{dv}{dt} = 0.$$

---

* So called because the arc emits a note whose pitch is the frequency of the alternating current through it.

4. ARC GENERATORS

Now
$$v = P + \frac{Q}{I - i_c};$$

$$\therefore \frac{dv}{dt} = \frac{Q \frac{di_c}{dt}}{(I - i_c)^2} = \frac{Q \frac{di_c}{dt}}{i_a^2} = \rho \frac{di_c}{dt};$$

whence
$$L \frac{d^2 i_c}{dt^2} + (R - \rho) \frac{di_c}{dt} + \frac{1}{C} i_c = 0.$$

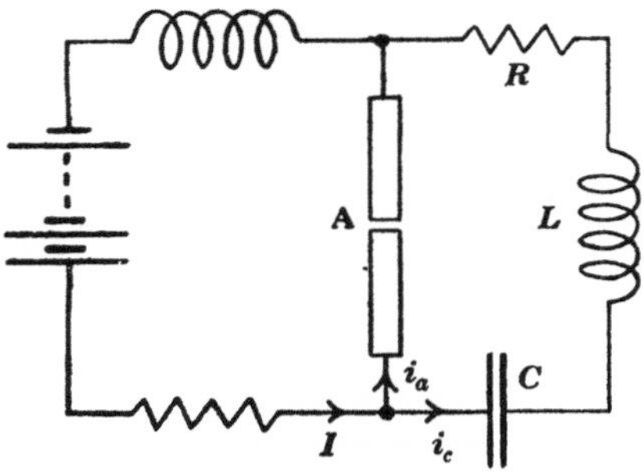

Fig. 75. Arc oscillator.

For small changes i_c of arc current, ρ is sensibly constant*. The equation then shows that the insertion of the arc in the CLR circuit has reduced the resistance of that circuit from R to $(R - \rho)$. If R is small enough, any incipient disturbance must grow as an oscillation of frequency approximately $\frac{1}{2\pi\sqrt{LC}}$, its amplitude increasing according to the exponential law $\epsilon^{(\rho - R)t}$ † until either our approximation of constant ρ no longer holds good or some new physical conditions supervene.

The curve of Fig. 73 is the arc characteristic for steady current. It obviously cannot hold for rapidly varying currents; for ions when formed take time to disappear by convection and recombination, and moreover the thermal condition of the arc at any instant must depend on the currents at preceding instants. At any value of the current i_a when it is decreasing, the temperature will be higher, and the P.D. v_a therefore lower, than the static values corresponding to that current. The kinetic P.D. will thus be higher than the static P.D. when the current is increasing, and lower when

---
* Unless $R \doteqdot \rho$, in which case the term in $\frac{di_c}{dt}$ is unimportant.

† See IV–5 (a).

it is decreasing. The kinetic characteristic of the Duddell oscillating arc therefore assumes some such form as Fig. 76; and it is obvious that the frequency of the oscillation must profoundly influence the behaviour of the arc. With an indefinitely great frequency there could be no negative slope at all, for there would be no time for the amount of ionisation to follow the changes of current.

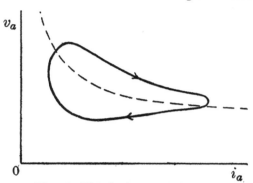

Fig. 76. Kinetic characteristic of arc.

For some such reasons as these, neither large power nor high frequency of oscillation can be got from an ordinary arc lamp as used by Duddell. The values in Duddell's experiments were of the order:

$L = 5\text{mH}$ $C = 1 \text{ to } 4\mu\text{F}$ $R = 0\cdot4\Omega$
R.M.S. current $= 3\text{A}$ Frequency $= 2000\text{–}4000$ c/s.

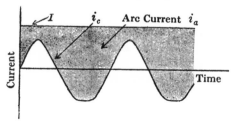

Fig. 77. Duddell arc oscillation.

4 (c). *The Poulsen arc.* The Duddell oscillations are characterised by the fact that the arc is never extinguished, the maximum value of the oscillatory current i_c being less than the steady feed current I, as in Fig. 77. Such arc oscillation is called *oscillation of the first type*, or here for short *α-oscillation*.

4. ARC GENERATORS

The arc became practicable as a generator of high-frequency currents for wireless telegraphy in the hands of V. Poulsen in 1903. The modifications made by Poulsen were threefold, and a Poulsen arc is illustrated diagrammatically in Fig. 78. These modifications are:

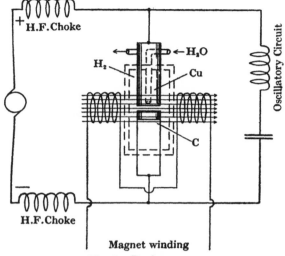

Fig. 78. Poulsen arc.

(i) The arc is enclosed in an atmosphere of hydrogen or coal gas or alcohol, kept as cool as possible by a water jacket.

(ii) The anode is of copper, cooled by internal water circulation.

(iii) A powerful magnetic field is maintained across the arc, acting as a magnetic blast driving the arc away from the electrodes.

Fig. 79 shows the marked effect of (i) and (ii) even on the static characteristic of the arc; and it is to be expected that the effect on the kinetic characteristics at high frequencies will be still more pronounced. The hydrogen atmosphere favours rapid de-ionisation, hydrogen of all gases possessing the highest thermal conductivity and the highest diffusion coefficient.

It is clear that α-oscillation can be much more easily produced with the steeper characteristic of Fig. 79, the limiting value of resistance of the oscillating circuit being proportional, as we have seen, to the steepness of the curve. An oscillating arc must always start up with the α-oscillation; but with the Poulsen arc, the oscillation once started builds up into what is called *oscillation of the second type*, or here for short *β-oscillation*, characterised by the oscillatory current having so large an amplitude that during part of the cycle the arc is actually extinguished. It is re-ignited later in the cycle at the moment when the P.D. between copper and carbon reaches the ignition or sparking value, which greatly exceeds the burning value—the more so the more completely de-ionisation has been effected during the period of extinction.

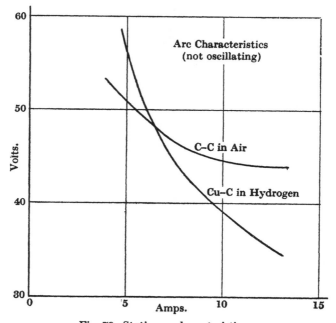

Fig. 79. Static arc characteristics.

It must be remembered that the presence of the (indefinitely large) chokes between the arc and the D.C. supply terminals removes any limitation imposed by the supply voltage on the instantaneous value of the arc P.D. The arc can call upon the chokes for an E.M.F. of any value by producing an appropriate (small) rate of change of current in them. Whereas in the α-oscillation there can be no abrupt changes in the current or P.D., since the representative point on the arc characteristic never leaves the smooth curve, in the β-oscillation abrupt changes do occur.

4. ARC GENERATORS

In Fig. 80 an attempt is made to trace the cycle of current and P.D. changes with β-oscillation. The complete cycle is divided into two parts: an epoch T_1 during which the arc is burning, and an epoch T_2 during which it is extinct. In the normal Poulsen working

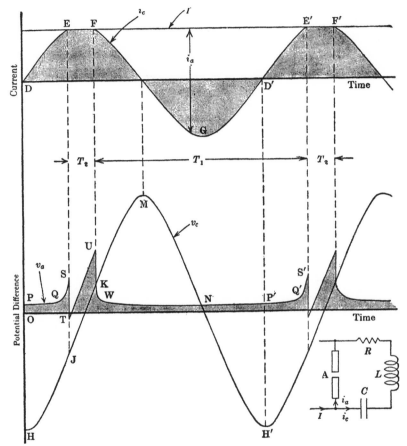

Fig. 80. Poulsen oscillation.

T_2 is very much shorter than T_1. Let us examine in succession (in Fig. 80) the course of the condenser current i_c and the arc current i_a, and the condenser P.D. v_c and the arc P.D. v_a; bearing in mind always that the chokes in the supply mains are supposed to make I sensibly constant, a condition approximately realised in practice.

128   V. THE PRODUCTION OF HIGH-FREQUENCY CURRENTS

Condenser current DEFGD' (*area shaded*). At D the condenser is fully charged in the negative direction, and along DE receives a current growing approximately sinoidally. At E this current equals the total invariable supply current I, so that no current is available

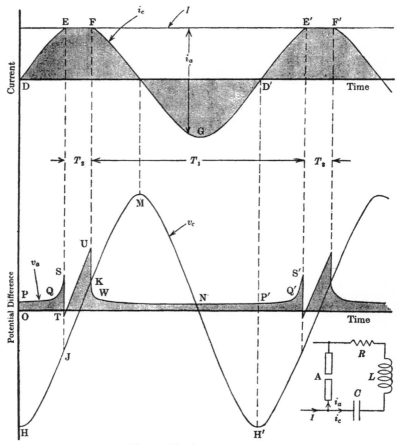

Fig. 80. Poulsen oscillation.

for the arc, which is therefore extinguished. Along EF the condenser receives the whole of the supply current I, and inevitably charges up more and more until its P.D. reaches such a value as at F to re-ignite the arc. This provides again a by-path to shunt away from the condenser some of the steady current I. The oscillatory circuit

4. ARC GENERATORS

CARL now being again closed through the arc, the condenser *C* discharges along FG with approximately sinoidal current, the arc tending to have less and less effect as the current through it rises. Except near the ends F and E', the condenser discharge FGE' is not very different from what it would be if an ordinary metallic conductor replaced the arc.

Arc current, i_a. This is the difference between the horizontal *I* line and the curve DEFGD'. i_a is always positive.

Condenser P.D., HJKMNH'. This is a very nearly sinoidal curve with its maxima opposite D and D', the portion JK being slightly changed from the sine shape owing to the horizontality of EF. If DEFGD' had been truly sinoidal, HJKMNH' would of course have been precisely another sinoidal curve lagging by $\frac{1}{4}$ period.

Arc P.D., PQSTUKWP' (*area shaded*). At P the arc current is large, so that the P.D., OP, is small. In the region Q the arc current is becoming small, so that the P.D. rises rather suddenly to the extinction voltage at S. Since i_c is now steady $(=I)$, v_a must be the P.D. of the condenser plus the ohmic drop in R; i.e. under EF, $v_a = v_c + RI$. v_a therefore rises along TU at a constant height TJ = UK = RI above JK to the striking voltage at U. U is generally higher than S on account of the progress in cooling and de-ionising during the extinct epoch T_2. On re-ignition at U and during the subsequent growth of i_a, v_a drops rapidly at U, passes through a sharply curving region at W corresponding to that at Q, gradually drops to a minimum height under G, and slowly rises again to P'.

What now of the power thrust into *RLC*, where energy is stored (in *L* and *C*) and lost or utilised (in *R*)? The arc P.D. v_a is the E.M.F. impressed on this circuit *RLC*, and the rate at which at any instant energy is passing into it is $v_a \times i_c$, i.e. the product of the ordinates of the curves below which the areas are shaded. During the middle portion of T_1 while i_c is negative, i.e. while the arc current exceeds the supply current *I*, this power is negative and proportional to the large-current P.D. of the arc. During all the rest of the cycle, except perhaps for a brief moment just after extinction at T, the power into *RLC* is positive, and plainly depends largely upon the heights S and U, i.e. upon the extinction and ignition voltages.

The following conclusions can be drawn from the consideration of these curves:

For large power delivered to the oscillatory circuit,

(i) I should be large;
(ii) extinction voltage large;
(iii) ignition voltage large;
(iv) resistance of oscillatory circuit large.

For small power taken back from the oscillatory circuit,

(v) I small;
(vi) large-current P.D. of arc small.

In the conflict between (i) and (v), (i) must obviously preponderate. For (ii) a long arc is wanted, well de-ionised; and for (iii) the same; but it must be noticed that whereas the ignition arc length is the shortest distance between the electrodes, the extinction length is much greater owing to the action of the magnetic field ("magnetic blast") which both bows the arc and separates the craters during the burning epoch T_1. The hydrogenous atmosphere and the coolness of the anode favour de-ionisation. (iv) is under control, as the oscillatory circuit can, within limits, be designed to suit the arc. For (vi) a short arc is required; this conflicts with (iii), but owing to the magnetic blast not with (ii).

These considerations confer some insight into the rationale of Poulsen's modifications of the Duddell arc, and into the conditions governing the design of Poulsen arcs.

A consequence of the brevity of the extinction epoch T_2 is that the feed current I is very nearly equal to the R.M.S. value of the oscillatory current i_c.

In assessing the practical utility of the Poulsen arc, a consideration of the first importance is the degree of steadiness, and approach to pure sine form, of the radiation emitted. In these respects the arc *per se* functions very imperfectly. The white-hot tip of the carbon cathode is slowly dissipated. It must be screwed forward towards the anode at intervals—perhaps every quarter of an hour—and it is maintained in slow rotation by a small motor, since the magnetic field drives the arc over to one side of the electrodes. Gross variations of this kind, however, are not the most serious. More troublesome are the slight irregularities from cycle to cycle, not measurable by an ammeter or wavemeter, but productive of sizzling and bubbling noises in a receiver, termed *mush*. And although the v_c curve is roughly sinoidal, we have seen that violent discontinuities occur within each cycle, so that the oscillation produced is necessarily rich in

4. ARC GENERATORS

harmonics of the fundamental frequency. All these defects have been countered with striking success by removing the arc from the antenna circuit—which formerly itself constituted the "oscillatory circuit" of Fig. 78—and letting the arc excite an intermediate circuit of low decrement loosely coupled to the antenna*.

To start a Poulsen arc it must be struck, as is an ordinary arc lamp. In keying, therefore, the oscillation cannot be started and stopped as in a spark transmitter. Instead, the Morse key—or, except in small powers, the set of relay contacts controlled by the key—is made to alter the wavelength slightly by short-circuiting an inductance inserted in, or coupled to, the oscillatory circuit. This is an undesirable feature of arc transmitters; for firstly the marking and spacing waves, although distuned but slightly†, together monopolise a wider band of frequencies than would otherwise be required; and secondly the power used on the spacing signal is wasted. Against these defects, however, the Poulsen arc, for long-wave transmitters of high power, has much to recommend it in respect of simplicity, robustness and economy.

Plate VIII shows one of the Elwell-Poulsen arcs in the Post Office station at Leafield, Oxfordshire; and Fig. 81 is a circuit diagram for this transmitter, the inductance values shown referring to the adjustments for wavelength 12,350 m‡. It is stated that in this arc as used, the extinction epoch (T_2 in Fig. 80) is only about 1 % of the cycle, and that the ignition voltage (at U in Fig. 80) is some 3300 V.

Working figures observed at a French station in an arc transmitter exciting the antenna without an intermediate coupled circuit were as follows:

Supply: 675 V × 300 $\sqrt{2}$ A = 285 kW.
Antenna: natural wavelength 6100 m;
loaded to work at 15,000 m;
capacitance 0·032 μF;
resistance 1·3 Ω;
current 300 A.

---

* See A. G. Lee and A. J. Gill, "The Leafield coupled arc," *Journ. I.E.E.* Vol. 63 (1925), p. 697.
† At Leafield, less than 1 %.
‡ See Lee and Gill, *loc. cit.* The condenser shunting some of the oscillatory current from the arc is found to be of service in stabilising the frequency and increasing the output. Certain subsidiary condensers, added to help in the elimination of high harmonics, are not shown in Fig. 81.

132 V. THE PRODUCTION OF HIGH-FREQUENCY CURRENTS

The efficiency in converting from D.C. power to aerial power was thus $\dfrac{300^2 \times 1\cdot3}{285,000} = 41\,°/_{\!\!o}$. The major portion of the lost energy is carried away in the cooling water of the arc.

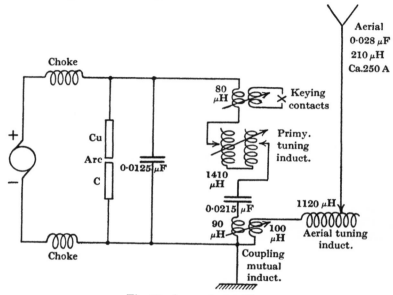

Fig. 81. An arc transmitter.

Plate IX shows a larger Poulsen arc, of American design. The magnetic field across the arc has here its axis vertical. and is produced by a single oil-cooled exciting coil seen below the arc chamber. Two rather larger arcs of the same pattern, each capable of taking 1000 kW, are installed at the Croix d'Hins station, near Bordeaux.

5. High-frequency alternators

5 (a). *General.* The alternating E.M.Fs. of low frequencies (25–100 c/s) used in power and lighting circuits are generated by rotating a coil of wire in a fixed magnetic field; or, more usually in large machines, by rotating a magnetic field about a fixed coil of wire. From quite early days, attempts were made to use this method—that of the dynamo—to produce currents of the very much higher frequencies required for wireless telegraphy. But the

5. HIGH-FREQUENCY ALTERNATORS

increase of frequency is of the order of 1000 times, and great practical difficulties had to be surmounted. Firstly, there is the obvious problem of mechanical strength at the enormous peripheral speeds required. Secondly, even with the highest practicable speeds, the number of poles necessary to give the required frequency is so great that leakage flux from the sides of the teeth can only be kept within bounds by providing an extremely small air-gap between rotor and stator. And thirdly, to moderate the eddy loss in the iron cores subjected to the rapidly changing fields, extremely fine laminations must be used. A fourth problem, that of maintaining the speed sufficiently constant—say to within 10 c/s*—seems to have been solved more easily than might have been expected; adequate constancy has been obtained with governors of mechanical (centrifugal), and of electrical (resonant circuit) types.

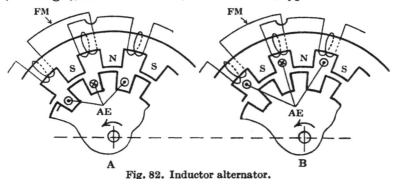

Fig. 82. Inductor alternator.

5 (b). *Inductor alternators.* The mechanical difficulties are much eased by abandoning the usual wound rotor in favour of the inductor type of alternator†, in which the rotor carries no winding at all. Fig. 82 shows diagrammatically at A an ordinary alternator‡, with stationary field magnet excited by a steady current in the winding FM, and producing an alternating E.M.F. in the rotating winding marked AE. At B the same alternator is shown modified by merely transferring the latter winding from the rotor slots to the stator slots. The machine is now an *inductor alternator*, and the mechanical difficulties attaching to high rotor speed are greatly reduced.

* See IX–7.
† This type is, indeed, commonly adopted even for frequencies such as 200–500 c/s, as required for synchronous spark transmitters. See Subsection 1 (b) of this Chapter.
‡ Except that the armature slotting is unusually coarse.

In Fig. 82, the whole field magnet flux must rise and fall as the rotor offers to each pole alternately a tooth and a slot. Hence the whole stator must be laminated (in the plane of the paper), and provision must be made to prevent a short-circuit for the alternating E.M.F. generated in the exciting winding FM. Indeed, there is now no need to provide the second winding AE, since FM may be made to serve simultaneously for D.C. and A.C., as does the single winding of a rotary A.C./D.C. converter.

A type of inductor alternator which offers advantages for very high frequencies is the homopolar type. In B of Fig. 82, let the field magnet winding FM, and the axial slots in which it lies, be replaced by a single coil coaxial with the rotor and lying in a circular

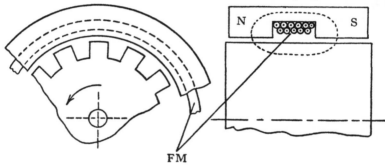

Fig. 83. Homopolar alternator.

recess in the stator, as in Fig. 83. The stator now has a complete ring of S pole at one end, and another of N pole at the other end. As the toothed rotor rotates, the positions of maximum air-gap flux-density move in the circumferential direction, but the total field magnet flux is constant. An alternating E.M.F. is produced in a winding appropriately disposed in narrow slots cut axially in the stator.

Ideally, these stator slots might be ever so narrow, in which case the constant-reluctance field circuit of Fig. 83 would be retained unimpaired. In practice the rotor slots are already so narrow, that it is desirable to make the stator slots wider rather than narrower. The substantial constancy of total flux is then made possible by the inequality of slot pitches in rotor and stator*. A suitable arrange-

* For a summary account of various possible slotting arrangements in inductor alternators, see M. Latour, "Radio frequency alternators," *Proc. I.R.E.* Vol. 8 (1920), p. 220.

5. HIGH-FREQUENCY ALTERNATORS

ment is shown in Fig. 84. The number of stator slots is here only two-thirds of the number of rotor teeth, so giving extra much needed winding space. $S_1, S_3 \ldots$ alternate with $S_2, S_4 \ldots$ in carrying the major part of the total air-gap flux. The E.M.Fs. in the AE conductors have the frequency $\dfrac{v}{l}$ and the instantaneous direction indicated in the figure.

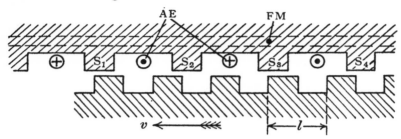

Fig. 84. Slot pitches in stator and rotor.

5 (c). *The Alexanderson alternator.* E. F. W. Alexanderson has constructed homopolar machines functioning electromagnetically as indicated in Fig. 84, but with rotors of disc (instead of drum) form, running at enormous speed. These machines are illustrated in Fig. 85. The stator pole-pieces are very finely laminated; but in

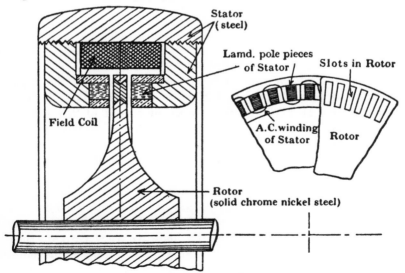

Fig. 85. Alexanderson alternator.

186 V. THE PRODUCTION OF HIGH-FREQUENCY CURRENTS

the rotor electrical efficiency is sacrificed to mechanical strength, and a solid steel disc, with slotted gaps, is used. The gaps are filled in with phosphor bronze, riveted over and finished off smooth, in order to avoid the additional windage loss which the irregularities would introduce. In a 2 kW, 100 kc/s machine of which particulars have been published, the rotor was about a foot in diameter, with 300 slots, and was driven at 20,000 revs. per min., at which speed every ounce of material at the periphery of the rotor demanded a radial force of 2 tons weight to hold it in place. The normal air gap was 0·015 inch.

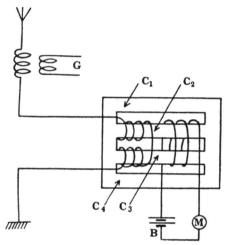

Fig. 86. Magnetic amplifier.

One of several more recent much larger machines of this type, installed at New Brunswick, U.S.A., is shown in Plate X. This machine runs at 2170 revs./min. with an output of 200 kW at 22 kc/s. The driving motor and gear box (ratio 1 : 2·97) are seen on the right; and above the alternator are two high-frequency transformers which serve to match the impedance of the alternator to the antenna it feeds.

The output of the alternator is controlled, for Morse keying (or for telephony), by a distuning process as usual with Poulsen arcs (Fig. 81). But in the Alexanderson outfit, the key contacts do not carry the whole or any part of the high-frequency current; they control a direct current which modifies the degree of magnetic saturation, and therefore the effective inductance, of a finely lamin-

5. HIGH-FREQUENCY ALTERNATORS

ated steel-cored coil included in the aerial or an associated high-frequency circuit. The principle will be clear from Fig. 86, in which G represents a C.W. generator. Modulation at M of the current from the battery B controls the M.M.F. round the core circuits C_1C_2 and C_3C_4. The alternating current in the other winding sets up an alternating M.M.F. round the core circuit C_2C_3—without producing any E.M.F. in the battery coil. The effective inductance of this winding depends on the degree of saturation in the cores C_2, C_3, and therefore upon the battery current. "A comparatively weak current of a few amperes controls as many hundreds of amperes in the antenna*."

5 (d). *Bethenod-Latour alternators.* Many high-frequency alternators have been constructed in France†, of inductor type, but with rotors of the ordinary drum shape. At the Centre Radioélectrique de Paris (Sainte Assise), alternators of 25 kW, 250 kW and 500 kW are in use, all generating directly at the radiation frequency. Interesting details of the 500 kW machines are as follows:

Rotor: 2500 revs./min.; peripheral speed 150 m/sec.; 360 teeth; laminations 0·01 cm thick.

Stator: two-thirds as many slots as in rotor.

Air-gap: less than 0·1 cm.

The complete machine is encased and partially evacuated to reduce air friction. If the rotor is run in air at atmospheric pressure it becomes too *hot*. The efficiency of the alternator is said to reach the remarkably high figure of 82 per cent.

Plate XI is a photograph of the group of machines for transcontinental transmission at Sainte Assise, and shows one of the 500 kW alternators in the middle. All the accessory apparatus, such as air pumps and oil pumps for the lubricating and cooling systems, are located in cellars below.

5 (e). *Frequency-multiplying alternators.* The mechanical difficulties inseparable from the project of an alternator for producing currents of very high frequency are reduced by the insertion of any frequency multiplying device between the alternator and the aerial. Various frequency multipliers are available. They may be classified

---

* E. F. W. Alexanderson, "Trans-oceanic radio communication," *Proc. I.R.E.* Vol. 8 (1920), p. 263. The device goes by the rather inappropriate name, *magnetic amplifier*.

† By the Cie. Générale de Télégraphie sans Fil, to the designs of M. Latour, *loc. cit.* p. 134.

as rotary and stationary. The rotary class has historical and theoretical interest, but the only working examples are probably obsolescent. Stationary frequency multipliers, depending on magnetic saturation in iron, are widely used.

The general principle of rotary multipliers may be illustrated by an elegant scheme due to M. Latour*. In Fig. 87, $R_1 \ldots R_4$ are four rotors rigidly joined together, with polyphase windings connected in pairs as shown. The stators S_1 and S_4 have single-phase windings; S_1 constitutes the D.C. field magnet of the prime generator $S_1 R_1$ supplying polyphase current at frequency n. Stators S_2 and S_3 have polyphase windings, and are connected together. The current of frequency n supplied by R_1 to R_2 produces in R_2 a

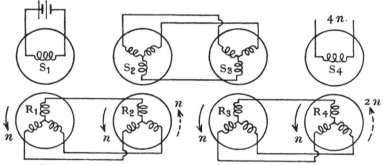

Fig. 87. Frequency multiplication by alternators in cascade.

field rotating (dotted arrow) at n relative to R_2, and therefore at $2n$ relative to S_2. The $2n$ current supplied by S_2 to S_3 produces in S_3 a field rotating at $2n$ relative to S_3, and therefore at $3n$ relative to R_3. And so on. Finally, with m machines a current of frequency $n \times m$ may be taken from the last machine.

No group of alternators of this kind seems to have found practical application; but R. Goldschmidt, acting upon ideas enunciated by P. Boucherot, developed a type of alternator, with a single stator and rotor but with multiple tuned circuits connected thereto, which is equivalent electrically to the Latour group with its several members superimposed. The principle of the Goldschmidt alternator is sufficiently indicated diagrammatically in Fig. 88. Here S is the stator, and R the rotor rotating at a speed corresponding with frequency n; and C_n, C_{2n}, C_{3n}, C_{4n} are shunts providing resonant paths for the currents of frequencies indicated by the subscripts.

* *Loc. cit.* p. 134.

5. HIGH-FREQUENCY ALTERNATORS

Alternators of this type are installed at Eilvese in Germany and at Tuckerton, U.S.A. At Tuckerton, the alternator is driven at 4000 revs. per min.; generates at a fundamental frequency of 10 kc/s; delivers about 120 kilowatts at 40 kc/s to the antenna; and has an overall efficiency of about 50 per cent. The core laminations are 0·002 inch thick; the rotor has 300 poles, is 3 ft. in diameter and weighs 5 tons; and the air-gap is less than 1 millimetre.

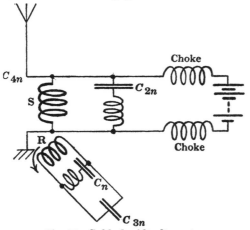

Fig. 88. Goldschmidt alternator.

5(f). *Saturated-iron doublers and triplers.* The mode of action of this type of frequency multipliers will be explained by reference to the form which has been most applied to wireless work, the Epstein-Joly doubler shown in Fig. 89. The iron cores of the two equal transformers A and B are magnetised by the steady battery current I to well over the bend of the magnetisation curve (i.e. they tend to saturation). A sine current, of the frequency n which is to be doubled, is passed through windings oppositely connected, as shown in the figure, so that the directions of the steady and alternating M.M.Fs. round the core are the same in A during epochs when they oppose in B. The two fluxes ϕ_A and ϕ_B, and E.M.Fs. e_A and e_B, are then as indicated by the several curves in the Figure. The sum of the E.M.Fs. in the third windings consists mainly of a sine curve of frequency $2n$. By applying this E.M.F. to a circuit tuned to the $2n$ frequency, a nearly pure $2n$ current is obtained.

The assumed sine current $I_1 \sin 2\pi nt$ would not be given by a sine P.D. of that frequency applied to each transformer winding, or to

140 V. THE PRODUCTION OF HIGH-FREQUENCY CURRENTS

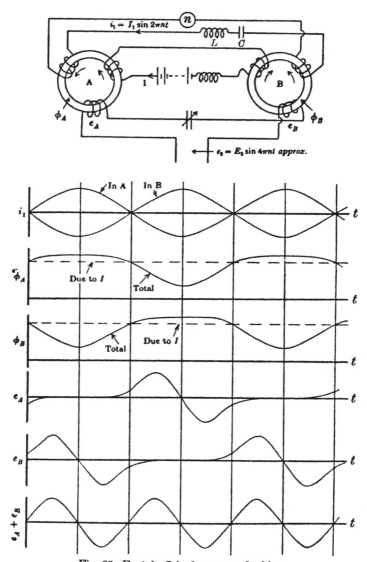

Fig. 89. Epstein-Joly frequency doubler.

5. HIGH-FREQUENCY ALTERNATORS

the two windings in series; but it can be got from a source of n-frequency sine E.M.F. with any closeness of approximation desired by the insertion of appropriate external inductance L and capacitance C.

We have examined the action of the doubler only qualitatively and for the open-circuit condition. The load current must modify the action seriously, since the $2n$-frequency M.M.F. introduced by it round each core cannot be countered by currents of the $2n$-frequency in the other windings. It is, indeed, extremely hard to calculate the load performance of such saturated-iron devices, even without allowing for hysteresis. Attempts have been made, but only with simplifying assumptions which diverge so far from the real conditions as to make the hardly won analytical results of small interest. From an inspection of the arrangement such as we have carried out with reference to Fig. 89, we can be sure that "it will work"; how well it works, and what are the optimum conditions as regards output and efficiency, are questions which would seem to be best treated experimentally.

Frequency doublers of this type are widely used, notably by the Telefunken Company* in the many high-power alternator transmitters installed by them, e.g. at Nauen in Germany, at Malabar in Java and at Assel in Holland. At the great transmitting station at Nauen, inductor alternators running at 1500 revs./min. deliver 400 kW at 6 kc/s to doublers of the type of Fig. 89 arranged two in cascade, so giving an antenna frequency of 24 kc/s ($\lambda = 12{,}600$ m). Most of the loss between input to alternator and input to antenna occurs in the cores of the doublers, which are cooled by powerful oil circulation.

5 (g). *Multipliers with very high saturation.* Other schemes are available resembling that of Fig. 89, and like it adapted to produce a total secondary E.M.F. whose form is a fairly close approximation to a sine curve of double or triple frequency. A different way of utilising the phenomenon of magnetic saturation for multiplying frequency depends upon pushing an alternating M.M.F. so far over the bend of the B-H reversal curve that during the greater part of the cycle the iron is sensibly saturated and magnetically inactive. The relation between B and H, or between flux ϕ and primary current i_1 (the secondary being on open circuit), then approximates to the rectilinear curve ϕ, i_1 of Fig. 90. The ϕ, t curve is the full-

---

* Gesellschaft für drahtlose Telegraphie m.b.H.

142   V. THE PRODUCTION OF HIGH-FREQUENCY CURRENTS

line truncated portion of the sine curve $\phi = ki_1$; and if τ is a smallish fraction $1/r$ of the half period, the E.M.F. (per turn of winding) is

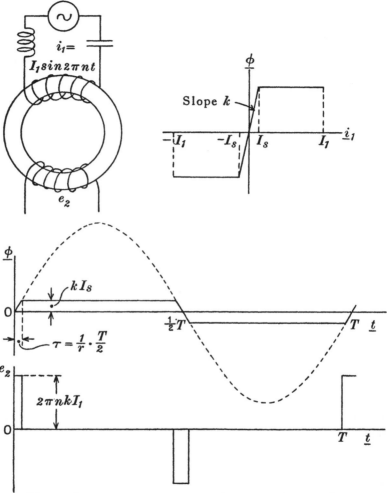

Fig. 90. Frequency multiplication by high magnetic saturation.

given by the sensibly rectangular curve e_2, t. Now this curve is representable by the Fourier series

$$e_2 = E \sin \frac{\pi}{r} \cdot \cos pt + \ldots + \frac{E}{m} \sin \frac{m\pi}{r} \cdot \cos mpt + \ldots,$$

5. HIGH-FREQUENCY ALTERNATORS

where $E \equiv 8nkI_1$. In this series, by choosing r to make $\sin\dfrac{m\pi}{r} = 1$ (i.e. $r = 2m$), the amplitude of the mth harmonic* is a maximum E/m, which is greater than $2/\pi$ times the amplitude of any other harmonic. Such an E.M.F., therefore, is well adapted to excite current of frequency $n \times m$ in a circuit tuned to that frequency.

Multipliers of this type have recently come into considerable use. Amongst the group of transmitters at Königs Wusterhausen† (German Post Office) are several operating in this way. The alternator supplies current at about 9 kc/s, producing intense saturation in the iron core of a choke. (The secondary winding of the transformer in Fig 90 is dispensed with.) The P.D. across the winding excites an oscillatory circuit tuned to the desired harmonic. By adjustment of this tuning and of the magnetising current I_1, any odd harmonic up to the 9th is selected for use; and the others are filtered out. For wavelengths shorter than about 4000 m, a second stage of frequency multiplication is provided.

The method seems to be practicable even up to some 500 kc/s. It has been adopted, for example, in the 12 kW broadcasting transmitter at Munich working on about 500 m. The following particulars refer to an installation of this character. An inductor alternator supplies 20 kW at about 9 kc/s to a frequency multiplier; the 9th harmonic is selected and is passed to a second multiplier, where again a 9-fold multiplication is effected. The two chokes contain respectively 4·4 lbs. of iron tape 0·05 mm thick wound with 60 turns of stranded wire, and 0·9 lb. of iron tape 0·008 mm thick wound with 50 turns. The efficiencies of the two transformations are said to be about 85 %   and 65 %.

The small size of the chokes is very striking. Plate XII shows one for an output of 10 kW at 600 kc/s. The core consists of a number of rings, spaced apart for cooling, each formed of tightly wrapped thin iron tape only a few millimetres wide. A cooling stream of paraffin oil is forced down the axial tube; it divides into four radial streams, and swirls round the core in the spaces between the rings and between the rings and the winding. It emerges at four peripheral openings, two of which are visible in Plate XII.

* Fig. 90 is approximately to scale for $m=9$.
† From information kindly supplied to the author at the station, and subsequently.

CHAPTER VI

THE DETECTION OF HIGH-FREQUENCY CURRENTS*

1. Methods of registering low-frequency currents

ALTERNATING currents are ordinarily registered by three methods—the thermal, the electro-magnetic, and the ferro-magnetic. In a hot-wire ammeter the current, of instantaneous value i, flowing through the wire produces heat at a rate proportional to i^2; and the consequent rise of temperature is registered by a mechanical indication of the expansion of the wire. In a dynamometer ammeter, the current i in one coil reacts with the same current i in a second coil; the forces between them are therefore proportional to i^2; and the mean value is registered by the flexure of a resisting spring. In a moving-iron ammeter, two pieces of soft iron are similarly magnetised by the current i; the reaction between them again is proportional to i^2 (or at any rate does not change sign with i), and is similarly registered.

Of these methods only the first is suitable for high frequencies. The large reactance of the coils of a dynamometer or moving-iron instrument, and the eddy currents in the iron of the latter, make these instruments highly inconvenient for wireless frequencies. The hot-wire ammeter is, however, used for measuring high-frequency currents of sufficient power, as in transmitters. But the hot-wire instrument, depending on the change of length of the heated wire, is a very insensitive instrument†.

If a thermo-couple and D.C. millivoltmeter are used to register the temperature of the heated wire, a much more sensitive instrument is obtained, especially when the thermo-couple and hot wire—now only a centimetre or less in length—are enclosed in a vacuous bulb. In Fig. 91, a full-scale deflection can be got in the millivoltmeter, if of the highly sensitive "Unipivot" pattern, with a consumption of power from the oscillatory circuit of about 500 μW‡.

---

* A fuller analysis of rectifying detectors will be found in Chapter XI.

† It consumes power of the order of 1000 times as much as a D.C. instrument of the d'Arsonval type of like size and robustness.

‡ But when the same millivoltmeter is used alone for a D.C. measurement, the power consumption is about $\frac{1}{1000}$ of this.

2. RECTIFICATION BY NON-OHMIC CONDUCTORS

In low-frequency practice, voltmeters are, of course, nothing but ammeters of high resistance, and their power sensitivity is necessarily even less than that of ammeters. In high-frequency practice, however, a type of voltmeter has been developed* which is not an ammeter in disguise, but in its best form consumes no measurable power from the circuit under examination. Such instruments are essentially calibrated thermionic rectifiers.

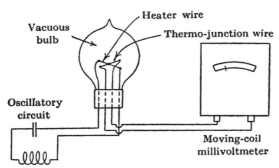

Fig. 91. A.C. measurement by thermo-couple.

2. Rectification by non-ohmic conductors

2 (a). *Without external resistance.* There is a form of converter whereby a direct current can be obtained from a feeble alternating current, which is at once more sensitive than the heater and thermo-couple and is devoid of the sluggishness inherent in thermal instruments. This is the *rectifier*, a conductor not obeying Ohm's law and exhibiting asymmetric change of conductance for rise and fall of P.D. When the problem is to detect small high-frequency currents for signalling purposes rather than to measure them, some form of rectifier is almost always employed. In speaking of receiving circuits, the word *detector* usually carries the significance of rectifier or rectifying detector†.

* In the first instance—as a calibrated portable instrument—by E. B. Moullin, whose elegant A.C. voltmeters made by the Cambridge Instrument Co. are now well known. See E. B. Moullin, "A direct-reading thermionic voltmeter and its applications," *Journ. I.E.E.* Vol. 61 (1923), p. 295; and "A thermionic voltmeter for measuring the peak value and the mean value of an alternating voltage of any wave-form," *Journ. I.E.E.* Vol. 66 (1928), p. 886. (The mean-value instrument is not suitable for high frequencies.)

† It will be convenient in this Chapter to use *rectifier* to signify the non-ohmic conductor itself, and *detector* to signify the rectifier in association with its output circuit. The diagrammatic symbol for any rectifier is seen in Fig. 92.

146   VI. THE DETECTION OF HIGH-FREQUENCY CURRENTS

The ideally "perfect" rectifier would be a conductor with a characteristic like that of Fig. 92. When the P.D. v is less than a critical value v_0 the resistance is infinite, and when the P.D. exceeds v_0 current flows as through a finite constant resistance r. (The smaller r is, the better.) Such a conductor would be a true non-return valve for the flow of electricity, like the valve of a pneumatic

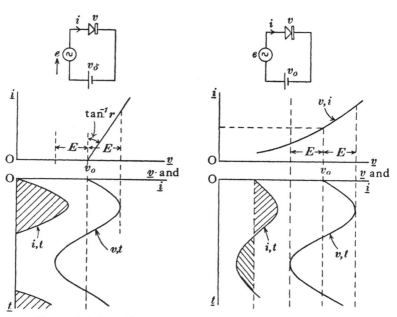

Fig. 92. Ideal rectifier.   Fig. 93. Real rectifier.

tyre for the flow of air. An alternating P.D. of amplitude E added to the steady P.D. v_0 would produce uni-directional, half-sine-wave pulses of current of amplitude $\dfrac{E}{r}$, as shown in Fig. 92; and the mean value of the current would be $\dfrac{E}{\pi r}$.

For the small signal P.Ds. dealt with in wireless reception no such "perfect" rectifier is known. Characteristics, which in the rough (i.e. for large changes of P.D.) appear to have the discontinuity of slope of Fig. 92, when examined on a finer scale show themselves to be more or less gently curved lines without discontinuities. Nevertheless an alternating P.D. whose mean value is zero, applied to a conductor with a curvilinear characteristic, does produce a current whose mean value is not zero; that is to say, *rectification*—though not "perfect" rectification—does occur.

This process is shown qualitatively in Fig. 93. Here v, i is the curved rectifier characteristic; v, t is the time curve of a sine fluctuation of P.D. about the mean value v_0; and i, t is the time curve of the current thereby caused to flow. The i, t curve is found by squaring across between the v, i and the v, t curves. It is clearly an asymmetric fluctuation above and below the value i_0 appropriate to $v = v_0$. Since the negative shaded areas are less than the positive, the effect of the E.M.F. $E \sin pt$ has been to change the mean current from i_0 to some larger value. If the v, i curve were convex over the working range instead of concave, the rectified current would have been negative—a decrease from i_0 instead of an increase. If the conductor obeyed Ohm's law, the v, i characteristic would be straight and there would be no rectification.

We proceed to derive a quantitative expression for the rectified current in the circumstances examined qualitatively in Fig. 93. Let the v, i characteristic be expressed quite generally as

$$i = f(v),$$

and let the alternating E.M.F. have the general form

$$e = E \sin pt + E_3 \sin 3pt + E_5 \sin 5pt + \ldots.$$

Then
$$i = f(v_0 + e),$$

which by Taylor's expansion is

$$i = f(v_0) + e f'(v_0) + \frac{e^2}{\underline{|2}} f''(v_0) + \frac{e^3}{\underline{|3}} f'''(v_0) + \ldots *.$$

In taking the time mean $\frac{1}{T} \int_0^T i\, dt$ over one period $T \equiv \frac{2\pi}{p}$ of the fundamental frequency, each of the terms in e, e^3 etc. is zero, and we find

$$\text{mean } i = f(v_0) + \frac{f''(v_0)}{4} (E^2 + E_3^2 + E_5^2 + \ldots)$$

$$+ \frac{f''''(v_0)}{64} (E^4 + E_3^4 + E_5^4 + \ldots + 2E^2 E_3^2 + \ldots) + \ldots.$$

* Where, in the usual notation, $f''(v_0)$ signifies $\frac{d^2v}{di^2}$ at $v = v_0$, and so on.

148 VI. THE DETECTION OF HIGH-FREQUENCY CURRENTS

If the values of the component E.M.Fs. E, E_3, E_5 etc., and/or the higher differential coefficients $f''''(v_0)$ etc., are small enough to make the f'''' and higher terms negligible, we have

rectified current $I_r \equiv$ change of mean i consequent on signal e,

$$\doteqdot \frac{f''(v_0)}{4}(E^2 + E_3{}^2 + E_5{}^2 + \ldots),$$

$$= \frac{\mathscr{E}^2}{2}\left[\frac{d^2i}{dv^2}\right]_{v=v_0},$$

where $\mathscr{E}$ is the virtual or R.M.S. value of the alternating P.D. applied to the rectifier. In most practical cases e is a nearly sinoidal E.M.F. $E \sin pt$, the harmonics $E_3, E_5 \ldots$ being negligible*; then $\mathscr{E}^2 = \frac{E^2}{2}$.

In practical rectification of the weak alternating E.M.Fs. constituting the received signals of wireless telegraphy, $\frac{E^2}{16}f''''(v_0)$ is often very small compared with $f''(v_0)$, so making $\frac{\mathscr{E}^2}{2} \cdot \frac{d^2i}{dv^2}$ a sensibly accurate expression for the rectified current; but whatever the shape of the curve $i = f(v)$, the rectified current tends more and more closely to the value $\frac{\mathscr{E}^2}{2} \cdot \frac{d^2i}{dv^2}$ as the signal strength $\mathscr{E}$ is reduced. The curve over the working range is found often to be very nearly a parabola

$$i = \alpha + \beta v + \gamma v^2.$$

Then $\frac{d^2i}{dv^2}$ has the constant value 2γ, and the above expression for the rectified current becomes accurate without restriction to weak signals. When the curve is not parabolic, the region giving the largest rectified current for a given signal P.D. E is the region where $\frac{d^2i}{dv^2}$ is greatest.

2 (b). *With external resistance.* To be of use, the rectified current must flow through some measuring or indicating instrument possessing resistance (and sometimes inductance). The pulses of electricity—the unequal positive and negative shaded areas of the i, t curve in Fig. 93—are accordingly collected and smoothed out in a condenser C (Fig. 94) shunting the indicating instrument, represented by the resistance R. We will assume that C is large enough to avoid any sensible high-frequency P.D. across it.

* As far as concerns the performance of the rectifier.

2. RECTIFICATION BY NON-OHMIC CONDUCTORS

The presence of R makes an accurate analysis difficult; for the change of the P.D. v across the rectifier consequent on the introduction of the signal e is compounded of the alternating E.M.F. e itself, and the change of steady opposing P.D. which appears across R. In Fig. 94, A is the state-point v_0, i_0 before the introduction of the signal E.M.F. e. Whereas under the conditions of Fig. 93 ($R = 0$) the mean P.D. retained its pre-signal value v_0 at A, in Fig. 94 ($R \neq 0$) the mean P.D. changes under the action of the signal to a lower value at A'. The v, t curve in Fig. 94 is consequently the full-line curve instead of the dotted curve. The change of mean current is no longer I_r, but is $I_r - (i_0 - i_0')$. This is the useful output of the detector. We may call it the output *signal current* I_s, retaining the term *rectified current* I_r for the rise of mean current attributable to the alternating component e of the total P.D. v across the rectifier. We have then

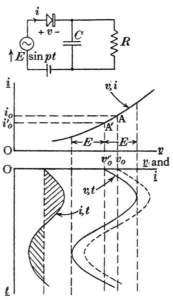

Fig. 94. Rectification with external resistance.

$$I_s = I_r - (i_0 - i_0').$$

The difficulty in calculating I_s from this expression—with given v, i curve and signal E—is that I_r depends upon $\dfrac{d^2 i}{dv^2}$ at A', which is in general different from its value at the pre-signal state-point A; and $(i_0 - i_0')$ depends upon the displacement AA'; while AA' itself depends upon I_s, since $v_0 - v_0' = RI_s$. For weak signals, however, AA' is a small displacement, so that

$$\left[\frac{d^2 i}{dv^2}\right]_{\text{at A'}} \quad \text{tends to} \quad \left[\frac{d^2 i}{dv^2}\right]_{\text{at A}},$$

and $\quad (i_0 - i_0') \quad$ tends to $\quad (v_0 - v_0') \left[\dfrac{di}{dv}\right]_{\text{at A}}.$

With sufficiently weak signals, therefore,

$$I_s \fallingdotseq \frac{\mathscr{E}^2}{2}\left[\frac{d^2i}{dv^2}\right]_{\text{at A}} - RI_s\left[\frac{di}{dv}\right]_{\text{at A}},$$

i.e.
$$I_s \fallingdotseq \frac{\mathscr{E}^2}{2}\frac{\dfrac{d^2i}{dv^2}}{1+R\dfrac{di}{dv}}.$$

This expression becomes the more accurate the weaker the signal. In practical operation it is sometimes sensibly accurate with signals much stronger than the actual signals.

Commonly the function of the rectifier is to produce an output signal P.D. E_s across R, rather than a signal current I_s in R*. Since

$$E_s = RI_s = \frac{\mathscr{E}^2}{2}\frac{d^2i}{dv^2}\frac{R}{1+R\dfrac{di}{dv}},$$

R should be made large in comparison with the slope resistance $\dfrac{dv}{di}$ of the rectifier. Then

$$E_s \fallingdotseq \frac{\mathscr{E}^2}{2}\frac{\dfrac{d^2i}{dv^2}}{\dfrac{di}{dv}}.$$

3. The detector in relation to the oscillatory circuit

3 (a). *Coefficients of performance of the detector.* The sensitivity of a detector may be expressed in two ways according to the aspect of its performance which is mainly in view. On the one hand, the ratio $C_{V/V} \equiv \dfrac{E_s}{\mathscr{E}}$ may be called the $\dfrac{\text{volts out}}{\text{volts in}}$ coefficient of performance, and expresses the sensitivity as the relation between the output rectified P.D. and the input high-frequency P.D. On the other hand, the quantity $C_{V/W} \equiv \dfrac{E_s}{W_i}$ may be called the $\dfrac{\text{volts out}}{\text{watts in}}$ coefficient of

---

* This is because the rectifier is used to actuate a potential-operated amplifier placed between it and the ultimate power-operated instrument (e.g. telephone or Morse recorder).

8. DETECTOR IN RELATION TO THE OSCILLATORY CIRCUIT

performance, and expresses the sensitivity as the relation between the output rectified P.D. and the input high-frequency power W_i.

We have seen that for a given input P.D. $\mathscr{E}$, in order to make the output P.D. E_s large, R is to be made large in comparison with the slope resistance $\frac{dv}{di}$; and that then

$$E_s \fallingdotseq \frac{\mathscr{E}^2}{2} \cdot \frac{\frac{d^2i}{dv^2}}{\frac{di}{dv}}.$$

Then
$$C_{V/V} \equiv \frac{E_s}{\mathscr{E}} \fallingdotseq \frac{\mathscr{E}}{2} \frac{\frac{d^2i}{dv^2}}{\frac{di}{dv}}.$$

In practice, the input P.D. applied to the rectifier is the high-frequency P.D. across the coil or condenser of an oscillatory circuit in which the available power is strictly limited (Fig. 95). The loading of that circuit by its association with the detector is therefore of primary importance. With weak signals the detector damps the circuit to the same extent as would a constant resistance of magnitude equal to the slope-resistance $\frac{dv}{di}$ of the rectifier at the operating region (A' of Fig. 94). The input power (A.C.) delivered to the rectifier from the oscillatory circuit is therefore

$$W_i = \mathscr{E}^2 \frac{di}{dv}.$$

The value of R is without influence because of the shunting capacitance C.

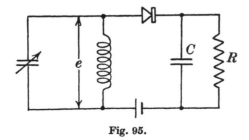

Fig. 95.

To obtain a large coefficient $C_{V/W}$, therefore, again R is to be made large in comparison with the slope-resistance $\frac{dv}{di}$; and then

$$C_{V/W} \equiv \frac{E_s}{W_i} \doteqdot \frac{1}{2}\frac{\frac{d^2i}{dv^2}}{\left(\frac{di}{dv}\right)^2}.$$

It is interesting to attach numerical values to these coefficients of performance of a detector. As a practical example we may take

$$\frac{d^2i}{dv^2} = 295\,\mu\text{A/V/V}, \qquad \frac{di}{dv} = 50\,\mu\text{A/V} = \frac{1}{20\,\text{k}\Omega}\;*.$$

For a weak signal the formulae for the two maximal coefficients of performance then give:

when $\qquad R \gg 20\,\text{k}\Omega, \qquad C_{V/V} = 29\,°/_\circ$ for $\mathscr{E} = 0\cdot 1$ V;

and $\qquad C_{V/W} = 0\cdot 06$ volt per microwatt.

3 (b). Connection to the oscillatory circuit. The optimum closeness of association of the rectifier with the oscillatory circuit depends on $\frac{di}{dv}$ —on whether the rectifier is of high or low resistance. In Fig. 96 the rectifier is, in effect, connected to the supply circuit through a variable-ratio, step-down transformer. The higher the sliding tap, the more power is taken by the detector per milliampere in the antenna.

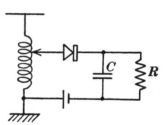

Fig. 96. Connection of detector to oscillatory circuit.

The resistance of the antenna may be divided into three parts:

r_r, the radiation resistance (see II–4 (c));
r_d, due to the flow of power to the detector;
r_h, due to heat production (earth resistance, eddy currents, etc.).

If a steady incoming signal produces in the antenna the high-frequency E.M.F. $\mathscr{E}_a$, the current in the antenna (if tuned) is

$$\mathscr{I}_a = \frac{\mathscr{E}_a}{r_r + r_d + r_h}.$$

* See p. 163, footnote, and p. 164.

3. DETECTOR IN RELATION TO THE OSCILLATORY CIRCUIT 153

The useful power is $\mathscr{I}_a^2 r_d = \mathscr{E}_a^2 \dfrac{r_d}{(r_d + r_r + r_h)^2}$.

For choice of r_d, this is a maximum when

$$r_d = r_r + r_h.$$

Accordingly, the correct position of the sliding tap in Fig. 96 is that which causes the antenna resistance to be doubled.

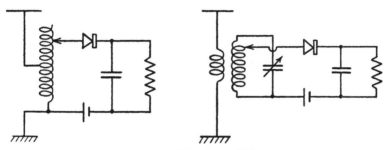

Fig. 97. Modifications of Fig. 96.

If the rectifier is of such high resistance that a step-up transformation is required, instead of the step-down of Fig. 96, it may be had by some circuit arrangement such as those in Fig. 97.

4. Practical detectors

We have been examining the effect in an A.C. circuit of a conductor possessing a curvilinear characteristic. Where are such peculiar conductors to be found? We have met one such conductor in the electric arc (Fig. 79); but this can occur only with heavy currents. Another is a fine wire of material with large temperature coefficient of resistance, like a lamp filament; but the heat capacity of even the finest practicable filament would preclude its use at high frequencies. But the *contact* between fragments of certain, usually crystalline, minerals lightly pressed together, or between the mineral and a metal point, is found to possess the non-ohmic curvilinear property for small currents in a marked degree; and the static characteristic appears to be closely followed even at high frequencies.

VI. THE DETECTION OF HIGH-FREQUENCY CURRENTS

A great variety of substances have been used to give the rectifying contact in crystal rectifiers. Two very common combinations are carborundum (SiC) and steel, and zincite (ZnO) and chalcopyrite (CuFeS$_2$). A fragment of the crystal is mounted in a small brass cup with solder or fusible alloy, and the other member of the pair is pressed against it by means of some spring adjustment. A good feature of the carborundum rectifier is that, presumably owing to

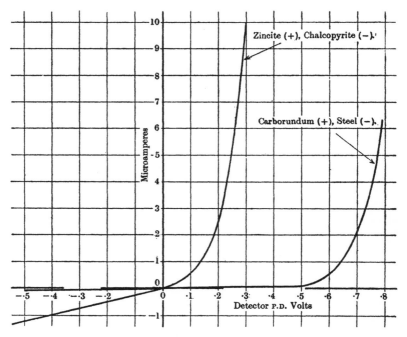

Fig. 98. Crystal rectifier characteristics.

its hardness, considerable pressure may be applied. Representative characteristics are given in Fig. 98. The curve for the carborundum rectifier shows that if it is to be sensitive to weak signals, a steady polarising P.D. of 0·6 volt or more must be applied. The best value varies widely with different samples of crystal, and may even be reversed in sign. The potential divider shown in Fig. 99 enables the operator to adjust v_0 by trial to the particular value giving greatest sensitivity. In the case of the zincite-chalcopyrite rectifier, although

4. PRACTICAL DETECTORS

a polarising P.D. of 0·1–0·2 V would improve sensitivity with very weak signals, there is fair curvature quite close to zero P.D.; so that with this rectifier, and others like it, the battery and potential divider are sometimes omitted for the sake of simplicity.

The simplest possible receiver, for spark telegraphy or for telephony, is obtainable by the use of a crystal rectifier as shown in Fig. 99. The course of events when receiving spark signals is shown in Fig. 100. Curve A shows the train of oscillation in the receiving antenna, or other high-frequency circuit to which the detector is

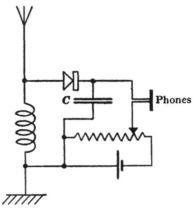

Fig. 99. Receiving circuit with crystal rectifier.

connected. Curve B represents the gradual rise of condenser P.D. as the rectified current passes into it, and the subsequent fall as the charge passes out through the telephone. Curve C shows the mean value (with respect to the high frequency) of the change of current through the rectifier due to the signal (the steady no-signal current i_0 not being shown). And finally curve D shows the resulting change of current in the telephone. The curves are not drawn to scale, but indicate, with the necessary exaggeration, the time relations subsisting between the quantities named. Owing to the back P.D. of the condenser, rectification ceases (i.e. the rectified current becomes zero) just before the end of the oscillation, and a small reverse current thereafter flows back through the rectifier from the condenser until the latter is quite discharged. The area of curve D is equal to the difference between

156 VI. THE DETECTION OF HIGH-FREQUENCY CURRENTS

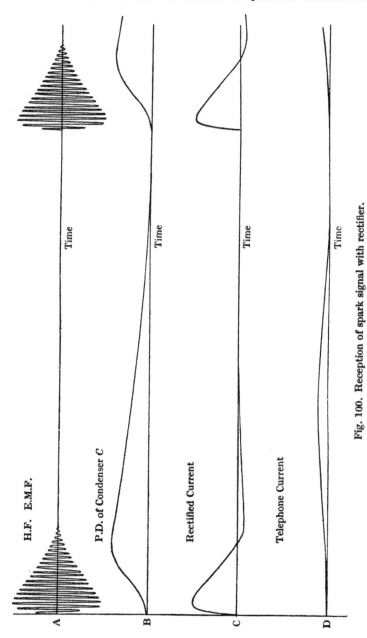

Fig. 100. Reception of spark signal with rectifier.

4. PRACTICAL DETECTORS

the positive and negative areas of curve C. The diaphragm of the telephone is attracted (or repelled) by the condenser discharge of curve D; so that the operator's ear-drum receives one little blow for every spark at the transmitter, and he hears a musical note whose pitch is the spark frequency. If the sparking is not regular (as in the spark transmitter of Fig. 66), he hears a sound of correspondingly scratching or buzzing character.

The condenser across the telephones (C, Fig. 99) is sometimes omitted in diagrams, and even in real circuits. But it is only the accidental self-capacitance of the telephone winding and connections which makes this permissible, some shunting capacitance across the telephone being absolutely necessary. The rectifier current due to the signal is in the form of brief impulses repeated at wireless frequency, and if the telephone winding were free from capacitance, it would oppose a back E.M.F. sufficient almost to cancel the high-frequency P.D. across the rectifier. As to the size of the capacitance C shunting the impedance fed by a rectifier, more will be found in Subsection 5(c) of this Chapter, and in XIII-5(d) with reference to telephonic reception. In the present connection it suffices to notice that C must be small enough not to produce perceptible blurring of the spark-frequency pulses of current traversing the telephone, and that it should preferably be large enough to accept the high-frequency pulses of current traversing the rectifier without setting up a sensible opposing P.D.

Although condenser C plays no appreciable part in the tuning of the oscillatory circuit, it is sometimes made adjustable for the purpose of giving some rough acoustic (tone-frequency) resonance in the telephone circuit; but the electrical dimensions of an ordinary telephone are such that electrical resonance is not very marked. The telephone-condenser circuit is, however, usually not actually aperiodic; so that curve D in Fig. 100 should accurately be shown as descending to negative values of current before finally becoming zero.

Crystal rectifiers are now seldom used except for testing purposes[*], and in broadcast reception by headphones quite near a transmitting station. They are, however, capable of excellent performance. Modern rectifiers made with certain synthetic crystals are more sensitive than the combinations whose characteristics have been given in Fig. 98,

---

[*] As the buzzer for generating trains of high-frequency oscillation (p. 117), so the crystal rectifier for observing them, is often fitted in portable wavemeters.

158 VI. THE DETECTION OF HIGH-FREQUENCY CURRENTS

and have not the inconveniently high resistance. Fig. 101 is the characteristic of a modern synthetic crystal rectifier*, plotted with the same relative scales for the coordinates as in Fig. 98, except that ordinates are milliamperes instead of microamperes.

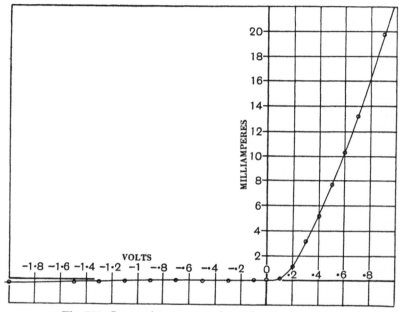

Fig. 101. Low-resistance crystal rectifier characteristic.

For the most part the curvilinear conductors now used are thermionic conductors. Such rectifiers have the great advantage over crystal rectifiers of being robust and constant. But the theory of the effects of curvilinearity of the characteristic applies equally well whatever the conductor which shows that characteristic. Thermionic rectifiers are treated in Chapter XI.

5. Rectification of strong signals

5(a). *The output of the detector.* Consider a detector operating as explained in Fig. 94 (p. 149), but with the ideal characteristic of Fig. 92, the mean rectifier P.D. being at A' (Fig. 102). The slide-back from the pre-signal P.D. to A' prevents current from flowing at all

* General Electric Co's "Gecosite," in fairly firm contact with a wire point: crystal positive, point negative where the P.D. is shown as positive.

5. RECTIFICATION OF STRONG SIGNALS

except during the epochs while $E \sin pt > E_s$. These epochs are represented by the angle 2ψ in Fig. 102*. Clearly if R is increased, ψ decreases towards zero, and E_s increases towards E, indefinitely.

The quantity of electricity passing per cycle is

$$Q_s = \frac{2}{p} \int_0^\psi \frac{E \cos \alpha - E_s}{r} d\alpha$$

$$= \frac{2}{pr}(E \sin \psi - E_s \psi);$$

$$\therefore E_s = RI_s = R \cdot \frac{pQ_s}{2\pi}$$

$$= \frac{R}{\pi r}(E \sin \psi - E_s \psi).$$

Now $E_s = E \cos \psi$;

$$\therefore \cos \psi = \frac{R}{\pi r}(\sin \psi - \psi \cdot \cos \psi),$$

i.e. $\dfrac{R}{r} = \dfrac{\pi}{\tan \psi - \psi}.$

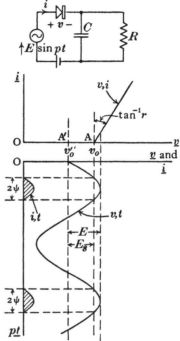

Fig. 102. Detection of strong signal.

The ratio $\dfrac{R}{r}$, whatever the strength of signal†, then determines the angle ψ, and therefore the ratio $\dfrac{E_s}{E} = \cos \psi$. With fixed $\dfrac{R}{r}$, the rectifier gives a strictly linear performance, $E_s = kE$ where k is a constant. Such a rectifier is sometimes called a "perfect" rectifier. The term does not imply that k is unity; it implies that k does not vary with the incoming signal strength. k is plotted as a function of $\dfrac{R}{r}$ in Fig. 103. E_s is nearly 98 °/₀ of E when $R = 1000r$, 90 °/₀ when $R = 100r$, and 64 °/₀ when $R = 10r$.

---

* It is therefore no longer the whole wave-form of the alternating E.M.F., but only the crest, which is significant. E. B. Moullin has designed a pattern of triode voltmeter operating on this principle for measuring peak values of alternating E.M.Fs. *Loc. cit.* p. 145.

† But if the pre-signal state-point A were to the left of O, there would be no rectification at all unless $E > OA$.

160 VI. THE DETECTION OF HIGH-FREQUENCY CURRENTS

Actual rectifier characteristics do not start up from zero current (or zero slope $\frac{di}{dv}$) discontinuously with initial finite slope $\frac{1}{r}$; they curve up more or less gradually. The actual régime approximates to the ideal régime just examined only when the slide-back is large enough to take the mean potential point A' (Fig. 94) off the characteristic. Then, with increasing signal strength, E_s rises less and less rapidly than as the square of E, and tends to become linearly proportional to E.

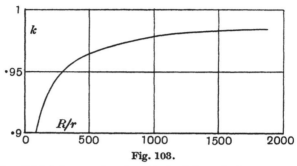

Fig. 103.

For the faithful reproduction desired in telephony, an approximately linear relation between E_s and E is necessary. The above described mode of operation is the basis of the means whereby approximate linearity is achieved.

5 (b). *The effective resistance of the detector.* The effective resistance of this detector obviously depends on the strength of the signal. For very weak signals, it would have the value of $\frac{dv}{di}$ at the pre-signal state-point on the characteristic; but when the signal is strong enough to make $E_s \doteqdot kE$ where k is a constant, it reaches a value which depends only on k and the resistance R. The pulses of current (shaded areas in Fig. 102) are taken from the A.C. source when the value of the E.M.F. lies between E and $E_s = kE$. The mean value with respect to current is nearer E than E_s, and with k not much less than unity it is sufficiently accurate to take it as E. The power taken from the source is therefore

$$\frac{p}{2\pi} Q_s E = \frac{E_s}{R} E, \quad \text{since } Q_s = \frac{2\pi E_s}{pR} \text{ (p. 159)},$$
$$= \frac{kE^2}{R}.$$

5. RECTIFICATION OF STRONG SIGNALS

But if r_d is the resistance which would absorb equal power from the A.C. source, this power is $\dfrac{E^2}{2r_d}$.

$$\therefore \frac{kE^2}{R} = \frac{E^2}{2r_d};$$

$$\therefore r_d = \frac{R}{2k}.$$

As R is increased and k tends to unity, r_d tends to the value $\dfrac{R}{2}$.

In this mode of rectification, therefore, the load on the oscillatory circuit can be reduced indefinitely by increasing R. There are, of course, practical limitations on the magnitude of R, but there is no inconvenience in making it exceed (say) a tenth of a megohm.

5(c). *The requisite shunting capacitance.* The condenser C shunting the load resistance R (Figs. 94 and 95, pp. 149 and 151) has throughout been assumed to be large enough to accept the pulses of electricity delivered to it without offering any sensible opposing P.D. The linear-law rectifier régime now under examination requires a larger capacitance to meet this condition than the square-law régime of the preceding Sections, because the rectifier delivers its output in the form of brief spurts while e only slightly exceeds E_s.

Our assumption of indefinitely large C implies that

$$\frac{1}{C} \cdot Q_s \ll E - E_s \quad \text{(Fig. 102)},$$

i.e. $C \gg \dfrac{Q_s}{E - E_s}$,

$$\gg \frac{2\pi E_s}{Rp \cdot E_s \left(\dfrac{1}{k} - 1\right)},$$

where $\dfrac{p}{2\pi} \equiv n$, the radiation frequency,

and $k \equiv \dfrac{E_s}{E}$, as before;

i.e. $CR \gg \dfrac{k}{(1-k)n}$;

i.e. if $k = 0\cdot 90$, $CR \gg \dfrac{9}{n}$.

There is no practical difficulty, for Morse signalling or even for telephony, in meeting this condition with short wavelengths (n large); but we shall see that with long waves difficulty may arise. If CR is not very great in comparison with $\frac{k}{(1-k)n}$, the performance of the rectifier must be somewhat modified in the direction of reduced output E_s. In the extreme case of C reduced to zero, it is easy to see that the rectified output (i.e. mean P.D. across R) falls to $\frac{1}{\pi} \cdot \frac{R}{R+r} \cdot E_{..}$

The obstacle to making C ever so large arises from the fact that, in signalling, the alternating E.M.F. applied to the rectifier is not sustained but is modulated at some acoustic frequency, say n_a. The E_s we have calculated is the rectifier output which would be given if n_a were ever so small; or, more precisely, if the time constant CR were negligible compared with the acoustic period $\frac{1}{n_a}$,

i.e. if $CR \ll \frac{1}{n_a}$.

If $k = 0.9$ and $n = 10^6$ c/s and $n_a = 900$ c/s, it is possible to select a value for C which meets both conditions well enough; with $CR = 10^{-4}$ sec., CR is more than 10 times $\frac{k}{(1-k)n}$ and is less than a tenth of $\frac{1}{n_a}$. Clearly, however, with longer waves, and with the higher values of n_a which must be met with in telephony, the two conditions we have assumed cannot be simultaneously fully satisfied.

6. Tests of actual rectifiers

The foregoing Sections have shown that the performance of a rectifier, with or without the capacitance-shunted load resistance R, should be calculable from the characteristic curve if the signal is weak; and that with a large load resistance, when the signal is strong, the automatic slide-back of the mean P.D. under the action of the signal tends to convert the square-law performance into a linear one. In this Section these theoretical conclusions are illustrated and checked by a comparison between the observed performance of an actual rectifier under various conditions with its performance as predicted from its observed static characteristic curve.

6. TESTS OF ACTUAL RECTIFIERS

In Fig. 104, the i,v curve (marked μA) is the observed characteristic of a certain conductor* for the range of P.D. 0 to 0·8 V. The $\dfrac{di}{dv}$ curve (marked $\mu A/V$) was obtained by plotting the slopes of the i,v curve; and the $\dfrac{d^2 i}{dv^2}$ curve (marked $\mu A/V/V$) was similarly obtained

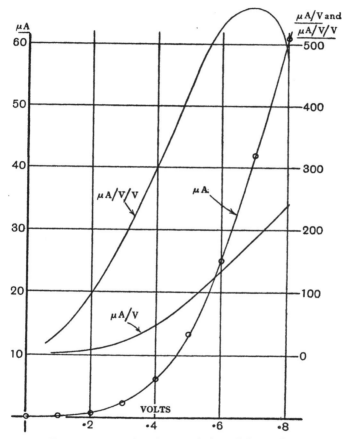

Fig. 104. A rectifier characteristic and derivatives.

* Actually a thermionic conductor, obtained in a triode disposition of considerable practical interest, referred to on p. 350. This is irrelevant in the present connection, except that it assures us that the static and the kinetic characteristics are identical. The signal E.M.F. in the measurements was approximately sinoidal of frequency 90 c/s. There is no ground for supposing that the performance of the rectifier would differ at the high frequencies with which it would be used in a wireless receiver.

from the $\frac{di}{dv}$ curve. The units throughout are microamperes and volts.

It will be seen at once that the i, v characteristic in this instance is far from being a parabola, since $\frac{d^2i}{dv^2}$ is far from constant; hence we cannot expect the formula

$$I_r = \frac{\mathscr{E}^2}{2}\frac{d^2i}{dv^2} \qquad \text{(p. 148)}$$

to hold except for weak signals.* Nor, of course, is the characteristic the ideal curve of Fig. 102, since its slope rises smoothly from zero; hence we cannot expect the relation

$$E_s = kE \qquad \text{(p. 159)}$$

to be reached until the signal strength is large.

The pre-signal state-point selected for the tests was $v = 0.4$ V, $i = 6\ \mu$A. The curves show that here

$$\frac{di}{dv} = 50\ \mu\text{A/V} = \frac{1}{20.0\ \text{k}\Omega},$$

$$\frac{d^2i}{dv^2} = 295\ \mu\text{A/V/V}.$$

Two values of the resistance R were used: one, 0.8 kΩ (i.e. the resistance of the microammeter alone); the other, 101 kΩ. The first is small and the second large compared with 20 kΩ, the rectifier resistance at the no-signal state-point.

Substituting these numerical values in the formula of p. 150, viz.

$$I_s = \frac{\mathscr{E}^2}{2} \cdot \frac{\frac{d^2i}{dv^2}}{1 + R\frac{di}{dv}},$$

we obtain, in microamperes and volts,

$$I_s = 142\mathscr{E}^2 \qquad \text{for } R = 0.8\ \text{k}\Omega;$$
$$I_s = 22.5\mathscr{E}^2 \qquad \text{for } R = 101\ \text{k}\Omega.$$

Fig. 105 shows, for small values of $\mathscr{E}$ and with $R = 0.8$ kΩ, the observed I_s plotted against $\mathscr{E}^2$. The points lie sensibly on a straight line, and this line ($I_s = 156\mathscr{E}^2$) is not far from the theoretical line $I_s = 142\mathscr{E}^2$.

* Unless further examination of the rectifier characteristic should show that, although $\frac{d^2i}{dv^2}$ is not constant, yet $\frac{d^4i}{dv^4}$, $\frac{d^6i}{dv^6}$, etc. are sufficiently small. See VI–2 (a).

6. TESTS OF ACTUAL RECTIFIERS

Fig. 106 shows the observed I_s, plotted this time against $\mathscr{E}$, for a wide range of $\mathscr{E}$ and with $R = 101$ kΩ. The points lie closely on the theoretical curve $I_s = 22\cdot 5 \mathscr{E}^2$ up to $\mathscr{E} = 0\cdot 25$ V. Here the slide-back is

$$E_s = 1\cdot 5 \,\mu\text{A} \times 101 \text{ k}\Omega = 0\cdot 15 \text{ V}.$$

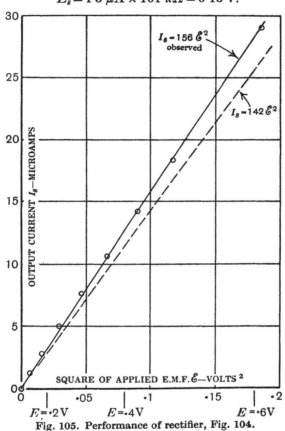

Fig. 105. Performance of rectifier, Fig. 104.

As $\mathscr{E}$ is increased, there ensues a transition stage extending to $\mathscr{E} = 0\cdot 4$ V ($E_s = 0\cdot 29$ V), after which the curve is sensibly the straight line

$$E_s = 0\cdot 83 E.$$

In this condition the coefficient of performance $C_{V/V}$ was 83% *; and the effective resistance of the detector must have been (p. 161)

$$\frac{101 \text{ k}\Omega}{2 \times 0\cdot 83} = 61 \text{ k}\Omega.$$

* Cf. 29% for the weaker signal, $\mathscr{E} = 0\cdot 1$ V, found on p. 152.

166 VI. THE DETECTION OF HIGH-FREQUENCY CURRENTS

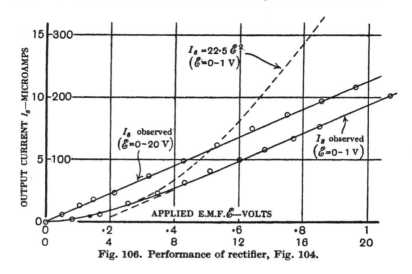

Fig. 106. Performance of rectifier, Fig. 104.

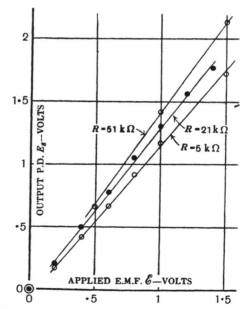

Fig. 107. Performance of crystal rectifier, Fig. 101.

If a rectifier like the crystal of Fig. 101 is used, whose characteristic approximates more closely to the "perfect" characteristic of Fig. 92, earlier inception of a sensibly straight line may be expected. Fig. 107 shows the values of E_s plotted against $\mathscr{E}$ observed with this crystal rectifier* with three different values of R, viz. 5 kΩ, 21 kΩ and 51 kΩ. They lie closely on straight lines nearly passing through the origin; and in the 51 kΩ case, E_s is sensibly equal to E.

7. Continuous wave reception

With spark transmitters, every spark produces a train of oscillation, with its accompanying change of current in the receiving telephones; and between the sparks the rectified current falls sensibly to zero. A sound is thus produced in the telephone as long as the transmitting key is held depressed. With C.W. transmitters, on the other hand (see Fig. 72, p. 119), the rectified current would rise at the beginning of the Morse dot or dash, and would remain substantially constant until the end; so that the telephone would indicate the occurrence of the signal only by a single click at the beginning and at the end. No summation of the received signal—psychological, electrical or mechanical—would take place, and the telephone would no longer be a suitable receiving instrument.

In order to retain aural reception in C.W. telegraphy some means must be provided for changing the steady rectified current into an unsteady one, varying preferably with a musical frequency. This might obviously be done by the use of an independently operated interrupter or switch connected somewhere in the receiving circuits. It was done in a better way in the earlier days of the Poulsen arc by P. O. Pedersen's *Tikker*, a form of vibrating or crazy-contact switch which combined the functions of interrupter and detector. These methods have, however, entirely given place to the *heterodyne* method, first proposed by R. A. Fessenden in 1902, but brought into common use only with the advent of the triode oscillator.

The principle of the heterodyne is the excitation of the receiving circuit by a steady, locally produced, oscillation of frequency differing slightly from that of the incoming signal. In the absence of a signal, the local oscillator produces a steady rectified current imperceptible in the telephone. During a signal, the two oscillations are combined, with the interference or beat effect familiar in acoustics

* Fig. 101 is the characteristic of the rectifier with its contact undisturbed after the measurements of Fig. 107 had been made.

168 VI. THE DETECTION OF HIGH-FREQUENCY CURRENTS

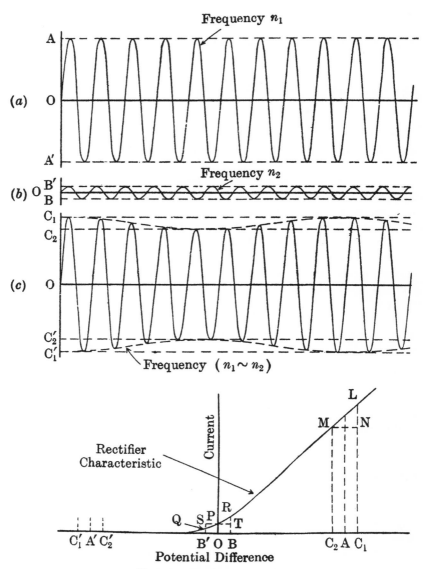

Fig. 108. Heterodyne reception.

7. CONTINUOUS WAVE RECEPTION

when two musical tones of slightly differing pitch are superimposed.

Let the E.M.F. in the detector circuit due to the local oscillation be $a \sin pt$; and that due to the incoming signal, when present, $b \sin qt$. The resulting E.M.F. is

$$a \sin pt + b \sin qt = a \sin (qt + \theta) + b \sin qt,$$

where $\theta \equiv (p-q)t$ is a phase angle which is not constant but changing with time—slowly if, as in actual heterodyne reception, $(p-q) \ll q$. These two terms are

syn-phased at $\qquad (p-q)t = 0, 2\pi, 4\pi\ldots,$
anti-phased at $\qquad (p-q)t = \pi, 3\pi, 5\pi\ldots.$

When syn-phased, the peak value is $(a+b)$; when anti-phased, it is $(a-b)$; the change from maximum $(a+b)$, through minimum $(a-b)$, and back to $(a+b)$, takes place $\dfrac{p-q}{2\pi} = (n_1 \sim n_2)$ times per second. The superposition of the two oscillations is portrayed graphically in Fig. 108. The rectified current rises and falls with the rise and fall of

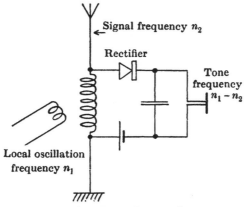

Fig. 109. Heterodyne receiver.

peak value of the alternating P.D. to which the rectifier is subjected. By adjusting the local oscillator so that n_1 differs suitably from n_2, a C.W. signal is made to declare itself by an audible tone in the telephone. For example, suppose the wavelength of the incoming signal is 3000 metres; $n_2 = 100$ kc/s. If we distune the local oscillator by (say) 1 per cent., making $n_1 = 101$ or 99 kc/s, the signal in the

VI. THE DETECTION OF HIGH-FREQUENCY CURRENTS

telephones is a shrill pure tone of a pitch (1,000 c/s) to which the telephone-ear combination is very sensitive, and which is quite distinct from the rustling or crackling sounds due to atmospherics or other causes. The simplest possible heterodyne receiver is shown in Fig. 109.

A comparison between continuous wave and spark telegraphy, as judged at the receiving station, may be summarised as follows:

Receiving apparatus. The same with both, except that with C.W. a local oscillation must be superimposed on the receiving circuits.

Sharpness of tuning. With C.W., if the receiving circuits are suitably proportioned, the sharpness of the resonance relation between wavelength of signal and tuning adjustment of the receiving circuits is much more pronounced. With spark, since the incoming signal itself has a large decrement, it is futile to strive after improvement by extreme reduction of receiver decrements.

Telephone note. Spark has the theoretical advantage that the identity of the sending station is defined by wavelength *and* note, whereas with C.W. the note is determined at the receiver. In practice, however, the purity of the C.W. heterodyne note, and the operator's ability to adjust it to any pitch which suits his apparatus or which aids him in dodging the particular interference he is experiencing at the moment, quite outweigh the advantage of the spark's characteristic note.

Sensitiveness of receiver. With equal distance and power of transmitter, the C.W. receiver is very much superior to the spark.

The superior sensitivity conferred by the heterodyne is due to an improved performance of the rectifier. The local oscillator can be made as powerful as desired, and the rectifier is therefore called upon to deal with a large instead of a very small alternating P.D. The consequent gain for the case when no large impedance is put in series with the rectifier—the case of Fig. 93 (p. 146)—may be easily grasped qualitatively by reference to the bottom diagram in Fig. 108. The pre-signal state-point is P, and the local oscillator causes the rectifier P.D. to sweep over the range AOA'. The advent of a weak signal causes the grid sweep to fluctuate at an acoustic rate between $C_1OC'_1$ (during syn-phase) and $C_2OC'_2$ (during anti-phase). LN, the difference between C_1L and C_2M, is a measure of the difference between the mean currents flowing at syn-phase and at anti-phase; and it is this difference which measures the acoustic current. If there had been no local oscillation, the effect of the signal would have been to produce a change of mean current which—quite apart from its undergoing no acoustic fluctuation—is measured by the relatively slight difference between RT and QS.

With the unrestricted alternating P.D. available from the local oscillator, the detector régime in a heterodyne receiver is easily made that of Section 5, illustrated by the observed performance curves of Figs. 106 and 107. We have seen that here, when the high-frequency E.M.F. is the sustained

$$e = E \sin pt,$$

the signal P.D. E_s is a constant k times the amplitude E. The heterodyned signal consists (Fig. 108) of a high-frequency E.M.F. whose amplitude is not constant, but which fluctuates at an acoustic rate between the extremes $(a + b)$ and $(a - b)$. Provided that the period of the fluctuation is large in comparison with the time constant CR^*, the signal output of the rectifier is an alternating P.D. of acoustic frequency and amplitude kb. However weak the incoming signal b, the signal output of the rectifier remains proportional to b; whereas in the absence of the local oscillator it would be proportional to b^2. The gain in sensitivity conferred by the heterodyne oscillator is therefore very great when the incoming signal is very weak.

A point of academic rather than practical importance is that although the output of the detector is a P.D. varying cyclically with frequency $(n_1 - n_2)$ between the limits $\pm kb$ above the no-signal value, this variation is not strictly sinoidal. For the dotted envelope labelled "Frequency $(n_1 \sim n_2)$" in Fig. 108 is not a sine curve unless b is vanishingly small compared with a. The acoustic output is an alternating P.D. of frequency $(n_1 - n_2)$ and amplitude kb; but the amplitude of the fundamental component is not in general kb.

* Discussed in Subsection 5 (c) of this Chapter.

CHAPTER VII

THERMIONIC TUBES: GENERAL PROPERTIES

1. Introduction

THE story of the discovery, theory and multifarious applications of the thermionic tube may be said to cover the period extending from 1883 till the present day and onwards. The story might be divided into chapters associated with the names of T. A. Edison, J. J. Thomson, O. W. Richardson, J. A. Fleming, L. de Forest and I. Langmuir. In 1883 Edison, engaged on an investigation of defects in the newly invented incandescent lamp, sealed a metal plate into the vacuous bulb in proximity to the filament (Fig. 110). He observed that the milliammeter A indicated a current when the battery B had the polarity shown, but none when the polarity was reversed; the conventional positive electricity could flow across the vacuous space in the direction from cold plate (*anode*) to glowing filament (*cathode*), but not from filament to plate. In 1899 Thomson proved that this current consisted of a stream of his "corpuscles"—our now familiar electrons—liberated from the hot filament; and in 1901 Richardson explained the physical process of their liberation. In 1904 Fleming utilised thermionic tubes (*diodes*) as rectifiers in wireless receivers*. In 1907 de Forest converted the diode into the *triode* by introducing between the filament and the plate a perforated electrode the potential of which was capable of influencing the plate current†. In 1913 Langmuir produced means of evacuating the bulb much more perfectly than had hitherto been possible.

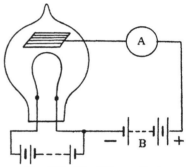

Fig. 110. The Edison effect.

The perfected vacuum—perfect in the sense that residual gas was so reduced that it had no sensible effect in the normal operation of

* See A, Plate XIII. † See B, Plate XIII.

1. INTRODUCTION

the tubes—removed the inconstancies and irregularities which had till then severely restricted their utility and impeded progress in research. It was obvious that a new instrument of limitless scope was waiting to be explored. The relative stagnation of the previous six years was succeeded by intense and world-wide activity. Not only in wireless, but in every branch of electrotechnics concerned with the detection or measurement of small effects, the thermionic triode has become an indispensable instrument of research in the laboratory and of public utility outside. The faithful reproduction at a distance of the fine acoustic detail of music and of the fine optical detail of a photographic picture, telephony across or under an ocean, telekinematography and the prospect of television—to name some of the more spectacular achievements—without the aid of the triode these could only be dreamed of.

As an artist can make a good picture without studying the chemistry of his paints, so a knowledge of the physics of the thermionic tube itself is not altogether essential to an investigator primarily concerned in discovering the uses to which it may be put. He can examine—mathematically as well as experimentally—the behaviour of divers electrical circuits connected in divers ways to the several terminals of the mysterious little bulb, knowing not *why* it works, but only *how* it works as disclosed by its several characteristic curves. Most of the striking potentialities of thermionic tubes— that is to say, their relations with circuits—have been discovered and developed in this way: and they would have been discovered (though not so widely applied) even if the tubes employed had remained unchanged from the earliest high-vacuum forms*.

In this book our more practical concern is with the outward applications of thermionic tubes, rather than their inward design. Nevertheless, as an appropriate introduction to a study of the uses of the instrument, a very summary account of the theory of its action is given in the following Sections of this Chapter†.

---

* Triodes of the pattern of Plate XIV, produced by the French early in the war of 1914–18, were used in enormous quantities during the war, and in Europe for years afterwards; until recently they have been superseded by others giving enhanced, rather than different, performance. All who were in touch with the developments during those early years must retain an affection for the old "R valve."

† The student who wishes to know more of the internal physics—of which much is known and much remains to be learnt—is referred to W. H. Eccles, *Continuous wave wireless telegraphy* (1921), Chap. V onwards; H. J. van der Bijl, *The thermionic vacuum tube* (1920).

2. THERMIONIC EMISSION

2 (a). *Rates of emission*. It used to be supposed that electricity is never found except in association with matter; that the current in the Edison effect (Fig. 110), therefore, *must* depend upon the presence of residual gas in the bulb or upon particles of matter leaving the carbon filament. We all know now that this doctrine was incorrect. But the phenomenon of free electricity is so foreign to the ordinary experiences of electrical engineering that a summary statement of fundamental conceptions of the relation between electricity and matter seems desirable here.

According to the electron theory, electricity is a kind of basic substance, a constituent of all matter, and has an atomic structure. The ultimate unit or atom of electricity is called the electron, and an atom of matter contains electrons in number and configuration according to the quality of the matter. The electron is the indivisible unit of electricity—of negative electricity according to the customary convention as to sign; and if an atom of matter contains an excess of electrons it is negatively charged, or if a deficit it is positively charged. Electrons have been marvellously measured and counted and weighed[*]. A current of electricity consists either of a flow of unattached electrons, or of a flow of atoms or groups of atoms containing an excess or deficit of electrons. Charged atoms or groups of atoms thus conveying, or capable by their motion of conveying, electricity are called ions.

In conducting materials there occurs a constant interchange of electrons between the atoms. At any instant, numerous electrons[†], temporarily free from atomic bonds, flit hither and thither across the wide inter-atomic spaces. In the absence of an external electric field, the velocities of the free electrons have random distribution; statistically, no electric displacement occurs in any direction. But under the influence of an electric field the motion ceases to be wholly random. A difference of potential maintained between two points in a conductor causes a steady net drift of dancing electrons from

---

[*] Mass $= 9 \times 10^{-28}$ gram $= \frac{1}{1850}$ of mass of hydrogen atom.
Charge $= 1 \cdot 6 \times 10^{-20}$ E.M. unit. 1 ampere $= 6 \cdot 3 \times 10^{18}$ electrons per sec.

[†] Probably the number per unit volume is of the same order as the number of atoms per unit volume at room temperature, and is proportional to $T^{\frac{3}{2}}$, where T is the absolute temperature.

2. THERMIONIC EMISSION

the region of lower towards the region of higher potential. There is no resultant transference of matter through the body, because one electron is precisely equal to another, and the electrons are fed into the one region of the body at the same rate as they are withdrawn from the other.

The free electrons must be in thermo-dynamic equilibrium with themselves and the molecules amongst which they move, their agitation increasing with the temperature of the material. It is assumed that they obey the laws of a perfect gas—a gas, it is true, of quite exceptionally small molecular weight $\left(\text{viz. } \frac{1}{2 \times 1850}\right.$ of that of the hydrogen molecule$\left.\right)$, and consequently of unusually high molecular speeds; but a gas the motion of whose molecules is described by the laws deduced in the dynamic theory of gases*.

Although in the interior of the body electrons exist in temporary freedom between the atoms, near the surface isotropic conditions obviously do not obtain. An electron approaching the surface to leave the body is no longer uniformly surrounded by atoms. As it penetrates the ill-defined boundary, the forces due to atoms behind become less and less balanced by the forces due to atoms ahead. The electron is accordingly subjected to restraining forces tending to hold it within the body; and it is well established that if an electron leaves the body, in so doing it must shed a definite quantity of energy specific to the material of the body. This energy may be written $q\phi$, where q is the charge of the electron and ϕ is a fall of electric potential. ϕ is variously termed the *work function*, the *electron evaporation constant*, and the *electron affinity* of the material. The last term (used by van der Bijl) seems to be the most descriptive.

The electron affinities ϕ of several materials are given in the Table †.

---

* O. W. Richardson used this hypothesis with great success in his pioneer work on thermionic emission; but he pointed out that although classical dynamics suffices for describing thermionic phenomena, its implications conflict with other observed properties of metals, e.g. the relations between their specific heats and optical properties. He indicated that the quantum theory must be brought in for a comprehensive treatment.

† The figures for the metals and carbon are taken from van der Bijl, *loc cit.* p. 173; and for the oxides from p. 188 of Richardson's book, *loc. cit.* p. 179.

They are expressed in volts, and the equivalent electron speeds*
are also given.

Material	Electron affinity volts	Equivalent speed cm/sec.
Tungsten	4·52	$1·28 \times 10^8$
Platinum	4.4	$1·26 \times 10^8$
Carbon	4·1	$1·21 \times 10^8$
Thorium	3·4	$1·11 \times 10^8$
Calcium oxide	Ca. 3.3	$1·09 \times 10^8$
Mixtures of barium and strontium oxides	Ca. 2·3	$9·1 \times 10^7$
Sodium	1·82	$8·1 \times 10^7$

An electron cannot escape from the surface of (say) tungsten unless it possesses a (velocity)² exceeding 4·52 equivalent volts with which to pay the toll demanded. It may obtain this velocity in three ways: by the incidence of radiation on the surface (*photo-electric emission*), by bombardment of the surface by high-speed electrons from without (*secondary emission*)†, and by thermal agitation within the body (*thermionic emission*). We are here examining the third phenomenon, thermionic emission.

According to the dynamic theory of gases, the speeds of the molecules—in our case, electrons—are distributed according to the curve $y = Ax^2 \epsilon^{-x^2}$. This curve is plotted in Fig. 111 with $A = \dfrac{4}{\sqrt{\pi}}$, and has the following significance. The shaded area $y\,dx$ below an element of the curve at any point P‡ is the fraction of the whole number of free electrons which at any instant possess speeds lying between x times and $(x + dx)$ times the *most probable speed*§. If we are interested in all electrons possessing speeds above some critical value, we have to find the point P corresponding to this value. Then the

---

* The kinetic energy $\tfrac{1}{2}mv^2$, or speed v, of an electron is customarily specified by the potential p through which it must rise to acquire that kinetic energy or that velocity on moving from rest. For, by the law of the conservation of energy, $\tfrac{1}{2}mv^2 = qp$; whence
$$v_{\text{cm/sec.}} = 60 \times 10^6 \sqrt{p_{\text{volts}}}.$$
We say that the energy or the (velocity)² of an electron is so many *equivalent volts*.

† See VIII–8 and IX–6(a).

‡ Or the ratio between the shaded area and the area beneath the whole curve; for the latter is unity, owing to our choice of the value of the constant A.

§ The most probable speed is represented in Fig. 111 by $x = 1$.

2. THERMIONIC EMISSION

proportion of electrons qualifying for our attention is the area below the curve from P to $x = \infty$, viz.

$$\frac{4}{\sqrt{\pi}} \int_x^\infty x^2 \cdot \epsilon^{-x^2} \cdot dx.$$

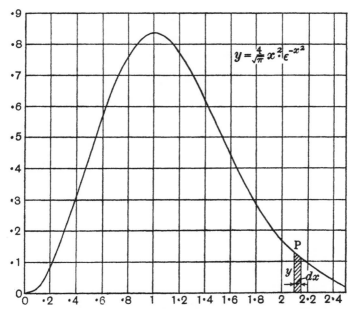

Fig. 111. Distribution of speeds of electrons within a conductor.

Let $C_{\text{m.p.}}$ be the most probable speed, and $\mathscr{C}$ be the root-mean-square speed. From the form of the curve, we find

$$C_{\text{m.p.}}^2 = \tfrac{2}{3} \mathscr{C}^2.$$

Now
$$\mathscr{C}^2 = \frac{3R}{M} T,$$

where R is the universal gas constant, $8 \cdot 3 \times 10^7$ ergs/gram molecule,
M is the molecular-weight number,
T is the absolute temperature (degs. K.)*.

Taking the molecular-weight number of the electron "gas" as $\frac{1}{1850}$, since $M = 2$ for hydrogen, we find

$$\mathscr{C}^2 = 46 \times 10^{10} T \text{ (cm/sec.)}^2,$$
$$C_{\text{m.p.}}^2 = 30 \cdot 5 \times 10^{10} T \text{ (cm/sec.)}^2.$$

* $T°$ K. = T degrees Kelvin = T centigrade degrees above absolute zero.

178   VII. THERMIONIC TUBES: GENERAL PROPERTIES

Hence at room temperature ($T = 288°$ K.), the most probable (speed)^2 is 0·0245 equivalent volt, and at $T = 2400°$ K. it is 0·205 equivalent volt.

Since 4·52 equivalent volts is the critical (speed)^2 for tungsten, only the electrons to the right of a point where

$$x = \sqrt{\frac{4·52}{0·0245}} = 13·6$$

possess super-critical speeds at room temperature, and to the right of $x = \sqrt{\frac{4·52}{0·205}} = 4·7$ at 2400° K. These areas are much too minute to be shown on the scale to which Fig. 111 is drawn.

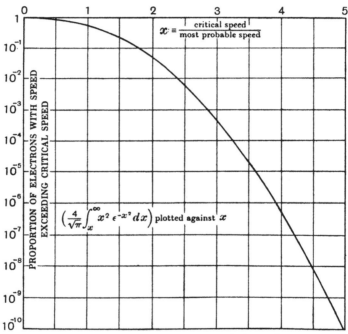

Fig. 112. Speeds exceeding a critical speed.

In Fig. 112, the areas $\int_x^\infty y\,dx$ below the curve of Fig. 111 are plotted to a semi-logarithmic scale. This curve shows how minute a proportion of the electrons possess super-critical speeds even when the critical speed is only a few times (x times) the most probable speed, and how rapidly the proportion rises as x is decreased. Thus the proportion of electrons with super-critical speeds is

1 in 10^{10} when $x = 5\cdot 0$,
1 in 10^{8} when $x = 4\cdot 4$,
1 in 10^{6} when $x = 3\cdot 9$.

With a small rise of temperature of the material, the most probable speed slightly increases ($C_{\text{m.p.}} \propto \sqrt{T}$); the value of x at the critical speed is slightly reduced $\left(x \propto \dfrac{1}{\sqrt{T}}\right)$—i.e. in Fig. 111 the point P moves slightly to the left; and the proportion of electrons with super-critical speed is much increased. Thus, on raising the tungsten from 2400° K. to 2600° K., x changes from 4·7 to (4·7 − 0·18), and Fig. 112 shows that the proportion of electrons with super-critical speed is multiplied by about 5. This illustrates the desirability, in a material to be used as a thermionic emitter, of the property of withstanding a high temperature without rapid volatilisation; for we want rapid evaporation of electricity with the least obtainable rate of evaporation of matter. Tungsten is such a material*.

Again, let us suppose the electron affinity of the material were reduced from 4·52 V (tungsten) to 3·4 V (thorium). The same proportion of electrons, viz. about 1 in 10^{9} ($x = 4\cdot 7$), as from tungsten at 2400° K. would have super-critical speeds with the value of $C_{\text{m.p.}}^{2}$ lowered in the ratio $\dfrac{3\cdot 4}{4\cdot 52} = 0\cdot 75$, i.e. at a temperature of

$$2400 \times 0\cdot 75 = 1800°\text{ K.}$$

This illustrates the desirability in a thermionic emitter of a low electron affinity. Thorium is one such material.

By applying in this fashion to assemblies of electrons the well-established results of the dynamical theory of gases, Richardson†

* If direct proof is needed that the thermionic current is not carried by matter, it is provided by the experimental fact that a tungsten filament during its life can emit a quantity of electricity equal to the charge of N electrons, where N times the mass of an electron exceeds the mass of the original filament. The departure of matter from the filament must reduce its mass; but electrons emitted from the filament are replaced by an equal number of electrons flowing to it through the wire connecting it to the external circuit.

† See O. W. Richardson, *The emission of electricity from hot bodies* (1921). Even after Richardson's work it seems to have remained for Langmuir to dispel all doubt that an electron current can be produced by the emission of electrons from a hot body without dependence on chemical or other reaction with a material atmosphere. See I. Langmuir, "The effect of space charge and residual gases on thermionic currents in high vacuum," *Physical Review*, Vol. II (1931), p. 450.

VII. THERMIONIC TUBES: GENERAL PROPERTIES

derived the following law for the rate of emission from a hot surface in vacuo:

$$I_s = \alpha T^2 \epsilon^{-\frac{\beta}{T}},$$

where α and β are constants for any one material, and T is the absolute temperature (degrees K.), I_s is the *maximum* rate of emission in the sense that it is the value of the current if none of the electrons which escape is allowed to return to the body; for if a proportion of the emitted electrons are pushed back, the net rate of emission is less than I_s.* I_s is called the *saturation current*, the *maximum emission*, or—more briefly, when there can be no doubt as to whether the gross or net emission is referred to—simply the *emission*.

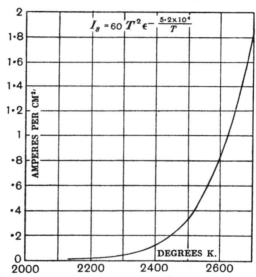

Fig. 118. Distillation of electricity from hot tungsten.

Expressed more specifically, the theoretical formula for the saturation current is

$$I_s = AT^2 \epsilon^{-\frac{b}{T}} \text{ amperes per cm}^2,$$

where $A = 60$,

T = absolute temperature in degs. K.,

$b = 1\cdot 16 \times 10^4 \phi$, ϕ being the electron affinity in volts.

---

* In the same way, the net rate of evaporation of a puddle in the road is greatest if a dry wind sweeps away the gaseous water molecules as soon as they are liberated from the liquid.

2. THERMIONIC EMISSION

In tungsten and certain other pure metals this formula seems to be fairly closely followed. It is plotted in Fig. 113 for tungsten, taking $\phi = 4\cdot5$ V ($b = 5\cdot2 \times 10^4$).

Tungsten, with its very high melting point of 3540° K. (cf. platinum, 1980° K.), can be run at a temperature high enough to give a usefully large emission while not so high as to cause too rapid volatilisation. Tungsten is better than any other metal (or than carbon) in this respect, and tungsten filaments are used as the emitter in thermionic tubes at temperatures in the neighbourhood of 2400° K.–2500° K. with a life of many hundreds of hours.

2 (b). *Low-temperature emitters.* In recent years a great deal of research has been carried out with the object of producing filaments for thermionic tubes which shall be superior to tungsten filaments. A larger emission per watt of heating power is desired, when the filament (of whatever material) is run at a temperature giving a long life, e.g. 2000 hours. Lime and some other oxides begin to emit appreciably at relatively very low temperatures; and in America oxide-coated filaments run at a low temperature have been used since the earliest days of high-vacuum tubes. The problem of producing a good coated filament, however, is not a simple one. Not only is the coating apt to disintegrate and fall away from the filament, but also the emission may be enormously affected by non-uniformity in the filament, by slight impurities in the coating, and by the presence of even minute quantities of residual gas in the tube. At the present time pure tungsten is still the only material used in high-voltage transmitting tubes; but during the last few years, for receiving tubes tungsten filaments have been gradually superseded by composite filaments of several types. These are run at lower temperatures than tungsten, and consume much less power for the same saturation current. Thus the filament of the triode of Plate XIV was of pure tungsten, and took about 2·5 W of heating power to give an emission of 20 mA; that of Plate XVI is of a coated type, and with about ¼th of the power gives the same emission and a longer life.

The Table gives values of A and b^* in the Richardson formula

$$I_s = AT^2 \epsilon^{-\frac{b}{T}}$$

as found for three representative types of filament. Fig. 114, plotted semi-logarithmically from these figures, shows the emission in am-

* See B. Hodgson, L. S. Harley and O. S. Pratt, "The development of the oxide coated filament," *Journ. I.E.E.* Vol. 67 (1929), p. 762.

182 VII. THERMIONIC TUBES: GENERAL PROPERTIES

peres per square centimetre over a wide range of temperatures. The three types of filament, in the order tabulated, appear to be run in practice at temperatures giving emissions of the order $0·1$ A/cm^2, $0·5$ A/cm^2 and 1 A/cm^2 respectively.

Material of hot surface	A	b
Tungsten	60·12	52 560
Thorium	100	53 100
Mixture of strontium and barium oxides	0·001	12 100

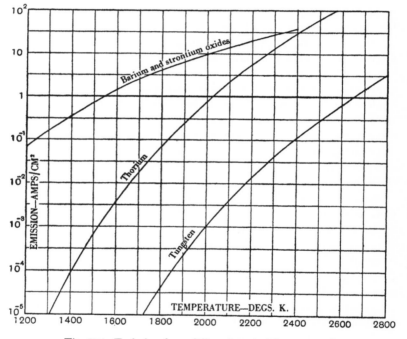

Fig. 114. Emission from different materials compared.

2 (c). *Speeds of emitted electrons.* We have been considering the number of electrons leaving a square centimetre of hot surface in a second. What are their speeds after they have paid the toll ϕ, and have passed the gate between matter and vacuum? The speeds of the electrons just after emission are known with more certainty

2. THERMIONIC EMISSION

than their speeds within the hot body; for the electron assembly outside is more certainly a "perfect gas" than is the assembly inside*.

The components of velocity after emission—say u normal to the hot surface and v, w in the other mutually perpendicular directions—are not equal. We will calculate the distribution and absolute values of the u components. Retaining our former notation, $U_{\text{m.p.}}$ is the most probable velocity and $\mathscr{U}$ the R.M.S. velocity.

The components, perpendicular to the hot surface, of the electron velocities just after emission are distributed according to the curve

$$y = Ax\epsilon^{-x^2}\dagger.$$

This is plotted in Fig. 115 for $A = 2$ $\left(\text{making} \int_0^\infty y/dx = 1\right)$, and $\int_x^\infty y\, dx$ is plotted in Fig. 116. These curves have precisely the same meanings as the corresponding somewhat different curves of Figs. 111 and 112. From the distribution of velocity components, it may be calculated that

$$U_{\text{m.p.}}^2 = \tfrac{1}{2}\mathscr{U}^2 = \tfrac{1}{3}\mathscr{C}^2 = 2\cdot 01 \times 10^{11}\, T\,(\text{cm/sec.})^2$$
$$= 5\cdot 6 \times 10^{-5}\, T \text{ equivalent volts.}$$

At $T = 2400°$ K., $U_{\text{m.p.}}^2 = 0\cdot 134$ equivalent volt.

So the most probable electron (velocity)² in the direction perpendicular to the hot surface just after emission is only $0\cdot 134$ equivalent volt at 2400° K.; and Fig. 116 shows that about 1 °/₀ of the electrons emitted have a normal component (velocity)² exceeding $0\cdot 63$ equivalent volt‡, and only 1 in 10^6 exceeding $1\cdot 9$ equivalent volts.

In most uses of thermionic tubes, even of the low-voltage reception patterns, the electrons are subjected after emission to fields producing velocities of the order of tens or hundreds of volts. The small emission

---

* See Richardson, *loc. cit.* p. 179.
† The distributions of the three velocity components are given by Richardson on p. 156 of his book (*loc. cit.* p. 179). They are of the form:

in the u direction $\qquad y = Ax\epsilon^{-x^2}$,
in the v and w directions $y = B\epsilon^{-x^2}$.

Calling $\mathscr{U}, \mathscr{V}, \mathscr{W}$ the R.M.S. values of the three component velocities, $\mathscr{U}^2 = 4\mathscr{V}^2 = 4\mathscr{W}^2$; so that if $\mathscr{C}^2 = \mathscr{U}^2 + \mathscr{V}^2 + \mathscr{W}^2$ is the mean square total velocity, $\mathscr{U}^2 = \tfrac{2}{3}\mathscr{C}^2$.

‡ These velocities are rather lower than those quoted for the same temperature by Eccles on p. 262 of his book, *loc. cit.* p. 173. His figures appear to refer to the total velocity $\mathscr{C}$ instead of our normal component velocity $\mathscr{U}$.

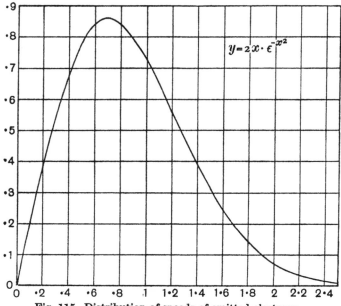

Fig. 115. Distribution of speeds of emitted electrons.

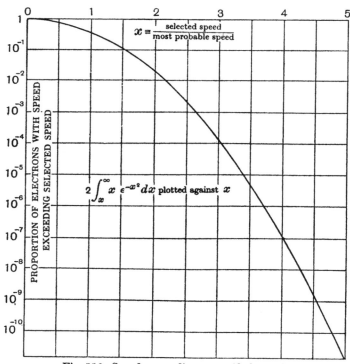

Fig. 116. Speeds exceeding a specified speed.

velocities are therefore seldom appreciable*, and for most purposes we regard the electrons as being merely liberated at, and not ejected from, the hot surface. With the cooler filaments of modern receiving triodes the emission velocities are still lower than those we have calculated for 2400° K.

3. THERMIONIC TUBES: DIODES

The thermionic tube is an instrument for utilising the phenomenon of the emission of electricity from hot bodies. To this end there are required:

(i) An emitter which will withstand the requisite temperature without rapid destruction.
(ii) Means for heating the emitter.
(iii) Means for protecting the emitting surface from the influence of other matter†; i.e. for maintaining a vacuum around the emitter.
(iv) Means for carrying away the emitted electrons.

Accordingly, the ordinary form of thermionic tube comprises

(i) a *cathode*, consisting of a material such as tungsten, which is usually
(ii) in the form of a filament, so that it may be conveniently heated by an auxiliary electric current.
(iii) The filamentary cathode is mounted in an evacuated glass bulb, in proximity to
(iv) a cold electrode, the *anode*. The anode serves as a landing ground for the flight of emitted electrons traversing the vacuum, whence they flow away from the tube through a wire sealed in the glass.

The anode is shaped so as to surround the cathode more or less completely, in order that as much of the hot emitting surface as possible may be utilised.

---

* A notable exception is in the grid rectifier. See XI–4 (b).

† An early "soft" triode is seen at C, Plate XIII, in which the gas pressure was regulated by warming the asbestos pellet sealed in the pocket of the bulb. Small quantities of various gases have sometimes been intentionally left for special purposes, but the practical interest of such tubes seems to be small. The study of the effects of residual gases, however, has much theoretical interest, and is a practical concern of the manufacturer in that his efforts are largely devoted to avoiding these effects by eliminating the imperfections of vacuum to which they are due. In this book we confine our attention to tubes with sensibly perfect vacuum.

VII. THERMIONIC TUBES: GENERAL PROPERTIES

The cathode is usually—not always—heated by an electric current flowing through it. But the heating current is purely an accessory*. The emission depends on the filament temperature and has no relation to the filament heating current except in so far as that current affects the temperature. Hence, although the filament must be provided with two terminals, the filament as a whole is to be regarded as one electrode of the thermionic tube. That it is not at a uniform potential is an accident, which, however, has sometimes to be taken into account. It is the custom to take the negative end of the filament as the datum line in specifying potentials. If we say, for example, that the anode potential is 100 V, we mean that it is 100 V above that of the negative end of the filament (and is therefore, perhaps, only 96 V above the potential of the other end).

Our thermionic tube is shown diagrammatically in Fig. 117. B is the highly evacuated bulb containing F, the hot filament or cathode, and A, the cold plate or anode. A is at the potential e; and the current i in amperes is $\frac{1}{6\cdot 3 \times 10^{18}}$ times the number of electrons passing across the vacuum from F to A in a second. There being only two electrodes, F and A, in the sense explained above, the tube is called a two-electrode thermionic tube or *diode*.

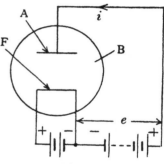

Fig. 117.

If the battery e were reversed in sign, no electron could arrive at A unless its velocity after emission had exceeded e; for in passing from F towards A it would be moving along a down-gradient of potential and so undergoing deceleration. We have seen that when the filament temperature is 2400° K., only 1 in 10^6 of the electrons emitted has a velocity exceeding 0·5 equivalent volt; so that if the anode potential were − 0·5 V, less than a millionth of the saturation current could flow across the vacuum. If, therefore, the battery e in Fig. 117 were replaced by an alternator of E.M.F. exceeding a few volts, the diode would act as a remarkably perfect uni-directional

* Tubes are now widely used in which the cathode is a hollow cylinder, containing an electrical heater but not connected therewith. Since the potential of the cathode in these "indirectly heated" tubes is independent of potential drop across the heater, the heating power (commonly 4 W) may be taken from A.C. mains, thus dispensing with a filament battery.

3. THERMIONIC TUBES: DIODES

valve, permitting flow of current when the E.M.F. is in one direction but resolutely maintaining sensibly zero current when the E.M.F. exceeds half a volt in the other direction*.

This non-return action of the diode can be demonstrated in a striking manner by the simple experiment illustrated in Fig. 118. A buzzer or electric bell vibrator sends an intermittent current from a feeble dry battery through the primary of a small step-up transformer, to the secondary of which a thermionic tube† and a condenser of 2 or 3 microfarads are connected in series. A spark gap set for 1000 volts or so is connected across the condenser. The latter gradually becomes charged up—it may take half a minute—until a P.D. is reached sufficient to crash across the spark gap.

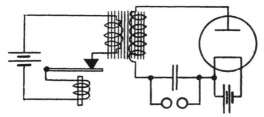

Fig. 118. Experiment with high-vacuum thermionic tube.

Diodes with tungsten filaments are used as valves in this way for obtaining a D.C. supply at high tension from an A.C. source. They are constructed in all sizes, giving outputs from (say) 30 mA at 200 V for domestic broadcast receivers, up to 100 times this current and 100 times this voltage (3 A × 20 kV). Such diode valves are very convenient for providing the D.C. supply at upwards of 1000 V required in high-power triode oscillators. Plate XV shows such a diode, suitable for providing a quarter of an ampere or so at several thousand volts.

* The presence of residual gas might declare itself here. Collision between an atom and an electron of sufficient speed ionises the atom, knocking off an electron and leaving a positive ion. Positive ions are carried in the direction filament-to-anode when the anode is of *negative* potential. Since *some* fraction—however minute—of the emitted electrons must possess the adequate speed (e.g., 10 equivalent volts for mercury vapour, 14 for hydrogen), the presence of residual gas must cause a current from filament-to-anode when the latter is at a negative potential. If sufficient gas is present, this current may be of appreciable magnitude; the phenomenon is used to test the perfection of vacuum in a tube. It sometimes goes by the inappropriate name of *backlash*.

† An ordinary receiving triode, with grid and anode connected together, may be used.

4. THE SPACE-CHARGE

When the anode battery is connected as shown in Fig. 117, so that the anode is positive, electrons pass across the vacuum from filament to anode; but in general the current is less than the saturation current I_s of the Richardson formula (p. 180). Of the electrons liberated at the cathode—we neglect their velocities of emission—some are carried across to the anode, but some return to the filament despite the positive potential of the anode. Langmuir showed* that the explanation lies in the electric field of the *space-charge*, the cloud of free electrons occupying the space between cathode and anode, of which the drift towards the anode constitutes the current traversing the vacuum. The number of electrons reaching the anode per second is a compromise between the influence of the positive anode in drawing electrons towards itself and of the negative space-charge in thrusting electrons back into the cathode.

As the anode potential is raised from zero, the current therefore does not leap to its saturation value, but rises smoothly up to the saturation value, after which no further rise of anode potential can increase the current. The curve relating anode current i to anode potential e is therefore some such curve as that of Fig. 119. We have to find the shape of the curve between $i = 0$ and $i = I_s$.

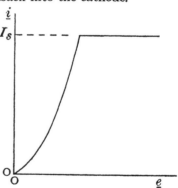

Fig. 119. Diode characteristic.

Langmuir* deals first with the simplest electrode system, viz. when cathode and anode are infinite parallel planes. In Fig. 120 these are seen in section; the lines with arrow-heads indicate the lines of electric field reaching out from the anode; and below are curves showing the corresponding relations between p, the potential at any point, and x, the distance of the point from the cathode. We follow the analysis given by Eccles†.

Let I_1 and E be the current per square centimetre and the anode

---

* *Loc. cit.* p. 179.
† P. 272 *et seq.* of book, *loc. cit.* p. 173.

4. THE SPACE-CHARGE 189

potential. At any distance x from the cathode, let p be the potential and F, D the electric field strength and flux-density; and let n be the number of electrons per unit volume, and u their velocity of translation towards the anode. Let m, q be the mass and charge of an electron. Obvious relations* between these quantities are:

$$I_1 = nqu, \quad F = \frac{dp}{dx}, \quad \tfrac{1}{2}mu^2 = qp.$$

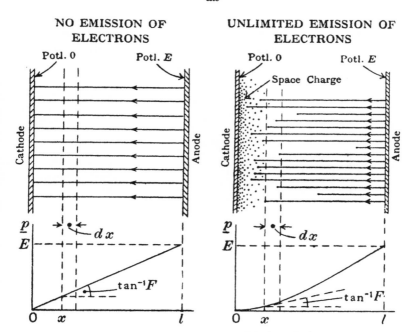

Fig. 120. Electric field in parallel-plate diode.

A fifth relation is given by equating the number of lines of electric flux-density ending in any element of volume to the charge of the electrons within this volume; i.e.

$$\frac{dD}{dx} = nq,$$

$$\text{or} \quad \frac{dF}{dx} = \frac{4\pi nq}{k},$$

where k is the dielectric constant of the vacuum.

* E.S. or E.M. units must, of course, be used throughout. See II–3.

To these four simultaneous equations must be added the terminal conditions:

at $x = 0$, $p = 0$ and at $x = l$, $p = E$; and

at $x = 0$, $F = 0$, because there is supposed to be an abundance of electrons emitted at the cathode, most of which are surplus and are reabsorbed by the cathode.

Eccles makes the trial solution

$$p = ax^b,$$

where a and b are to be determined by substitution in the equations. This leads to the solutions, expressed in amperes, volts, centimetres and seconds:

$$I_1 = E^{\frac{3}{2}}(2 \cdot 33 \times 10^{-6}\, l^{-2}),$$
$$p = E\,(l^{-\frac{4}{3}})\, x^{\frac{4}{3}},$$
$$n = E\,(2 \cdot 50 \times 10^5\, l^{-\frac{2}{3}})\, x^{-\frac{2}{3}},$$
$$F = E\,(1 \cdot 33\, l^{-\frac{4}{3}})\, x^{\frac{1}{3}},$$
$$u = E^{\frac{1}{2}}(5 \cdot 9 \times 10^7\, l^{-\frac{2}{3}})\, x^{\frac{2}{3}}.$$

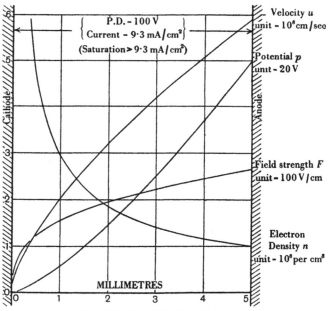

Fig. 121. Parallel-plate diode.

It is interesting to attach numerical values to these expressions. Let us take electrodes 0·5 cm apart at a P.D. of 100 V; i.e. $l = 0.5$,

4. THE SPACE-CHARGE

$E = 100$. The current density I_1 is then $0\cdot093$ A/cm² $= 9\cdot3$ mA/cm². Fig. 121 shows the values of n, F and u at any point between cathode and anode. It is noteworthy that the region of rapid change of field strength F and electron density n is confined to the near neighbourhood of the cathode. While the electron density is in-

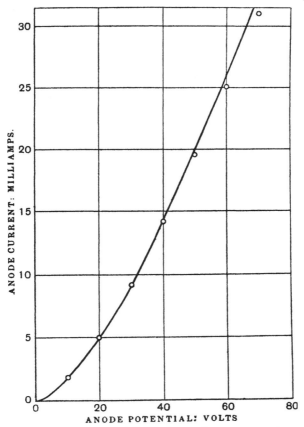

Fig. 122. Foot of diode characteristic. Curve calculated for anode 1·9 cm long, 0·5 cm radius. Points marked observed with Army "U 3" valve.

definitely large ever so close to the cathode, at one-tenth of the distance to the anode (0·5 mm in our example) it has only 4·6 times its value at the anode. The field strength F is $0\cdot77\ F_0$ at one-tenth the distance to the anode, and is $1\cdot33\ F_0$ at the anode, where F_0 is the field strength which would have obtained throughout if there had been no space-charge. ($F_0 = 200$ V/cm in our example.) The space-charge and its effects are mainly concentrated in a layer very close to the cathode.

The analysis has shown that the shape of the curve in Fig. 119 is $I = A \cdot E^{\frac{3}{2}}$. Langmuir also obtained a solution* of the more difficult but more practical electrode arrangement in which cathode and anode are concentric cylinders. When the cathode is a filament of diameter small compared with the diameter D of the anode, and end effects are neglected, for the same conditions of unlimited emission and zero emission velocity, the current per unit length of cylinders is

$$I_{\text{amps.}} = \frac{2 \cdot 9 \times 10^{-5}}{D} E_{\text{volts}}^{\frac{3}{2}},$$

and the potential at x from the axis is

$$p = E \left(\tfrac{1}{2} D\right)^{-\frac{2}{3}} x^{\frac{2}{3}}.$$

This formula for I is plotted against E in Fig. 122 for an actual diode† in which the anode was 1·5 cm long and 0·5 cm in radius. To allow for fringing, 0·2 cm has been added at each end, bringing the effective length to 1·9 cm. The points shown on the graph were experimentally observed with this diode, and are seen to follow the $\tfrac{3}{2}$-power law very closely.

Langmuir showed further that the $\tfrac{3}{2}$-power law holds not only for plane and cylindrical electrodes, but for electrodes of any shape whatever.

5. Introduction of a Third Electrode

We pass now to the effect on the space-charge of the introduction into it, or near it, of a third electrode whose potential is determined by conditions outside the tube. The tube then becomes a three-electrode tube or *triode*. The third or control electrode is commonly called the *grid*.

The control electrode may be of any shape and in any position where it can exert an influence on the space-charge determining the flow of current between cathode and anode. The most effective arrangement is a perforated plate located between anode and cathode, near to the latter where the space-charge is dense. The grid is visible

---

* *Loc. cit.* p. 179.
† Substantially the triode of Plate XIV with grid omitted.

5. INTRODUCTION OF A THIRD ELECTRODE

in B of Plate XIII as a zigzag wire situated between the filament and the plate-form anode. In C and D of the same Plate, and in Plate XVII, it is a fine wire meshwork surrounding the filament, the whole being surrounded by the cylindrical anode; and in Plates XIV and XVI it is clearly visible as a cylindrical or flattened open wire helix. The introduction of the grid was the work of L. de Forest in 1907, and although the rationale of the triode does not appear to have been fully understood at the time, this addition to the Fleming Valve marks the conversion of the thermionic tube from a mild convenience (as a substitute for the crystal rectifier) into a potent and indispensable instrument with a wide field of divers applications.

The reader may be helped to some physical picture of the action of the grid by studying the diagrams in Fig. 123. Here C, G and A are sections of the cathode, grid and anode, which may be regarded for simplicity as portions of large plane surfaces perpendicular to the paper. The diagrams are intended to show the lines of electric stress in the space between anode and cathode which would act upon the electrons if any were present, urging them to move in the direction against the arrow-head. The presence of the electrons modifies the field after the fashion investigated in the last Section, but by first considering the field in their absence we can obtain some insight into the action of the grid. To arrive at the electron currents, which are stated on the right of the diagrams, we have to picture a cloud of electrons congregated near the cathode, repelling freshly emitted electrons back into the cathode, but ready and waiting themselves to be dragged away towards grid and anode by an electric field in the direction A to C.

Below saturation, i.e. as long as more electrons are emitted than are carried away from the cathode, no lines of force from anode or grid terminate on the filament, but each line ends on a free electron constituting part of the space-charge. In Fig. 124, an attempt is made to show the lines of force urging the free electrons towards the anode in the case of the diode, and towards grid and anode in the case of the triode. The grid is supposed to be somewhat positive (i.e. with respect to the cathode), and the condition unsaturated. It will be seen that the grid to some extent shields the space-charge from the anode, though only slightly if the grid mesh is coarse; but from its proximity to the cathode it exercises a powerful influence over the dense portions of the space-charge, drawing out electrons and impelling them in the direction filament-to-grid. Many of these

194   VII. THERMIONIC TUBES: GENERAL PROPERTIES

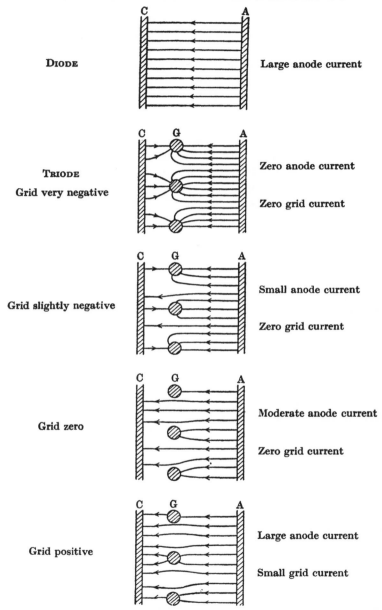

Fig. 123. Electric field between anode A maintained at high potential, and cathode C at zero potential, when grid G is varied in potential. Cathode not heated.

6. TRIODE METRICAL RELATIONS

electrons shoot through the interstices of the grid and are seized by the anode. If the grid is of very fine mesh, even when positive in potential its insertion between anode and filament may decrease the anode current. But in any case the anode current is obviously influenced by the grid potential, and may be very acutely influenced if the grid is of suitable form and suitably situated. When the grid is below the cathode in potential, its current is zero*, and no power is absorbed by it; and even when its potential is positive, if the grid wire is fine the grid current is small. In the practical use of triodes, the grid current is sometimes very small and often zero.

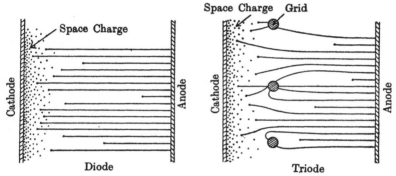

Fig. 124. Lines of force from anode and grid when electrons are emitted from cathode.

6. Triode metrical relations

6(a). *Anode current a function of anode and grid potentials.* When the high-vacuum triode was first produced, Langmuir discovered empirically that the anode current (well below saturation) followed a $\frac{3}{2}$-power law as in the diode. But whereas in the diode the potential e_a (with respect to the cathode) of only one electrode is involved, in the triode the potentials of the anode e_a and of the grid e_g are both effective in controlling the anode current. In the diode

$$i_a = A e_a^{\frac{3}{2}},$$

* If the electrons were released from the cathode with infinitesimal velocity, this would be strictly true; actually some of them have appreciable velocities of emission, so that the grid current does not become zero until the grid potential is a volt or so negative. Strictly speaking, it never becomes *quite* zero. See VII–2 (c).

VII. THERMIONIC TUBES: GENERAL PROPERTIES

and in the triode the law was found to be

$$i_a = B(e_a + \mu \cdot e_g)^{\frac{3}{2}},$$

where μ is a constant of the triode determined by the dimensions and relative positions of the electrodes.

The relationship between anode current and grid and anode potentials—when the current is well below saturation and the grid is at much lower potential than the anode—has since been deduced theoretically for both plane and cylindrical electrodes*. For the triode with cylindrical electrodes the formulae are given in Fig. 125, in which is included for comparison the formula for the cylindrical diode already given on p. 192. i_a is the anode current per unit length, neglecting end effects; and the units are amperes and volts.

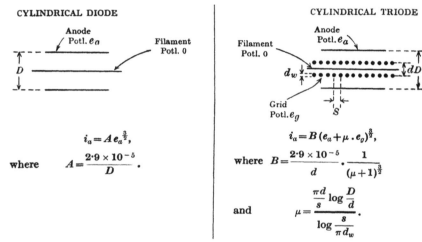

Fig. 125. Cylindrical diode and triode.

Now the revolutionary possibilities of the triode are wrapped up in this determination of anode current by a function of

$$(e_a + \mu \cdot e_g).$$

This expression implies that the current in the anode circuit is affected as much by a volt of grid potential as by μ volts of anode potential; so that if μ can be made large, changes of current can be effected in the anode circuit by E.M.Fs. introduced into the grid

* By W. H. Eccles, *Radio Review*, Vol. 1 (1919), p. 67. The student should consult his book, *loc. cit.* p. 173. The formulae quoted above are those deduced by Eccles.

6. TRIODE METRICAL RELATIONS

circuit of much smaller magnitude than would be needed if they were applied in the anode circuit directly. This condition alone, however, does not imply that there is here an amplifier necessarily any more remarkable than (say) the ordinary alternating-current step-up transformer (Fig. 126). The same change in i_s could be effected by introducing at X a much smaller E.M.F. than would be needed at Y; but we should have to pay, in terms of power, just as high a price at X as at Y. In the triode, on the other hand, we have seen that there is no necessary relation between the effect in the anode circuit and the work done on the grid in producing that effect. The grid current is *not* determined by the anode current; it is indeed often maintained sensibly zero, in which case large anode circuit powers may be controlled by changes of grid potential involving the expenditure of no power*.

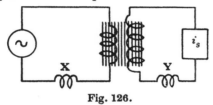

Fig. 126.

In the diode when e_a, and in the triode when $(e_a + \mu . e_g)$, is raised enough to bring i_a to the saturation value, no further rise of potential can produce further increase of current. The theoretical curve between i_a and e_l†$\equiv (e_a + \mu . e_g)$ is therefore the full-line curve of Fig. 127, which resembles the full-line curve in Fig. 119 (p. 188) for the diode.

Curves found experimentally agree well with the theoretical curve when the current is far from saturation. We have had an illustration of this in Fig. 122. But the sharp transition from the $\frac{3}{2}$-power law to saturation shown in Figs. 119 and 127 is never realised. There are obvious reasons for this, the chief being the fringing effects at the free edges of the electrodes and the non-uniform temperature of the filament along its length. The temperature of the filament is a maximum at the middle, falls gently for some

---

* The control over energy transformations exercised by the negatived grid of a high-vacuum triode provides perhaps the most perfect realisable physical analogue of the function of the will in deciding whether a man shall do this or that with the energy within him; neither the grid potential nor, we may suppose, the will entering into the energy equations.

† $e_l \equiv e_a + \mu . e_g \equiv$ *lumped potential*—a useful term coined by Eccles.

way on each side of the middle, and then near the supports falls rapidly. A millimetre of filament near a cool end will be delivering its whole emission when e_l has not been raised nearly far enough to draw away all the electrons emitted from a millimetre near the middle of the filament. Since i_a is the sum of the currents taken from all the millimetres of the filament, the onset of saturation of the whole filament is not sudden. If the full line of Fig. 127 is the curve which would be given with a uniform filament at a certain temperature, the dotted curve indicates what may be expected with a filament which has this temperature only at the middle and tapers off in temperature towards the ends.

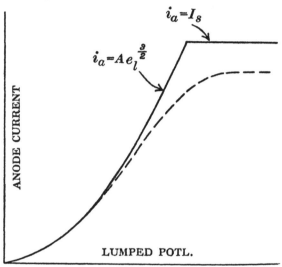

Fig. 127. Effect of temperature gradient along filament.

Other causes of departure from the curves predicted by the theory of the ideal triode are the finite and not uniform velocities of emission; contact E.M.F. between the cathode and other electrodes, especially the grid; and non-uniformity of cathode potential. The emission velocities have been discussed in VII-2 (c). The *contact E.M.F.* between two materials is the difference of their electron affinities (p. 175), and may amount to a volt or so, depending on the materials of the grid (or anode) and of the surface of the filament. The P.D. across the ordinary filamentary cathode may be from 2 V to 6 V in receiving tubes and up to 30 V in high-power transmitting

6. TRIODE METRICAL RELATIONS

tubes. The effects of these departures from the ideal conditions assumed in the theory become appreciable when the potentials are such as to draw only a very small current to the grid (or anode).

A disturbing force not contemplated in the simple theory is the force $B_y qu$ along the z direction experienced by an electron of charge q moving with velocity u along the x direction in a magnetic field of flux-density B_y along the y direction. In high-power tubes with large heating current I in the filament, the magnetic flux-density $\frac{2I}{r}$ at a distance r from the axis of the filament may have a sensible effect in curving the paths of electrons.

Although the real conditions within a triode and the ideal conditions postulated in the simplified theory differ enough to make the real characteristics depart widely from the ideal curves

$$i_a = B\,(e_a + \mu \cdot e_g)^{\frac{3}{2}},$$

the current does approximately follow the equation

$$i_a = f(e_a + \mu \cdot e_g).$$

That the function f is not the $(\frac{3}{2})^{\text{th}}$ power does not affect the fundamental intrinsic property of the triode referred to on p. 196. It is still true that a volt at the grid has the same effect on the anode current as μ volts on the anode, although it is not the effect specified by the $\frac{3}{2}$-power law. The *gradual* incidence of saturation as i_a increases tends to make slope $\frac{di_a}{de_l}$ decrease, and offsets the increasing $\frac{di_a}{de_l}$ of the $\frac{3}{2}$-power law; so that in practice a long middle region of the characteristics is ordinarily nearly rectilinear. For many purposes this is very advantageous.

6 (b). *Characteristic curves.* We have then

$$i_a = f(e_a + \mu \cdot e_g),$$

where f is a function which cannot be readily calculated, and μ is a function of the geometry of the electrodes more or less accurately calculable for particular geometrical forms.

To ascertain the properties of any given triode in relation to external circuits, f and μ must be found. This is readily done experimentally by plotting i_a against e_g for two (or more) constant values of e_a. If μ were strictly a constant, only one point on the second curve would be necessary*, since the curve for $e_a = E_2$ would be an exact copy of the curve for $e_a = E_1$ shifted back through a distance $\frac{1}{\mu}(E_2 - E_1)$ (Fig. 128). This becomes obvious on writing

$$f[E_1 + \mu e_g] = i_a = f\left[E_2 + \mu\left(e_g + \frac{E_2 - E_1}{\mu}\right)\right].$$

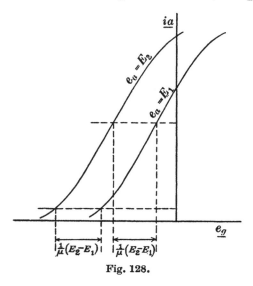

Fig. 128.

An equivalent procedure is to plot curves of i_a against e_a for constant values of e_g.

In view of the simplifying approximations made in the theoretical analysis, it is not certain, or even likely, that f and μ will remain entirely unchanged over a wide range of values of e_a and e_g. It is prudent, therefore, to plot a family of such characteristic curves, extending over the range which may be covered in the practical operation of the triode. Fig. 129 is such a set of characteristics, belonging to the triode of Plate XIV.

A glance at these curves suffices to show that the function f is at least approximately independent of the potentials, except when the grid potential is positive and not much less than the anode potential.

* If f and μ were wholly independent of the anode and grid potentials, for any given filament current there would be a unique lumped characteristic (i_a, e_l). This curve and the value of μ would together completely specify the performance of the triode at that filament current.

But under these conditions the grid current—as the lowest curve in Fig. 129 shows—is not very much smaller than the anode current. Now the theory which led us to expect

$$i_a = f(e_a + \mu \cdot e_g)$$

was based on the assumption that sensibly all the electrons reaching

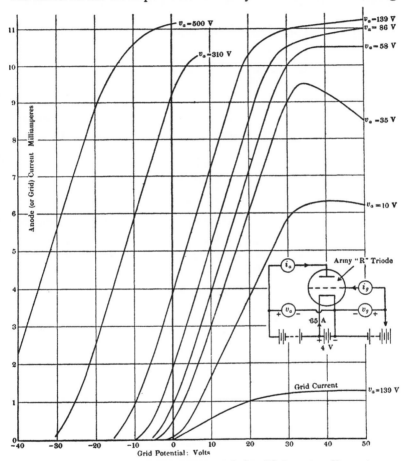

Fig. 129. Typical small high-vacuum triode with tungsten filament: observed characteristic curves.

the grid passed forward to the anode; so the change of shape of the curves at potentials giving considerable grid current need cause no surprise. Nor is it of much practical import, since the triode

VII. THERMIONIC TUBES: GENERAL PROPERTIES

is not used under these conditions*. It must be remembered that, whatever the projected area of the grid wires, the grid current is nil when the grid potential is negative.

A direct test of the constancy of μ is made on picking out from the family of characteristics pairs of values of e_a and e_g corresponding with the same value of i_a. This is done for $i_a = 5$ mA in the Table.

e_a	500	310	139	86	58	35	10
e_g	-32	$-12\frac{1}{2}$	$+3\frac{1}{2}$	$+9\frac{1}{2}$	$+13\frac{1}{2}$	$+16\frac{1}{2}$	$+26$
$e_a + 10e_g$	180	185	174	181	193	200	270

The third line of the Table shows that $(e_a + 10e_g)$ is nearly constant until the grid current becomes large, proving that here μ is a constant, and equal to 10†.

A feature of the characteristics is the remarkable straightness of a large part between the foot, where we may expect

$$i_a \propto (e_a + \mu . e_g)^{\frac{3}{2}},$$

and the knee, where saturation becomes very apparent. This straightness is a very fortunate property, since it gives a wide region where the relations between i_a, e_a and e_g are very nearly linear. In this linear region we may write

$$i_a = P + Q(e_a + \mu . e_g),$$

where P and Q are constants.

At any value i_a, the slope g of an i_a, e_g characteristic found at constant anode-potential is

$$g \equiv \frac{\partial i_a}{\partial e_g} = \mu . f'(e_a + \mu . e_g);$$

and the slope a of an i_a, e_a characteristic found at constant grid potential is

$$a \equiv \frac{\partial i_a}{\partial e_a} = f'(e_a + \mu . e_g).$$

These two slopes have the dimensions of conductances, and may be called respectively the *slope* (or *differential*) *anode/grid conductance*

---
* Except in oscillators working at high efficiency, in which case the exact shape of the curves has little interest.
† See also p. 209.

6. TRIODE METRICAL RELATIONS

and the *slope* (or *differential*) *anode/anode conductance**. They are usually given the less specific but briefer names, the *mutual conductance* and the *anode conductance*. Their ratio is μ; and this, with obvious appropriatenes, is called the *amplification factor* of the triode. The reciprocal of the anode conductance a is the *anode resistance*

$$\rho \equiv \frac{\partial e_a}{\partial i_a}.$$

For currents in the curvilinear regions, g and a vary with the current, although their ratio $\mu = \frac{g}{a}$ is nearly constant. For currents in the rectilinear regions, μ is constant, and g and a are also individually constant. In the triode whose characteristics are given in Fig. 129, over the linear regions—say between 2 mA and 8 mA—

$$g = 0\cdot 33 \text{ mA/V}, \quad \mu = \frac{g}{a} = 10,$$

$$a = 0\cdot 033 \text{ mA/V}, \quad \rho = \frac{1}{a} = 30 \text{ k}\Omega.$$

6 (c). *Simple equivalent circuit.* Consider the triode circuit in Fig. 130.

$$\delta i_a = \frac{\partial i_a}{\partial e_a} \cdot \delta e_a + \frac{\partial i_a}{\partial e_g} \cdot \delta e_g$$

$$= a \cdot \delta e_a + g \cdot \delta e_g$$

$$= \frac{1}{\rho}(\delta e_a + \mu \cdot \delta e_g).$$

This transformation justifies the equivalence expressed in the Figure. In words: the change of anode current flowing through Z produced by the change δe_g of grid potential is the same as would be produced by a P.D. $\mu \cdot \delta e_g$ applied to Z if the resistance of Z were augmented by ρ.

For anode currents in the rectilinear region of the triode characteristics, ρ is a constant despite finite changes δi_a. In the curvilinear

---

* The British Engineering Standards Association has adopted "A.C." to signify "slope" or "differential," but this nomenclature is often awkward and is apt to mislead. In dealing with triodes it is, in fact, seldom necessary to state explicitly that the *slope* resistance $\frac{\partial e}{\partial i}$ is meant. "Resistance" signifying $\frac{e}{i}$ is a term no one is tempted to apply to triode characteristics, since it exhibits no sort of constancy. Confusion does not arise, because, in the metallic parts of circuits which include triodes, the slope resistance $\frac{\partial e}{\partial i}$ and Ohm's resistance $\frac{e}{i}$ are equal.

regions, ρ is a function of the current, and can be treated as constant only for sufficiently small changes.

This conception of the triode simple equivalent circuit is often very helpful.

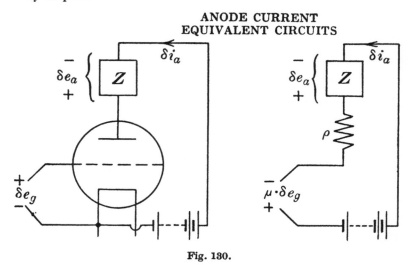

Fig. 130.

7. Modern patterns of triodes

7 (a). *Low power*. In most applications of triodes, linearity of the anode current characteristics over the working range is unobjectionable, and it is often highly desirable. The only important exception is in their application as anode rectifiers, when it is the curved foot of the characteristic which is operative. The suitability of a triode for a particular service is determined, therefore, primarily by the two slope conductances a and g in the straight region of the characteristics. These are often most conveniently specified by the amplification factor μ and the anode resistance ρ. Taken separately, high μ and low ρ are always to be regarded as virtues in assessing the practical effectiveness with which a triode can be made to function if provided with circuits dimensioned to suit it. There are, of course, other considerations, such as the maximum output, and the power drawn from the requisite filament and anode batteries.

Since μ and ρ are not independent variables in the design of a triode, without further investigation, and without specification of

7. MODERN PATTERNS OF TRIODES

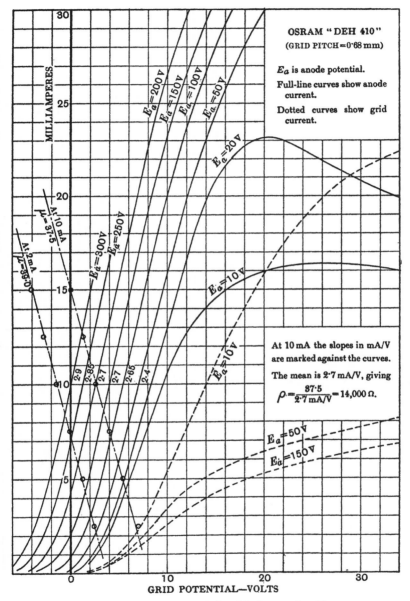

Fig. 131. A receiving triode with fine-mesh grid.

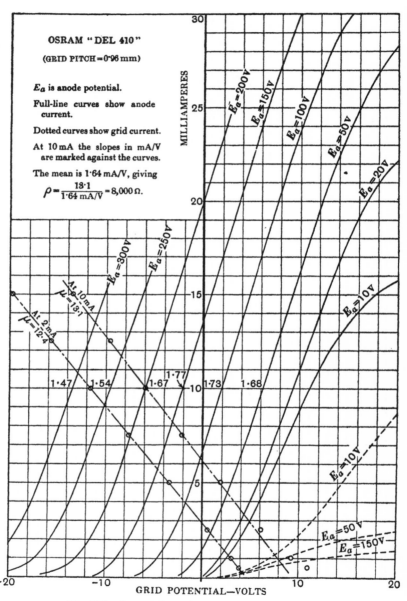

Fig. 132. A receiving triode with medium-mesh grid.

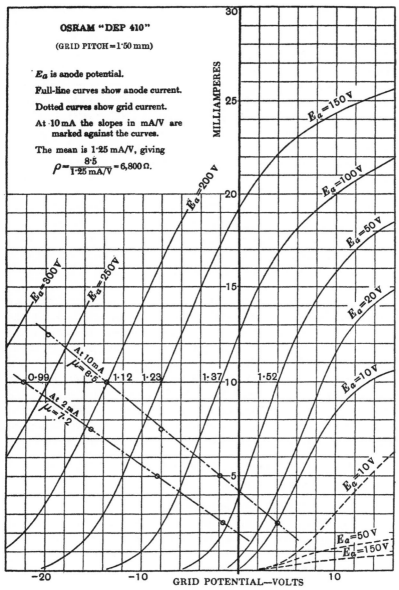

Fig. 133. A receiving triode with coarse-mesh grid.

the particular application, it is not possible to balance the advantage of an increase in μ against the disadvantage of an accompanying increase in ρ. Nevertheless, a comparison of the characteristics of a series of modern small triodes having divers μ values with those of their predecessors, such as the "R" triode of Plate XIV and Fig. 129, shows at once that great improvements have been made. As typical of the latest practice we take the M.O. Valve Co's graded series of three triodes of one class designated "DEH 410," "DEL 410" and "DEP 410." These three triodes differ only in the grids, the pitches of the grid windings being respectively 0·68, 0·96 and 1·50 mm. The filaments are rated at 0·1 A × 4 V (max.); they are of a coated type giving a very high emission per watt with long life. Anode potentials up to 150 V are recommended by the makers. These triodes are shown in Plate XVI, and very complete sets of characteristic curves are given in Figs. 131, 132 and 133. The curves are the accurate plots of observations on specimen triodes. To facilitate comparison, they are all drawn with the same scales, and with the same set of anode potentials for each triode. The filament P.D. was 3·8 V throughout.

The amplification factor μ at any particular anode current (say 10 mA) is best found from such a family of characteristics as the slope of the straight line which most nearly passes through the set of points x, y, where x is the grid potential and y is the anode potential of a 10 mA point on a characteristic. The construction is carried out for 10 mA and for 2 mA on each of the three Figures by the dot-and-dash lines ruled through the point indicated by small circles*. Thus in Fig. 133, the 10 mA line cuts the 250 V characteristic at $e_g = -19·8$ V, the 50 V characteristic at $e_g = +4·0$ V, and so on. The x, y points giving the dot-and-dash line are accordingly $-19·8$ V, 250 V; 4·0 V, 50 V; and so on. The slope $\frac{dy}{dx}$ of this line is 8·5 V/V, so that $\mu = 8·5$.

It will be seen from the Figures that:

(i) If we except any points for which e_g is positive and not small compared with e_a, the points do lie very evenly on a straight line. This indicates that μ is indeed approximately constant at any one value of i_a.

* The y-scale taken for this purpose is 20 V per unit of the marked (milliamperes) scale.

7. MODERN PATTERNS OF TRIODES

(ii) Although 2 mA is on the curved foot of each characteristic—where the slopes $\dfrac{\partial i_a}{\partial e_g}$ have fallen to about half the slopes at 10 mA—the value found for μ is much the same at 2 mA as at 10 mA. This indicates that μ is indeed approximately independent of i_a.

As further illustration of the approximate constancy of μ at all anode currents, the following set of direct determinations in the triode of Fig. 131 may be cited. At anode currents of 4000, 2000, 300 and 50 microamperes, μ was found to be 36·5, 38, 36·5 and 34 respectively. The theoretical constancy of μ (in the absence of grid current) is thus experimentally confirmed.

We find illustrated in this series of triodes that a geometry making for high μ also makes for high ρ. For different purposes one or another of these three triodes will be best. We shall have occasion to illustrate this in subsequent numerical examples of various triode applications. Each of these triodes is markedly superior to the old "R" triode of Plate XIV with its pure tungsten filament, either in respect of μ, or of ρ, or of both μ and ρ. This is brought out in the Table, where the μ and ρ values are collected.

Name of triode	"R"	"DEH 410"	"DEL 410"	"DEP 410"
μ	10	37	13	8
ρ	30 kΩ	14 kΩ	8 kΩ	7 kΩ*

The use of a filament emitting at a much lower temperature than tungsten, besides conferring the direct advantage of reduced filament power for a given emission, has also made it practicable to construct triodes with smaller spacing between the cathode and the grid. The production of the high vacuum has been very much facilitated in low-power tubes by the eddy-current heating of the electrodes during evacuation, and by the "gettering" process. The grid and anode used to be heated, while the tube was on the pumps, by heavy electron bombardment from the filament, which was very damaging to the filament; but they are now heated by the application from outside

* The value of ρ in the useful working range is so little less than that of "DEL 410" as to make that excellent triode preferable for any purpose, except in so far as rather lower anode potentials for equal current are required in "DEP 410."

the tube of an alternating magnetic field of very high frequency. The gettering process consists in volatilising a fragment of magnesium (or other material) mounted within the tube, which deposits itself on the glass and takes with it, in chemical combination or otherwise, most of the residual molecules of gas left after the pumping. The silvery deposit on the glass so noticeable in most modern receiving triodes is the result of this process.

7 (b). *High power.* Whereas in the small receiving tubes we have been considering, the power supply to the anode circuit is of the order of a watt, transmitting triodes are made in which it is measured in kilowatts. In developing high-power triodes, three classes of problems have had to be faced, viz.: (i) how to dispose of the large fraction (20 % to 80 % in practice) of the anode circuit power which remains as heat in the anode; (ii) how to lead into the bulb through vacuum-tight seals the large filament current required to give large emission; and (iii) how to maintain the very perfect vacuum in the bulb necessary to avoid catastrophic discharges of the spark or arc type* under the great electric fields accompanying the high voltages.

These difficulties have restricted the use of glass bulbs to anode dissipation powers of a kilowatt or two, and have led to the development of diodes and triodes (i) with envelopes of silica†; (ii) with envelopes partly of glass and partly of metal, with a vacuum-tight fused junction between them‡; and (iii) with composite metal and glass envelopes, fitted together with demountable ground joints permitting easy renewal of the filament, but necessitating constant pumping during operation§.

Plate XVII shows a glass triode for anode supply at 10 kV and

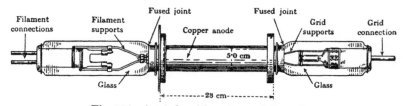

Fig. 134. A triode with water-cooled anode.

* I.e. not pure electron currents.
† See Morris-Airey, Shearing and Hughes, "Silica valves in wireless telegraphy," *Journ. I.E.E.* Vol. 65 (1927), p. 786.
‡ See W. J. Picken, "Cooled-anode valves, and lines of transmitting valves," *Journ. I.E.E.* Vol. 65 (1927), p. 791.
§ See C. F. Elwell, "The Holweck demountable type valve," *Journ. I.E.E.* Vol. 65 (1927), p. 784.

7. MODERN PATTERNS OF TRIODES

dissipation up to 200 W; and Plate XVIII a single-ended metal-glass triode capable of dissipating 7·5 kW at 15 kV.

The main features of a metal-glass, cooled-anode triode* representing modern high-power practice are shown in Fig. 134. The metal portion of the envelope is itself the anode, and is surrounded by a cooling jacket containing water or oil in rapid circulation; 5 or 10 kW can be removed in this way without troublesome rise of temperature of the anode or glass seals. The filament is of pure

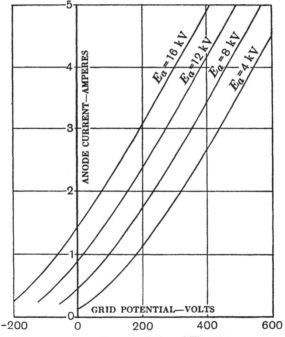

Fig. 135. Characteristics of Fig. 134.

tungsten, as in all high-voltage diodes or triodes; it is 1 mm in diameter and 4×10 cm long. Its heating power is $19 \text{ V} \times 49 \text{ A}$, and the saturation current about 8 A. The grid is a gauze woven of fine tungsten wire. Characteristics of this triode—viz. i_a, e_g for constant e_a, as in Figs. 129–133—are plotted in Fig. 135.

Triodes of this pattern are specially suitable for short-wave work. They are used in the Marconi beam stations at wavelengths of

* Developed and manufactured by the M.O. Valve Co. and designated "C.A.T. 2." The particulars are taken from the paper by Picken, *loc. cit.* p. 210.

14-2

some 15 m. At such very high frequencies (20×10^6 c/s), capacitance currents between the electrodes reach very high values. The double-ended construction allows the inter-electrode capacitance to be kept to a minimum, and gives room for a heavy-current grid lead and seal. In this triode the grid lead can carry a current of 100 A without overheating.

Plate XIX shows what is probably the largest model of metal-glass triode yet constructed*. The overall dimensions are 108 cm × 17 cm. The permissible continuous anode dissipation is 50 kW at about 12 kV. The filament is heated with 30 V × 225 A, and the total emission is then about 40 A. The amplification factor is 45; and with anode and grid potentials around 12 kV and 0, the anode resistance is 3500 Ω.

8. Grid current

If the velocities of emission of the electrons at the cathode were strictly zero, the grid current would be zero for all grid potentials below zero†. This appears to be so in Figs. 129–133, but that is only because the scale of current is too coarse. The theory of the distribution of emission velocities discussed in Subsection VII–2 (c) indicates that the current cannot be *quite* zero however great the field urging emitted electrons back to the cathode. The more sensitive the current indicator used, the larger is the negative potential of the grid (or anode) required to make the current appear to be zero.

In Fig. 136 grid current curves are given for a bright-emitter "R" triode like that of Plate XIV and Fig. 129, and for the dull-emitter "DEL 410" triode of Plate XVI and Fig. 132. In both the filament P.D. was 3·8 V and the anode potential 10 V. In both triodes it is seen that grid current which formerly seemed to be zero is actually finite though small. By using a reflecting galvanometer instead of a relatively coarse milliammeter we have pushed back the grid potentials at which grid current seemed to vanish from 0 and $+1$ V to $-1·7$ V and $+0·2$ V. The more microscopic the measurements the farther to the left would the vanishing point appear to be. Strictly speaking, there is no vanishing point; practically speaking, the current may be called zero when the slope $\dfrac{\partial i_g}{\partial e_g}$ of the curve is small com-

* "C.A.T. 10" of Marconi's Wireless Telegraph Co., to whom I am indebted for the photograph and dimensions.
† This assumes perfect vacuum. See footnote on p. 187.

pared with the smallest conductance significant in the external circuit. We can see from the two curves in Fig. 136 that at $e_g = -1\cdot 7$ V and $e_g = +0\cdot 2$ V respectively the slope $\dfrac{\partial e_g}{\partial i_g}$ exceeds 10 MΩ; if the external circuit is such that a conductance of $\dfrac{1}{10\,\text{M}\Omega}$ is insignificant, the curve has for all practical purposes vanished at those potentials.

Large grid currents such as those visible in Figs. 129–133 may occur in triode oscillators. The small but finite grid currents such as those visible in Fig. 136 are important when the grid-filament path in a triode is used as a rectifier—a very common arrangement. In amplifiers and the modulators of telephone transmitters it is usually desired to avoid all grid current. This is practically achieved in triodes with dull-emitter filaments like those of our "DE-410" series if the grid potential is never allowed to rise above that of the negative end of the filament; and with a pure tungsten filament if the grid potential is kept at least 2 V lower.

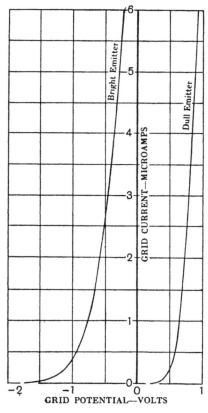

Fig. 136. Grid current.

The displacement of the curve to the right (Fig. 136) on changing from a bright emitter to a dull emitter is, qualitatively, what would be expected from a reduction of the most probable velocity of emission on reducing the temperature of the emitter. But a quantitative enquiry shows that, large as the temperature difference is*, its effect in shifting the foot of the grid current characteristic must be very small. We found (p. 183)

* While the tungsten filament is at a bright white temperature, the dull emitter filament has a temperature too low to be visible.

that the most probable velocity $U_{m.p.}$ for the glowing tungsten is only 0·134 equivalent volts, and for the dull emitter it must be between that and zero. Hence temperature could account only for a shift of less than 0·134 V. The shift is probably to be attributed almost entirely to the inequality in the two triodes of the contact E.M.F. between cathode surface and grid. The contact E.M.F. is the difference of the electron affinities (see p. 176), and the electron affinity of the coating of the dull-emitter filament is a volt or more lower than that of tungsten.

9. Multi-grid tubes

9 (a). *Two grids as control electrodes*. Familiarity with the triode has shown forcibly how a grid placed in an electron current is able, while retaining a large measure of transparency to the electron stream, to control the magnitude of the current. If one grid can thus be interposed between cathode and anode, so obviously can a second and a third, and so on. Whether a series of grids can in practice be advantageously employed is therefore a question worth investigation.

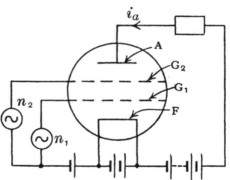

Fig. 137. Two control grids.

Now in wireless practice it often happens that the signal dealt with contains two frequencies; one, the high frequency of the radiation; another, a slightly different high frequency (as in heterodyne reception), or a much lower key or acoustic frequency of variation of amplitude. A double-grid thermionic tube or *tetrode* can be used to deal with one of these frequencies on one grid, and simultaneously with the other on the second grid. Each grid is then functioning in a similar way, viz. as a control grid on which some form of signal

9. MULTI-GRID TUBES

is impressed with the object of producing a corresponding fluctuation of anode current. Thus in Fig. 137 it is clear that the anode current i_a would contain components of both the frequencies n_1 and n_2*.

Tetrodes functioning in this manner, viz. with both grids used as control electrodes, are not of much practical interest; for the same result can be obtained—and in practice better obtained—by the use of two triodes associated with circuits suitably linked.

9 (b). *Fixed-potential space-charge grid.* In the triode the duty of the anode battery may be described as that of producing a field between cathode and anode sufficiently strong to urge electrons across the vacuum at an adequate rate; and of the grid, as that of impeding the operation of the anode battery to an extent depending

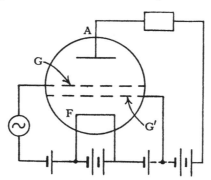

Fig. 138. Fixed-potential space-charge grid.

on the grid potential. In receiving triodes an anode battery of the order of 100 V is called for. We are now concerned with efforts to economise in the voltage of that battery.

* Authors have taken mutually inconsistent views as to how the triode theory should be extended to cover two grids. In the one view A and G_2 are regarded as forming an equivalent anode located at G_2 and of potential $\left(\dfrac{e_a}{\mu_2}+e_{g_2}\right)$. This leads to
$$i_a = f\left[\left(\dfrac{e_a}{\mu_2}+e_{g_2}\right)+\mu_1 e_{g_1}\right].$$
In the other view, the two grids are regarded as qualitatively on a par in affecting the space-charge near the filament. This leads to
$$i_a = F\left[e_a+\mu_2 e_{g_2}+\mu_1 e_{g_1}\right].$$
Probably the triode theory is overstrained in either treatment, and which view is nearer to the truth depends on the electrode configuration. If the two grids were spatially intermingled, the second view would be correct.

216   VII. THERMIONIC TUBES: GENERAL PROPERTIES

The kinetic energy of an electron on reaching the anode is proportional to the anode potential, and is lost as heat at the anode. Now we require an adequate number of electrons arriving at the anode per second, but we have no wish that they should arrive with high velocity. A larger number of electrons in motion across the vacuum at a proportionately smaller speed would be equally effective in the external circuit. In the triode FGA of Fig. 138 (G' for the moment being supposed absent), reduction of the anode battery voltage would result in a reduction of the anode current—which, by hypothesis, is intolerable. Now provided that the requisite number of electrons per second travelling with small velocity reach a position where their further course (forward or backward) is strongly influenced by the potential of G, it matters not how they get there. In the triode FGA (G' absent) they are dragged from F to G by the anode, which, being unfavourably situated to influence the space-charge between F and G, must be given the high potential of (say) 100 V. But a positively charged electrode at G' is in a much more favourable position to draw out electrons from the space-charge near F; and if this electrode is an open grid, most of them will pass through it to G, and so be available as anode current controlled by the potential of G.

Reverting to animistic metaphor, whereas the duty of the anode battery was formerly to produce a field reaching from A right through G to the cloud of electrons waiting at F, it now has to reach only from A to a cloud of electrons waiting at G; another battery (in practice, part or all of the anode battery) connected to G' having undertaken the responsibility of forwarding the supply of electrons as far as G*. With any given geometry of F, G and A, the anode voltage needed to provide any specified anode current is thus much reduced.

Illustrative characteristics are given in the next Subsection.

9 (c). *Fixed-potential screen grid.* Roughly speaking, the tetrode of Fig. 138 at small anode potentials gives the same performance (for weak signals) as would a triode constructed with the same electrodes F, G and A at large anode potentials. At the same i_a, neither $\dfrac{\partial i_a}{\partial e_a}$ nor $\dfrac{\partial i_a}{\partial e_g}$ has been much altered by the insertion of G'.

---

* The effect is almost the same as if G' and F were replaced by an emitting surface—of larger area and smaller emission density than F—situated at or just below G.

9. MULTI-GRID TUBES

But let the auxiliary fixed-potential grid be located above G (Fig. 139) instead of below. G' now serves as a more or less perfect electrostatic screen, more or less preventing changes of anode potential from affecting the space-charge conditions below G'. $\frac{\partial i_a}{\partial e_a}$ is reduced by the presence of G', and tends towards zero as G' is made of finer mesh.

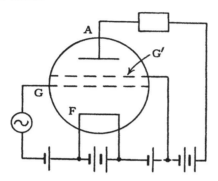

Fig. 139. Fixed-potential screen grid.

If G' absorbed none of the electrons reaching it, its presence would have no effect on the value of $\frac{\partial i_a}{\partial e_g}$ at any specified i_a. But unless G' were given a positive potential, the potential of A would have to be made inconveniently high in order to draw out electrons from F at this rate. Hence in practice G' is given a positive potential as indicated in Fig. 139. G' then robs A of a proportion of the electrons arriving from F, and thereby decreases $\frac{\partial i_a}{\partial e_g}$*. The net result to be expected from the introduction of G', therefore, is an increase of anode resistance $\rho \equiv 1 \Big/ \frac{\partial i_a}{\partial e_a}$, and a not very different increase in the amplification factor $\mu \equiv \frac{\partial i_a}{\partial e_g} \Big/ \frac{\partial i_a}{\partial e_a}$. No economising in anode battery voltage is to be looked for.

---

* Actually this effect may be more than compensated by secondary emission from G' under bombardment by the primary electrons arriving from F. This is referred to in Subsections 9 (d) and 9 (f) below.

The effects of an auxiliary fixed-potential grid used as in Figs. 138 and 139 are illustrated in Fig. 140. In this Figure stand contrasted the characteristic curves for the same tetrode used in the two different ways illustrated in Figs. 138 and 139*.

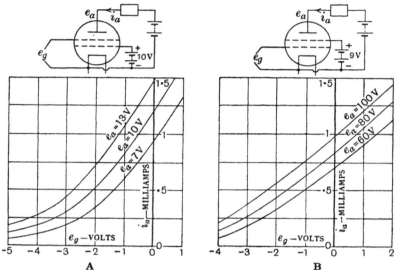

Fig. 140. Effects of fixed-potential grids compared.

Examination of the curves at $i_a = 0.75$ mA yields the following figures:

In A, with $e_a \fallingdotseq 10$ V,

$$\frac{\partial i_a}{\partial e_g} = 0.33 \text{ mA/V},$$

$$\mu = \frac{6 \text{ V}}{1.4 \text{ V}} = 4.3;$$

$$\therefore \rho = 13 \text{ k}\Omega.$$

In B, with $e_a \fallingdotseq 80$ V,

$$\frac{\partial i_a}{\partial e_g} = 0.21 \text{ mA/V},$$

$$\mu = \frac{40 \text{ V}}{1.32 \text{ V}} = 30;$$

$$\therefore \rho = 140 \text{ k}\Omega.$$

9 (d). *Screen-grid tetrode for preventing retroaction.* The screening effect of the auxiliary grid in Fig. 139 and B of Fig. 140 was to raise

* The tetrode is Mullard's "DEDG"; filament $1.8 \text{ V} \times 0.4 \text{ A}$. The curves are prepared from those published by the maker. It should be noticed that in comparing the two arrangements of Figs. 138 and 139 by reversing the connections to the two grids of a single tetrode, we are not accurately following the procedure contemplated in the foregoing argument. There we supposed that the control grid G remained unchanged, and that the auxiliary grid G' was first below and then above G.

the amplification factor by raising the anode resistance. Results at least as good could have been obtained in the ordinary triode by appropriate design of its single grid*. But there is another effect of screening of much greater practical importance.

With a triode, however carefully the external circuits Z_g and Z_a (A, Fig. 141) joined to grid and anode may be isolated from each

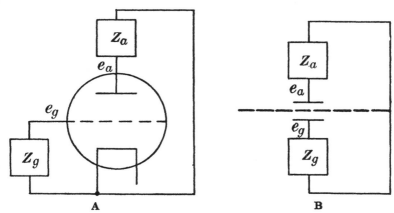

Fig. 141. Prevention of retroaction by screen grid.

other, they are coupled together by the mutual capacitance† subsisting between anode and grid within the tube. In virtue of this coupling, and whether electrons are being emitted or not, a change of potential e_a produces a flow of current through Z_g and therefore a change of potential e_g (B, Fig. 141). Now the action of e_g on e_a, through the electron stream passing across the vacuum, is the essential property of the triode; but an internal reaction of e_a on e_g is often very undesirable (as we shall see in the next Chapter). This reaction is completely prevented in B, Fig. 141 by the insertion of a screening plate whose potential is held fixed, as indicated by the dotted line. The auxiliary grid G' in Fig. 139 is an approach to such a screen, and the finer its mesh the more nearly perfect a screen it is.

Tetrodes of the type of Fig. 139 are accordingly manufactured‡,

---

* For example, see the triode of Fig. 131.
† At least 2 or 3 $\mu\mu$F.
‡ The idea seems to have originated with W. Schottky in 1919; but effective screen-grid tetrodes have only recently come into general use.

with a fine-mesh outer or screen grid carefully disposed to shield the anode and its internal connections from the grid and its internal connections, and are used with a supplementary external screen, as shown diagrammatically in Fig. 142. Owing to the fine mesh of the screening grid, to obtain sufficient anode current without excessively high anode potential the screen grid must be given a high positive potential.

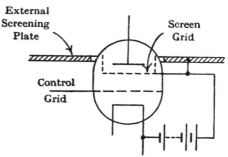

Fig. 142. Usual screen-grid disposition.

Characteristic curves (i_a plotted against e_a with constant e_g) for a screen-grid tetrode* are shown in Fig. 143. At the operating anode potential of about 130 V, we read from these curves

$$\frac{\partial i_a}{\partial e_a} = 0\cdot 0045 \,\text{mA/V} = \frac{1}{220\,\text{k}\Omega},$$

$$\frac{\partial i_a}{\partial e_g} = 0\cdot 85\,\text{mA/V};$$

$$\therefore \mu = \frac{0\cdot 85}{0\cdot 0045} = 190,$$

$$\rho = 220\,\text{k}\Omega.$$

The residual anode-grid capacitance in this tube has been found to be $0\cdot 014\,\mu\mu\text{F}$†.

Screen-grid tetrodes are very valuable devices for high-frequency amplification. They would be much more widely applicable if the anode resistance could be reduced without detriment to the

* The M.O. Valve Co.'s "S 215." The curves are those published by the makers.

† N. R. Bligh, "Measurement of the grid-anode capacity of screen-grid valves," *Experimental Wireless*, Vol. vi (1929), p. 299.

effectiveness of the screening. Their use in amplifiers is referred to in VIII-7.

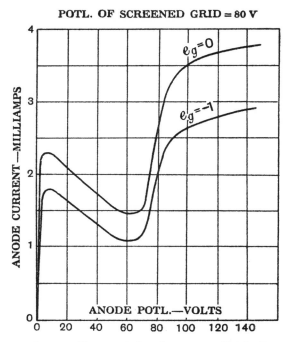

Fig. 143. Characteristics of a screen-grid tetrode.

9 (e). *Negative conductance with screen-grid tetrode.* The downward sloping portions of the characteristics in Fig. 143, indicating negative slope conductance (and resistance*), are due to the phenomenon of *secondary emission*, i.e. the emission by a cold electrode under the action of bombarding electrons, called *primary*, of other electrons, called *secondary*. The number of secondary electrons emitted per primary electron increases with the speed of impact of the latter; as many as 20 has been observed by A. W. Hull.

---

* The phenomenon calls for a rather difficult inversion of one's customary habit of thought. While both negative conductance and its reciprocal, negative resistance, are equally amenable to algebraic treatment, the former is apt to be the less confusing to think about. The larger the magnitude of a negative conductance, the more potent it is for practical applications. As a component of electric circuits, a very large negative resistance is as unserviceable as a very large positive resistance.

Secondary emission occurs at the anode of the ordinary triode, but is not noticed under the usual conditions because the secondary electrons are drawn back into the anode. The effect is, however, observable if the grid is made higher in potential than the anode, an arrangement developed by Hull in 1915 and called by him the *dynatron*. The dynatron effect is well illustrated by the curve of Fig. 195 (p. 296) obtained with an ordinary small triode*.

In the normal use of the screen-grid tetrode the anode is at a higher potential than the screen, and consequently re-absorbs the electrons it emits. But at anode potentials lower than that of the screen (viz. 80 V in Fig. 143) the secondary electrons leave the anode and are absorbed by the screen. Let n_1 electrons per second from the filament reach the anode, and let n_2 secondary electrons be emitted from the anode. Owing to the efficient screen, n_1 is nearly independent of e_a. The anode current is

$$i_a = n_1 - n_2,$$

$$\therefore \frac{\partial i_a}{\partial e_a} = -\frac{\partial n_2}{\partial e_a}.$$

Since $\frac{\partial n_2}{\partial e_a}$ is positive, $\frac{\partial i_a}{\partial e_a}$ is negative.

In Fig. 143 the downward slopes are about

$$-0\cdot167 \,\text{mA/V} = \frac{1}{-60\,\text{k}\Omega}.$$

Accordingly this tube, with the electrode potentials adjusted to bring the representative point to this region of the curves, would have the property of reducing by $60\,\text{k}\Omega$ the total resistance of a circuit comprising the anode-filament path.

Various applications of such a negative-conductance device are feasible. Two are examined in VIII-8 and IX-6 (a).

9 (f). *Pentodes.* In the screen-grid tetrode (VII-9 (d)), the screen is given a very fine mesh and the anode resistance ρ as well as the amplification factor μ is consequently very high. When perfect screening is not desired, the screen can be given a coarser mesh, resulting in a lower μ but with the compensation of a lower ρ, which

* With an Osram "LS 5" triode (thoriated filament), the grid being at 220 V, it was easy to make the number of secondary exceed the number of primary electrons, causing the anode current actually to reverse in direction. But see footnote on p. 297.

9. MULTI-GRID TUBES

is in general highly desirable. In order to obtain a large anode current the auxiliary grid is given a high fixed potential—in practice, the whole of what is available from the anode battery. Since in use the anode potential undergoes wide fluctuations of potential, this would introduce the complications of secondary emission discussed with reference to Fig. 143.

But the loss from the anode of the secondary electrons emitted by it is completely avoidable by the simple expedient of introducing yet one more grid, situated next to the anode and maintained at a lower potential. The direction of the electrical field near the anode is then as in the normal triode, viz. such as to urge electrons towards

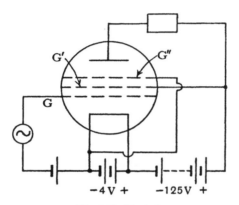

Fig. 144. Pentode.

the anode. A five-electrode tube, or *pentode*, is thus arrived at (Fig. 144). The innermost grid G is the control electrode to which the signal is applied; the intermediate grid G′, at a fixed high potential, is a partial screen as in Fig. 139; and the outermost grid G″, at a fixed low potential, causes the anode secondary emission to return to the anode.

Characteristic curves for a pentode* operating in this fashion are given in Fig. 145.

It will be noticed that the characteristics, although nearly straight over a useful range, have not the triode's virtue of parallelism; the anode resistance at $e_g = -10$ V is seen to be nearly twice the value

---

* Mullard's "P.M.24." Filament, 4 V × 0·15 A. The characteristics are drawn from the maker's published curves.

at $e_g = -2$ V. Near $e_g = -6$ V, $e_a = 125$ V, it is seen that approximately

$$\frac{\partial i_a}{\partial e_g} = 2\cdot 05 \text{ mA/V},$$

$$\mu = \frac{100 \text{ V}}{2\cdot 1 \text{ V}} = 50;$$

$$\therefore \rho = 24 \text{ k}\Omega.$$

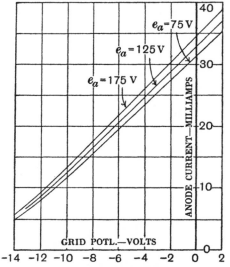

Fig. 145. Characteristics of a pentode.

Comparison of this pentode with ordinary triodes shows that a triode (e.g. that of Fig. 131) would be capable of giving much the same performance as regards ρ and μ only with a positive control-grid potential, which would cause large grid current to flow; or with much higher anode potential, which would destroy the triode.

CHAPTER VIII

THE TRIODE AS AMPLIFIER

NOTE. *In this Chapter frequent reference is made by way of numerical illustration to the typical modern receiving triode whose characteristics are given in Fig. 132 (p. 206). For brevity this triode is denoted " our reference triode."*

1. OUTPUT OF THE TRIODE

THE fundamental property of the triode is that a change of potential of the grid, or control electrode, effects a change in the circuit of the anode, or repeat electrode. The input to the device is a change δe_g of grid potential (Fig. 146); the significant output may be change δi_a of anode current, change δe_a of anode potential, or change $\delta(e_a\, i_a)$ of the product of these. The term *amplifier* should, perhaps, strictly be applied only to dispositions in which the significant output is δe_a, the *amplification* then being the pure number $\dfrac{\delta e_a}{\delta e_g}$; but an extension to cover uses in which the anode current is significant is both convenient and customary.

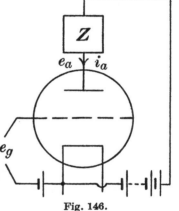

Fig. 146.

According to its function, the triode is conveniently designated an *amplifier triode* or an *output triode**.

Let us suppose that Z in Fig. 146 is an ammeter of resistance R whose function is to measure changes of grid potential. Then $\delta e_a = - R \cdot \delta i_a$. But

$$\delta i_a = \frac{1}{\rho}(\delta e_a + \mu \cdot \delta e_g)$$

$$= -\frac{R}{\rho}\delta i_a + \frac{\mu}{\rho}\delta e_g,$$

$$\therefore\ \delta i_a = \frac{\mu}{\rho + R}\delta e_g.$$

* Cf. the commercial term *power valve* for the last tube in a broadcast receiver.

226   VIII. THE TRIODE AS AMPLIFIER

If the P.D. δe_g had been applied directly to the ammeter, the change of ammeter current would have been $\frac{1}{R}\delta e_g$. The insertion of the triode has multiplied the ammeter current by $\frac{\mu R}{\rho + R}$. This may be greater or less than unity. In the case of our reference triode and a high-resistance "Unipivot" microammeter, $\mu = 13$, $\rho = 8000\,\Omega$, $R = 800\,\Omega$, so that $\frac{\mu R}{\rho + R} = 1\cdot 2$; the interposition of the triode has produced a slight increase of sensitiveness. If it were practicable to wind the ammeter to a much higher resistance, this gain would be raised from 1·2 towards the limiting value μ. But the interest of the arrangement lies rather in the feature that, with the triode interposed, the power taken from the source of the E.M.F. δe_g is nil, however high or low the resistance of the ammeter may be.

When the output instrument Z is of the nature of a microammeter, it is necessary to connect it to the triode in such a way that it carries only the small *changes* of anode current consequent on the changes

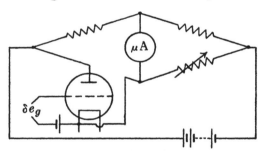

Fig. 147. Balancing the no-signal current.

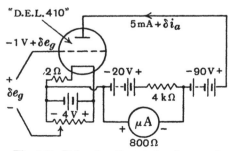

Fig. 148. Balancing the no-signal current.

1. OUTPUT OF THE TRIODE

of grid potential. Thus, the particular microammeter of our example undergoes a full-scale deflection with 25 μA, so that the several milliamperes of total anode current must not be passed through it. Some bridge or balance circuit is therefore employed, such as those of Figs. 147 and 148. In Fig. 148 appropriate values for our reference triode are suggested as a numerical illustration of a convenient disposition for laboratory measurements. Every millivolt increase of δe_g would deflect the microammeter through about 1·5 μA.

The current-operated apparatus Z in the anode circuit is usually not an ammeter, but an A.C. instrument such as a telephone, the input signal being then a sinoidal E.M.F. of acoustic frequency. By providing a suitable winding in the telephone, or by inserting a suitable

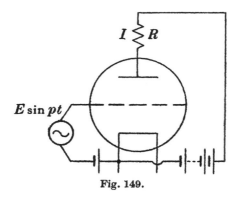

Fig. 149.

transformer between it and the anode, the impedance Z can be given any desired magnitude; for example, that producing the largest output for a given input signal. To take a simple case, if Z is a resistance R (Fig. 149), and if E and I are the amplitudes of the alternating components of grid potential and anode current, we have

$$I = \frac{\mu}{\rho + R} E;$$
$$\text{output power} = \tfrac{1}{2} I^2 R = E^2 \frac{\mu^2}{2} \frac{R}{(\rho + R)^2}.$$

This is a maximum for choice of R when $R = \rho$. Hence to secure maximum output power with a given input E.M.F., the instrument should be given a resistance equal to the anode-filament slope resistance of the triode. The power falls off only slowly as R is made greater than ρ, being $\frac{8}{9}$ths of the maximum when $R = 2\rho$. Since the greatest permissible input E without passing off the rectilinear region of the triode characteristics is inversely proportional to I, it is often advantageous to make $R > \rho$.

It can be shown that when the available input E.M.F. is unlimited, for rectilinear operation without grid current the maximum output power is obtained when $R = 2\rho$, and is $\dfrac{(B_a - P_{\text{min.}})^2}{16\rho}$, where B_a is the P.D. of the anode battery and $P_{\text{min.}}$ is the value of the lumped potential at the lower end of the substantially straight portion of the characteristic. This is proved in IX-2 (c) for the linear régime in a self-oscillating circuit; and a self-oscillator is merely a steadily functioning amplifier in which the grid excitation happens to be provided by some coupling with the anode circuit (retroaction)*.

In amplifier triodes (in the restricted sense), the desired output is in the form not of current or power, but of an anode potential fluctuation which is a magnified copy of the grid potential fluctuation. With Z a pure resistance R, the amplification is then:

$$\text{Amplification} \equiv \frac{\text{anode potential change}}{\text{grid potential change}},$$

$$= \frac{-R \cdot \delta i_a}{\delta e_g},$$

$$= \frac{-R \dfrac{\mu}{\rho + R} \delta e_g}{\delta e_g},$$

$$= -\mu \frac{R}{\rho + R}.$$

This increases (negatively) as R is increased, tending to the limiting value μ, the amplification factor of the triode, as R becomes indefinitely greater than the anode resistance ρ.

Such an arrangement may be regarded as an amplifier of P.D., in which a rise of 1 unit in the grid potential produces a fall of $\mu \cdot \dfrac{R}{R + \rho}$ units in the anode potential. There is no appreciable time lag in this amplifier. Owing to the small mass of the moving parts,

* See also IX-2 (d).

1. OUTPUT OF THE TRIODE

the electrons, their velocity is very great: any change of grid potential produces the corresponding magnified change of anode potential almost instantaneously*. Hence this simplest form of triode amplifier, where the only impedances of the circuits are plain resistances, constitutes a truly aperiodic arrangement available for amplifying steady or alternating E.M.F. of any frequency. Its action can be tested very directly with steady E.M.Fs. by observing the change of reading of an electrostatic voltmeter connected between anode and filament when the grid potential is changed a known amount (say a volt or two), as indicated in Fig. 150.

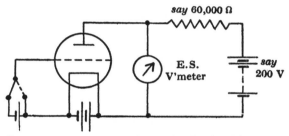

Fig. 150. Measurement of amplification by voltmeter.

A serious practical objection to making R large enough to give an amplification approaching μ is the no-signal voltage drop occurring across R, which necessitates a corresponding increase in the voltage

R	Amplification	Anode battery
1 kΩ	10 % of μ	155 V
10 kΩ	53 % of μ	200 V
100 kΩ	92 % of μ	650 V

of the anode battery. Thus, suppose a no-signal anode current of 5 mA with a no-signal grid potential of -5 V is demanded; with our reference triode, an anode potential of 150 V must be provided. The anode battery must therefore have a P.D. of $(150 + 5R)$ volts

---

* The velocity acquired by an electron in passing from the filament to the anode at (say) +70 volts is 5×10^8 cm. per sec. Hence if the anode is (say) 0·5 cm. from the filament, the time taken for the electron to travel across the vacuum is of the order of 2×10^{-9} sec.

if R is in kilohms; while the amplification obtained is $\mu \dfrac{R}{9+R}$ (since at 5 mA $\rho \fallingdotseq 9$ kΩ). The Table on p. 229 shows the values of amplification obtained and the battery voltage required with various values of R.

On the other hand, suppose R is made 100 kΩ while the anode battery is restricted to 150 V. On interpolating between the curves of Fig. 132, it is found that the anode current is about 0·55 mA. At such small current values the characteristics are very curved, and ρ has risen from 9 kΩ to about 25 kΩ. The amplification (for weak signals) has therefore fallen from 0·92 μ to 0·80 μ.

The reduction of anode battery from 650 V to 150 V with $R = 100$ kΩ, while appreciably reducing the amplification for weak signals, has a much more serious consequence: the strength of the strongest signal which can be amplified without serious distortion is much reduced. We have seen (p. 227) that if E is the amplitude of grid potential, the amplitude of anode current is $I = \dfrac{\mu}{\rho + R} E$. With the grid bias of -5 V, E might reach 5 V without appreciable grid current; then, using the 650 V battery,

$$I = \frac{13 \times 5}{9 + 100} \text{ mA} = 0\cdot6 \text{ mA}.$$

The anode current accordingly fluctuates between $(5-0\cdot6)$ mA and $(5+0\cdot6)$ mA. Over this range ρ changes so little that the amplification—which is proportional to $\dfrac{1}{\rho + R}$—is sensibly constant and distortion is absent. But with the 150 V battery, if ρ remained 25 kΩ throughout the signal cycle, we should have

$$I = \frac{13 \times 5}{25 + 100} \text{ mA} = 0\cdot5 \text{ mA}.$$

Now over the range $(0\cdot55 \mp 0\cdot5)$ mA the value of ρ changes widely. It is obvious that severe distortion would occur if 5 V input signals were applied.

If the input signal to the amplifier is an alternating E.M.F., the large resistance R in the anode circuit can be replaced by an inductive impedance, with marked economy in the anode battery. This disposition is analysed in the next Section.

2. Analysis of Amplifier with Mixed Impedance

Let the external path between anode and filament possess inductance L as well as resistance R; and let i, e_a, e_g now signify the *changes* of anode current, anode potential and grid potential accompanying the input signal E.M.F. e_g. Fig. 151 shows the actual and the equivalent simple circuits*. From the latter we can read off

$$i = \frac{\mu e_g + e_a}{\rho};$$

$$e_a = -Ri - L\frac{di}{dt};$$

$$\therefore \rho e_a = -\mu R e_g - R e_a - \mu L \frac{de_g}{dt} - L \frac{de_a}{dt};$$

i.e. $L\dfrac{de_a}{dt} + (\rho + R)e_a = -\mu\left(Re_g + L\dfrac{de_g}{dt}\right).$

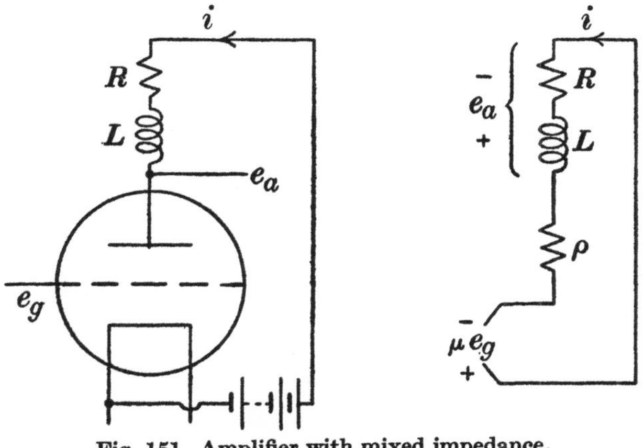

Fig. 151. Amplifier with mixed impedance.

In general, e_g is some function of t, say $e_g = Ef(t)$. Then the output e_a is given by the solution of the differential equation

$$L\frac{de_a}{dt} + (\rho + R)e_a = -\mu E[Rf(t) + Lf'(t)].$$

* Cf. Fig. 130.

VIII. THE TRIODE AS AMPLIFIER

Commonly the input signal is a sinoidal alternating E.M.F. $E \sin pt$. Then

$$L \frac{de_a}{dt} + (\rho + R) e_a = -\mu E (R \sin pt + pL \cos pt),$$
$$= -\mu E \sqrt{R^2 + p^2 L^2} \sin (pt + \phi),$$

where $\phi \equiv \tan^{-1} \frac{pL}{R}$.

The complete solution is easily found. It is

$$e_a = \alpha \sin (pt + \phi - \theta) + \beta \epsilon^{-\frac{\rho + R}{L} t},$$

where
$$\alpha \equiv -\mu E \sqrt{\frac{R^2 + p^2 L^2}{(\rho + R)^2 + p^2 L^2}},$$

$$\theta \equiv \tan^{-1} \frac{pL}{\rho + R}.$$

We investigate further only the steady state, when $t = \infty$. Then

$$e_a = -\mu E \sqrt{\frac{R^2 + p^2 L^2}{(\rho + R)^2 + p^2 L^2}} \sin \left(pt + \tan^{-1} \frac{pL}{R} - \tan^{-1} \frac{pL}{\rho + R}\right).$$

If $R \gg pL$,

$$e_a \doteqdot -\mu E \frac{R}{\rho + R} \sin pt;$$

i.e. the amplification is $\mu \frac{R}{\rho + R}$, and the anode potential lags $180°$ on the grid potential, whatever the value of R*.

On the other hand, if $R \ll pL$,

$$e_a \doteqdot -\mu E \frac{pL}{\sqrt{\rho^2 + p^2 L^2}} \sin \left(pt + \frac{\pi}{2} - \tan^{-1} \frac{pL}{\rho}\right);$$

i.e. the amplification is $\mu \frac{pL}{\sqrt{\rho^2 + p^2 L^2}}$, but the anode potential lags less than $180°$ on the grid potential unless $pL \gg \rho$.

If I, E_a are the amplitudes of the anode current and potential in the steady state,

$$I = \frac{E_a}{\sqrt{R^2 + p^2 L^2}},$$
$$= \frac{mE}{\sqrt{R^2 + p^2 L^2}},$$

---

* This has already been found more directly on p. 228.

where m stands for the magnitude of the amplification, viz.

$$m \equiv \frac{E_a}{E} = \mu \sqrt{\frac{R^2 + p^2 L^2}{(\rho + R)^2 + p^2 L^2}}.$$

The practical advantage of a large reactance over a large resistance in economy of anode battery is illustrated by an extension of a former numerical example (p. 229). The formula shows that the same amplification ($m = 0.92\,\mu$) given by the former 100 kΩ of pure resistance is given by 21 kΩ of pure reactance; alternatively, if 108 kΩ of pure reactance is provided, the current amplitude for the same input signal (and therefore the permissible strength of input signal) is the same as with the 100 kΩ of pure resistance, and the amplification is raised to $m = 1.00\,\mu$. While 650 V is required with the 100 kΩ of pure resistance, only 150 V is required with the pure reactance.

Against this, however, it must be remembered that when signals covering a range of frequencies are to be amplified, resistance has the advantage over reactance of having a value which is independent of the frequency.

3. Construction of the Impedance

If the P.D. to be amplified is of high (radiation) frequency, it is difficult to obtain either resistive or inductive impedances of the desired large values in which the shunting by stray capacitance is negligible*.

The minimum of self-capacitance in resistors is obtained by using a conductor of very high resistance per unit length. Accordingly the resistance element of a high-frequency amplifier takes the form (i) of a short rod of some composition of very high (equivalent homogeneous) resistivity, such as a compressed mixture of graphite and a non-conducting powder, or a partially carbonised cellulose filament; or (ii) of an extremely thin metal or carbon film, such as a heavier form of the grid leak shown at A in Plate XX. The latter is preferable as being more constant, since the former is—or, at any rate, is apt with age to become—a series of crazy contacts between particles of high conductivity.

For acoustic frequencies, a length of fine resistance wire can readily be wound in various ways so as to possess very small shunting

* The effect of capacitance shunting is investigated in Sections 5 and 6 of this Chapter.

admittance. A convenient form is shown at B in Plate XX. The conductor is bare chromium-nickel wire of diameter 0·03 mm with slightly oxidised surface, wound in a nearly close helix round a cotton core; the helix is cotton-covered, and forms a cord of diameter about 1 mm and resistance about 50 kΩ per metre. A length of this cord

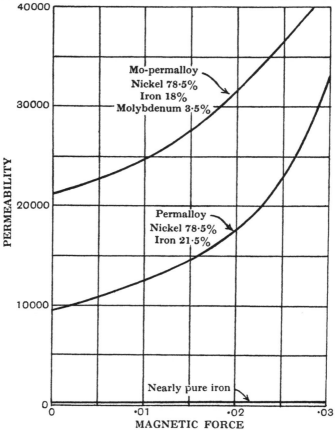

Fig. 152. Magnetic alloys for weak fields.

is enclosed in a protecting tube of insulating material closed by a brass terminal cap at each end*.

High-frequency chokes are usually air-core coils of fine copper wire, wound as single-layer solenoids for the smaller inductance values, and in narrow slots in a former for the larger values.

Chokes for acoustic frequencies are constructed as fine-wire coils

* See also p. 263.

3. CONSTRUCTION OF THE IMPEDANCE

wound on closed cores of laminated steel. At 500 c/s the 21 kΩ of our example (p. 233) is given by an inductance of 6·7 H. But at 5000 c/s the reactance would be 210 kΩ, and at 50 c/s it would be 2·1 kΩ; and the corresponding amplifications in our example would be 1·00 μ and 0·22 μ (cf. 0·92 μ with 21 kΩ of pure reactance or 100 kΩ of pure resistance). For telephonic—especially musical—signals, therefore, where all frequencies between (say) 50 c/s and 5000 c/s should be amplified without discrimination, if reactance is employed it should be large enough to give an amplification nearly equal to μ even at the lowest relevant frequency. For our practical numerical illustration, we might specify an amplification not less than 0·71 μ* at 50 c/s. With $\rho = 9$ kΩ as before, a reactance not less than 9 kΩ at 50 c/s is required, i.e. an inductance of not less than 29 H.

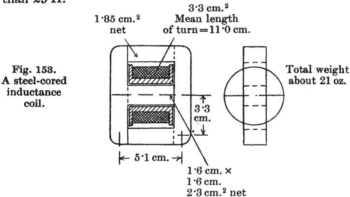

Fig. 153. A steel-cored inductance coil.

This inductance can be provided by a small steel-cored coil. But in designing inductance coils for such purposes it is necessary to allow both for the magnetic saturating effect of the steady component of anode current (5 mA in our example), and for the very small slope or differential permeability $\dfrac{dB}{dH}$ of the core for the small alternating magnetic force produced by weak signals†. An idea of construction and dimensions may be had from Fig. 153, showing a coil with closed "Stalloy" core which might be used to give the

* Two such stages would amplify signals of 50 c/s half as much as those of the higher frequencies.
† The new alloys of the "permalloy" and "mumetal" type, which have done so much for the loading of submarine cables, may be of great assistance here. See G. W. Elmen, "Magnetic alloys of iron, nickel and cobalt," *Bell System Tech. Journ.* Vol. VIII (1929), p. 435. Fig. 152 is taken, by permission, from Fig. 20 of this paper.

29 H of our example. Experiment has shown* that a 5000-turns winding on this core, when carrying 5 mA of steady current, would show an inductance of about the required 29 H† when subjected to a P.D. of 13 V at 50 c/s, i.e. when the input signal on the grid of our reference triode is 1·4 V at 50 c/s. The 5000 turns of copper wire might be of S.W.G. 38 D.W.S., with a resistance of some 400 Ω, which would have negligible effect on the performance. But a finer guage would probably be preferable, the space thus freed being devoted to a more thorough sectionalising of the winding in order to reduce its self-capacitance. An example of a very well-sectionalised winding is seen in the transformer in Plate XXI.

With input signals much weaker than the 1·4 V at 50 c/s of our example, the inductance would be appreciably smaller. A core of somewhat larger dimensions would therefore be advantageous; but quite apart from economic and mechanical considerations favouring small bulk, there is a practical limitation on inductance value imposed by the unavoidable stray capacitance, which may become important before the frequency reaches 5000 c/s.

General principles governing the design of amplifier circuits, which emerge clearly from considerations such as the foregoing, are:

(i) To obtain high amplification for signals of any one frequency without resorting to excessive anode battery voltage, inductive is superior to resistive impedance.

(ii) If inductance is employed, to obtain uniform amplification over a range of frequencies, its reactance *at the lowest frequency* must be so much larger than the triode resistance ρ that sensibly full amplification μ is given at the lowest frequency (and, *a fortiori*, at higher frequencies). On the other hand, when pure resistance is employed, the amplification, whether a large or small fraction of μ, does not fall off at low frequencies.

(iii) The effect at the upper frequencies of stray capacitance imposes an upper limitation on the permissible inductance value;

---

* See L. B. Turner, "Some measurements of a 'Stalloy' core with simultaneous D.C. and A.C. excitation," *Experimental Wireless*, Vol. IV (1927), p. 594. The beneficial effect of a small air-gap in the Stalloy circuit has been investigated by A. A. Symonds: "Loop permeability in iron, and the optimum air-gap in an iron choke with D.C. excitation," *ibid.* Vol. v (1928), p. 485.

† In the absence of the 5 mA D.C. excitation, the inductance would be 2 or 3 times as great. The inductance values quoted by manufacturers must always be suspect unless both the relevant D.C. and A.C. values are specified.

while a lower limitation is imposed by the necessity of keeping the reactance large compared with ρ even at the lowest frequency. Consequently good telephonic performance in reactance amplifiers can be got only with triodes of sufficiently low ρ (and correspondingly low μ).

4. CASCADE AMPLIFIERS

By making the grid much finer (and/or the anode distance larger), the amplification factor of a triode may be increased without limit. For example, the $\mu = 8.5$ in Fig. 133 has been raised to $\mu = 37.5$ in Fig. 131. But the practical objections of the accompanying increase in ρ and in the requisite anode battery voltage stand in the way of obtaining a very large amplification by the use of a single triode. The more convenient course is to use a chain of two or more triodes in cascade, the anode fluctuations of one triode being impressed on the grid of the next. Then with n triodes, each amplifying m times, the total amplification is m^n. For example, if $m = 10$ and $n = 3$, the amplification is 1000.

In Fig. 154 two triodes, each amplifying as in Fig 151, are arranged in cascade. The battery GB_2 has a P.D. slightly larger

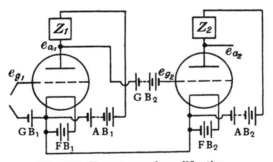

Fig. 154. Two stages of amplification.

than the potential of anode 1, and so serves to maintain grid 2 at a suitable small negative mean potential. It has no effect on the fluctuations of potential which occur when e_{g_1}, and therefore e_{a_1}, fluctuate; and it carries no current. The amplification obtainable with this two-triode combination is the product of the amplifications of the two triodes separately.

It is an obvious step* to simplify the circuits of Fig. 154 by using a common filament battery FB and a common anode battery AB as in Fig. 155. An extension to three or more triodes, still with common filament and anode batteries, but with separate high-voltage grid batteries GB_2, GB_3, etc., is also obvious.

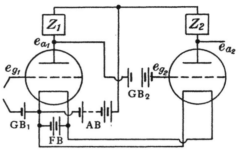

Fig. 155. Common batteries.

If the impedances Z_1, Z_2 in Figs. 154 and 155 are pure resistances, these circuits are capable of amplifying signals (changes of potential e_{g_1}) of every frequency, including quasi-steady changes. But in wireless telegraphy and telephony the signals to be amplified are

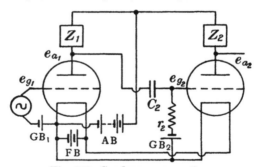

Fig. 156. Condenser connection.

usually in the form of alternating currents, either of radiation frequency such as tens or hundreds of kilocycles per second, or of acoustic frequency such as hundreds or thousands of cycles per second. For such signals, by the use of transformers and condensers, the

* Familiarity breeds contempt. One may recollect, with feelings of rueful amusement, that this step was not always obvious, and that instruments intended to be portable have been issued with separate filament and anode batteries for the separate triodes.

4. CASCADE AMPLIFIERS

steady currents can be separated from the signal currents superposed on them, with very marked practical simplifications. Thus for amplifying ordinary telephonic signals, suitable circuits are shown in Fig. 156, where, it will be seen, the necessity for the high-voltage battery GB_2 of Fig. 155 has been avoided.

Ignore, in the first place, the path provided by the high resistance r_2. Grid 2 cannot be permanently positive, because until it became slightly negative electrons would continue to reach it from the filament, removing its positive charge. It is, therefore, at a negative potential of indeterminate value, the difference between its potential and that of anode 1 being taken up across the condenser C_2. Any change in e_{a_1} is then accompanied by a precisely equal change in e_{g_2}. In order that this change of e_{g_2} may be effective, e_{g_2} must be of a value to give the representative point a suitable location on the anode current characteristic, as already explained. It is to ensure this condition that the *grid leak* r_2—commonly a megohm or more—is fitted. The grid leak keeps the mean grid potential at the desired value, but, owing to the largeness of its resistance, does not appreciably reduce the fluctuations of grid potential. There is no upper limit to the permissible value of C_2, since its only function is that of the battery GB_2 in Fig. 155, viz. to protect grid 2 from the steady component of potential of anode 1 while allowing the signal fluctuations to pass. Hence any capacitance will do whose reactance at the signal frequency is small compared with the resistance of the grid-filament path and its shunt r_2—i.e. substantially r_2 if the grid potential is held negative.

An example of a grid leak is shown at A in Plate XX. The conductor is a very thin film of colloidal carbon sprayed on to a glass rod. This is mounted within an evacuated glass tube to protect the film from mechanical and chemical damage.

The advantage of this self-adjustment of mean grid potential becomes very great when more than two or three triodes are connected in cascade. With it the voltages of the batteries may vary widely without any ill effect; whereas in a multi-triode amplifier of the pattern of Fig. 155, the effects of very tiny changes of voltage in AB, GB_2, GB_3,... would be so magnified as to bring the grid potentials of the later triodes of the chain to quite unsuitable values. Multi-triode amplifiers have been constructed containing seven or more triodes linked by this capacitance-and-resistance connection. An advantage of this type of amplifier is its

240 VIII. THE TRIODE AS AMPLIFIER

simplicity and compactness; for anode circuit resistances R, grid leaks r, and grid condensers C can be easily constructed in forms occupying very little space, and they need not be adjusted to any precise values. The grid batteries GB_1, GB_2, GB_3,... may of course be united.

5. Transformer connection between triodes

5 (a). *General.* In general terms, the function of the impedance Z in the anode circuit of the amplifiers so far discussed is, by opposing change of anode current, to make the signal at the grid reproduce itself in the form of the largest obtainable fluctuation of anode potential. This fluctuation is applied to the grid of the next triode in the chain. When the impedance Z is that of a choke coil, the addition of a secondary winding (Fig. 157) not only obviates the need of the grid condenser C_2 and leak r_2 of Fig 156, but also offers a prospect of obtaining a step-up of signal strength in the transformer, thereby removing the theoretical limit μ to the amplification obtainable per stage.

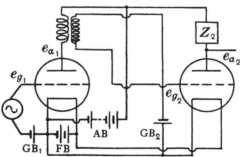

Fig. 157. Transformer connection.

Since the transformer secondary is not called upon to supply current to the grid which it feeds, if Fig. 157 represented completely the real circuits, there would be no theoretical limitation on the step-up ratio of the transformer. A secondary circuit devoid of current is without influence on the primary circuit, and its resistance is of no significance. Hence, with a given primary, by merely filling the space available for the secondary winding with ever more turns of finer wire*, an ever larger ratio $\frac{e_{g_2}}{e_{g_1}}$ would be obtained. As so

---

* In practice the guage of wire used is the finest that mechanical—not electrical—considerations will permit; e.g. about S.W.G. 47.

5. TRANSFORMER CONNECTION BETWEEN TRIODES

often happens, however, the undesired but unavoidable stray capacitances impose severe restraints upon the designer; the secondary *is* loaded with the capacitance current.

The foregoing remarks in this Section, and much of what follows, apply to transformer connection between triodes without respect to frequency. But the term transformer connection is ordinarily intended to connote—as it does here—connection of anode 1 to grid 2 through the mutual inductance between two coils which have as little magnetic leakage, and as little capacitance shunting, as can be contrived, and which provide a step-up in voltage by reason of the number of secondary turns exceeding the number of primary. Now without the aid of steel cores it is impossible to get small magnetic leakage between the two coils unless they are wound together, spatially intermingled. The capacitances then existing between contiguous parts of the primary and secondary would make this procedure futile except in the special case of a 1 : 1 transformer, where there need be no P.D. between any portion of the primary winding and the corresponding portion of the secondary lying against it. Accordingly, transformers for high frequencies are restricted to 1 : 1 transformers*. The connection between triodes which such transformers provide is indistinguishable from simple choke connection (i.e. 1 : 1 auto-transformer). For this, the analysis, including the effects of resistance and stray capacitance, is given in Sub-section 6 (b) of this Chapter. The secondary winding merely serves to permit the change of circuit from Fig. 156 to Fig. 157.

Other forms of connection between high-frequency triodes necessarily involve more or less loose coupling, and great frequency discrimination. The circuits of both anode 1 and grid 2 may be tuned circuits, loosely coupled together. The behaviour of such circuits *per se* has already been examined in Chapter IV (e.g. in Sub-section 8 (b)); an extension to cover their use between triodes presents no difficulty. If the secondary circuit alone is provided with a tuning condenser, and a moderate coupling is employed—a common arrangement—the analysis of the next Section may be followed; for, although in that Section it is very non-selective acoustic amplifiers which are especially held in view, the algebraic treatment is of general application.

* In winding, the primary and secondary wires may be run together and simultaneously into the grooves of the former. The insulation between them is for D.C. only.

The desiderata in acoustic inter-triode transformers are large primary inductance, large step-up ratio of turns, small self-capacitances of windings, and small magnetic leakage between them. An example of a very good construction is shown in Plate XXI. The winding on each of the three formers is well sectionalised, with a good deal of air spacing; and the secondary, occupying the middle former, is sandwiched between the two parts of the primary*.

Each winding of a transformer for acoustic signals must possess self-capacitance lying perhaps between 10 $\mu\mu$F and 100 $\mu\mu$F according to the disposition of the turns; and it is hard to avoid another 5 or 10 $\mu\mu$F between grid and filament, and between anode and filament, within the triode and from the connecting leads. If the total stray capacitance is represented by C across the secondary, and if the inductance of the primary is L, the step-up ratio of turns s, and the magnetic leakage negligible, the transformer—so far from being in the open-circuit condition suggested in Fig. 157—is actually a closed oscillatory circuit which resonates at the frequency $\dfrac{1}{2\pi s \sqrt{LC}}$.

If, for example, $L = 29$ H (as on p. 235), $s = 2$, $C = 25$ $\mu\mu$F†, the resonant frequency is 2950 c/s. At frequencies much below the resonant frequency, the introduction of the secondary winding has merely multiplied the amplification by s; but at frequencies much above the resonant frequency a short-circuited condition is approached. At intermediate frequencies magnetic leakage must be taken into account and detailed analysis is necessary.

5 (b). *Dependence of amplification upon frequency (acoustic).*
The circuit under analysis is shown in Fig. 158. A sustained alternating E.M.F. e_1 of frequency $\dfrac{p}{2\pi}$ is the input signal to the grid of a triode whose parameters are ρ, μ. The anode circuit comprises the transformer primary of inductance L_1 and negligible resistance. The transformer has a secondary of inductance L_2 with mutual inductance M, the coefficient of coupling $k \equiv \dfrac{M}{\sqrt{L_1 L_2}}$ being appreciably less than unity. Stray capacitances across primary and secondary

---

* With the steel core removed, the coefficient of coupling between primary and secondary is about 70 %. It would be a good deal lower but for the sandwiched disposition.

† Made up, say, of 20 $\mu\mu$F across the primary and 20 $\mu\mu$F across the secondary.

5. TRANSFORMER CONNECTION BETWEEN TRIODES

are represented by an equivalent single capacitance C across the secondary*. The signals are assumed to be weak enough to permit ρ and L_1 to be taken as constants. e_2 is the output E.M.F. available for application to the grid of the next stage in the amplifier chain.

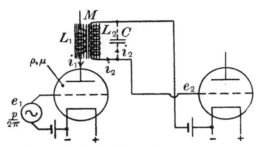

Fig. 158. Analysis of transformer connection.

Letting the symbols e, i stand for the vector quantities, and using $j \equiv \sqrt{-1}$, we have the equations:

$$\mu e_1 - (\rho + jpL_1) i_1 - jpMi_2 = 0,$$

$$e_2 = jpMi_1 + jpL_2 i_2,$$

$$i_2 = -jpCe_2.$$

On eliminating i_1 and i_2 from these three equations, we find

$$\frac{e_2}{e_1} = \frac{\mu k \sqrt{L_1 L_2}}{L_1[1-(1-k^2)p^2L_2C] - j\frac{\rho}{p}[1-p^2L_2C]}.$$

Now $p_0^2 L_2 C = 1$, where $\frac{p_0}{2\pi}$ is the resonant frequency of the transformer secondary with its equivalent shunting capacity C;

$$\therefore \frac{e_2}{e_1} = \frac{\mu k \sqrt{L_1 L_2}}{L_1\left[1-(1-k^2)\left(\frac{p}{p_0}\right)^2\right] - j\frac{\rho}{p}\left[1-\left(\frac{p}{p_0}\right)^2\right]}.$$

This complex quantity expresses the ratio between the output and input E.M.Fs. completely, i.e. in magnitude and phase. We are

---

* This, of course, is strictly permissible only if $k=1$. However, actually k is nearly unity; and in any case when the step-up ratio is upwards of 2, stray capacitance on the primary side has relatively small influence.

interested in the amplification m^*, which is the modulus or magnitude of this ratio, and is

$$m = \mu \sqrt{\frac{k^2 L_1 L_2}{\left[L_1\left\{1-(1-k^2)\left(\frac{p}{p_0}\right)^2\right\}\right]^2 + \left[\frac{\rho}{p}\left\{1-\left(\frac{p}{p_0}\right)^2\right\}\right]^2}}.$$

In any satisfactory amplifying system for the reproduction of music, constancy of amplification over a wide range of frequencies —say at least 50 c/s to 5000 c/s—is required. Let us examine the formula for m to see how far this desirable constancy can be obtained. In the first place, whatever the values of k and p_0, it is clearly necessary to have

$$(pL_1)^2 \gg \rho^2$$

at the lowest significant frequency†. In the second place, it is necessary to have

$$(1-k^2)\left(\frac{p}{p_0}\right)^2 \ll 1$$

at the highest significant frequency. If k^2 is not nearly unity, i.e. if magnetic leakage is serious, this condition is met only if $p_0^2 \gg p^2$ at the highest frequency: say numerically, if $\left(\frac{p_0}{2\pi}\right)^2 \gg 5000^2$. That this is no easy condition is shown by our former example‡. It is for this reason that magnetic leakage is so objectionable. Moderate magnetic leakage would not matter if there were no stray capacitance; moderate stray capacitance would matter less if there were no leakage; the designer is forced to compromise between the two ills.

A complicated formula such as that for m (above) does not convey very much until numerical values are attached. It is illustrated in Fig. 159 by four curves showing the amplification calculated from the formula at frequencies between 0 and 4000 c/s. The same triode, of anode resistance 30 kΩ, is used with four different trans-

---

* The hearer, fortunately, is nearly or quite unable to appreciate phase relationships between the Fourier components of a complex sound. If he were sensitive to phase, a shift of his position by a few inches with respect to the performers in an orchestra would alter the music he hears; and a change of a few miles in the length of a telephone line would change the quality of the speech and music received.

† This is the same condition as that already found for the choke connection in Section 2 of this Chapter. In the numerical example in Section 3 it led to $L_1 \not< 29$ H with our reference triode.

‡ P. 242. There $\frac{p_0}{2\pi} = 2950$.

5. TRANSFORMER CONNECTION BETWEEN TRIODES

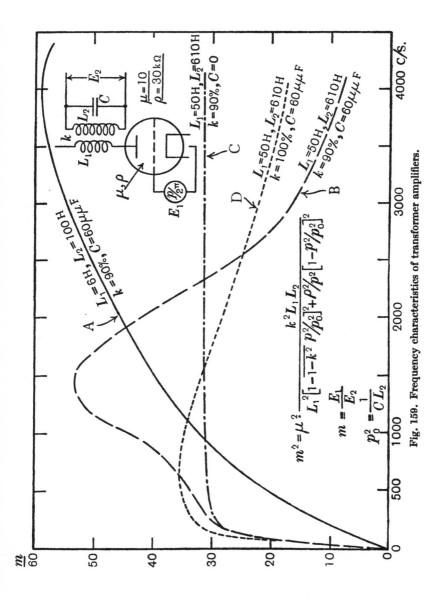

Fig. 159. Frequency characteristics of transformer amplifiers.

formers A, B, C, D of dimensions tabulated below. A and B represent real transformers with somewhat exaggerated magnetic leakage, while C and D exhibit the effect of ideal conditions, viz. no capacitance and no magnetic leakage respectively.

Transfr.	L_1	Turns ratio s	k	C	$\dfrac{p_0}{2\pi}$	$s\mu$
A	6 H	4	0·90	60 $\mu\mu F$	2050 c/s	40
B	50 H	3·5	0·90	60 $\mu\mu F$	830 c/s	35
C	50 H	3·5	0·90	0	∞	35
D	50 H	3·5	1·00	60 $\mu\mu F$	830 c/s	35

The inductance and capacitance in transformer A are those of an early pattern, designed primarily for small bulk* and for use with Morse signals. Its outstanding fault for telephony is clearly the smallness of the primary inductance, which makes the amplification very poor at low frequencies; but the stray capacitance causes the amplification to exceed $s\mu$ at frequencies above 1500 c/s. B represents a transformer planned much more generously for telephonic service; it has nearly the same turns ratio as A. Its superiority over A at low frequencies is pronounced; but its performance is marred by the large stray capacitance and/or large magnetic leakage which produce a well-marked peak in the curve, and cause the amplification to fall below $s\mu$ at frequencies exceeding 2500 c/s. Curve C shows the much superior performance—a sensibly uniform amplification of 31 above 500 c/s—which would be given by transformer B if there were no stray capacitance. Curve D, in comparison with curve B, shows how much the ill effect of stray capacitance would be reduced if there were no magnetic leakage. These performances would, of course, all be much better if the triode were our reference triode with its much lower ρ and slightly larger μ.

5 (c). *Measurements of a transformer, and of a three-stage amplifier, for acoustic frequencies.* The high-class performance which can be obtained with a well-designed transformer in conjunction with a triode of low anode resistance is illustrated in Fig. 160. These amplification-frequency curves refer to the Ferranti "AF 5" transformer shown in Plate XXI, when used with our reference triode

* To economise in fine wire, during the war, when supplies were short.

5. TRANSFORMER CONNECTION BETWEEN TRIODES

taking 4 mA of anode current. Measured values of the triode and of the transformer* are as follows:

$$\rho = 8700 \, \Omega; \quad \mu = 12\cdot 7;$$
$$L_1 = 80 \text{ H (a mean value)}; \quad s = 3\cdot 50; \quad \therefore \; L_2 \doteqdot 80 \times (3\cdot 5)^2 = 980 \text{ H};$$
$$1 - k^2 = 0\cdot 006.$$

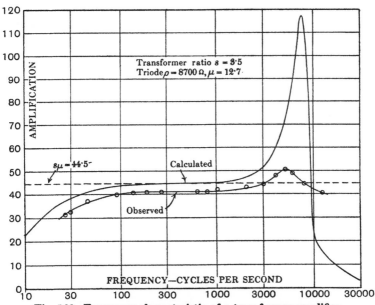

Fig. 160. Frequency characteristic of a transformer amplifier.

Including a condenser of 270 $\mu\mu$F across the primary (embodied by the makers within the housing of the transformer), and an additional 30 $\mu\mu$F connected across the secondary,

$$p_0 = 2\pi \times 590 \text{ c/s}; \quad \therefore \; C = 75 \, \mu\mu\text{F}.$$

Inserting these figures in the formula on p. 244, we obtain

$$m^2 = \frac{2050}{(1 - 0\cdot 0172 N^2)^2 + \dfrac{0\cdot 000300}{N^2}(1 - 2\cdot 87 N^2)^2},$$

where $N \equiv \dfrac{n}{1000}$ is the frequency in kilocycles per second. The upper

* The author is indebted to Mr C. Gordon Smith for much assistance in obtaining the experimental data for Figs. 160 and 161.

curve in Fig. 160 is plotted from this formula over the range 10–30,000 cycles per second.

This theoretical curve is to be compared with the actual observations of amplification shown on the lower curve. These points are plotted from the measured input signals e_1 (Fig. 158) required to produce an output signal e_2 maintained at about 2 V (R.M.S.) at all frequencies*. It will be seen that the calculated and observed curves agree fairly well, except that the high peak at about 5000 c/s in the former was not realised in the latter. This discrepancy is probably attributable mainly to one or both of the following inaccuracies in the theoretical treatment:

(i) the self-capacitance distributed throughout the windings is not adequately representable at high frequencies by a concentrated capacitance across the terminals;

(ii) the resistance of the transformer is not negligible†.

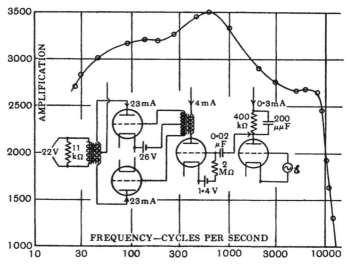

Fig. 161. Frequency characteristic of a 3-stage amplifier.

The overall performance of a complete three-stage acoustic amplifier is depicted in Fig. 161. The three stages are (i) a "DEH 410"

* The capacitance introduced across the transformer secondary by the measuring instrument constitutes the additional 30 $\mu\mu$F referred to above.

† Although the primary resistance with D.C. was 2·4 kΩ, at 256 c/s, the secondary being on open circuit, the resistance had risen to some 70 kΩ (and even higher in other specimens which did not appear to be defective).

5. TRANSFORMER CONNECTION BETWEEN TRIODES

triode (Fig 131) with a resistance anode circuit; (ii) a "DEL 410" triode (Fig. 132) with a Ferranti "AF 5c" transformer*; and (iii) two output triodes (Osram "P 425," $\rho = 2 \cdot 3$ kΩ, $\mu = 4 \cdot 5$) arranged in "push-pull" association, feeding a load circuit of 11 kΩ through a 1 : 1 transformer (Ferranti "OP 3c"). The amplifications to be expected at middle frequencies, in the absence of undesired stray capacitance and retroactive effects, are in the several stages respectively about 25, 45 and 3; i.e. an overall amplification of $25 \times 45 \times 3 = 3400$. In the measurements, at each frequency the input P.D. $\mathscr{E}$ was adjusted to produce a P.D. of about 22 V (R.M.S.) across the output terminals. The overall amplification $\dfrac{22}{\mathscr{E}}$ observed is seen to lie between 2500 and 3500 over the whole important acoustic range of frequencies.

It is easy to underestimate the really very remarkable triode amplifier performance here exemplified. Over the wide range of frequencies, 25 c/s to 5000 c/s, and without any adjustment for change of frequency, the amplification retains the large value of 3100 subject to a variation not exceeding about 12 %; and this in an instrument in which no fine adjustment of any sort is required. Moreover, since the instrument is designed to deliver a power output, the magnitude of its performance is more fairly assessed by reference to power levels. If the input P.D. $\mathscr{E}$ is supposed to occur across a resistance equal to that of the output apparatus (11 kΩ), the ratio of output power to input power is $(3100)^2$, i.e. about 10 million.

The chief advantages of the "push-pull" arrangement embodied in this amplifier are the avoidance of a D.C. component of magneto-motive force round the core of the output transformer, and the constancy of the total feed current to the anodes of the output triodes. If the same two triodes had been used in parallel (with the transformer primary correspondingly reduced from T to $\dfrac{T}{2}$ turns), the magnetomotive force would have been $46 \times \dfrac{T}{2}$ mA turns instead of $\left(23 \times \dfrac{T}{2} - 23 \times \dfrac{T}{2}\right) = 0$; and the mean feed current of 46 mA would have been subject to a large fluctuation at the signal frequency

---

* This is precisely similar to the "AF 5" transformer of Fig. 160 except that a centre tap is provided in the secondary winding.

(viz. about $\pm 8\sqrt{2}$ mA when the output P.D. is 22 V). If then all the anodes were fed from the same source (as was the condition in the measurements of Fig. 161), retroaction from the output triodes to the anode of the input triode due to impedance in this common source could be avoided only with special precautions.

6. High-frequency amplifiers

6 (a). *Stray capacitance in resistance amplifiers.* The unavoidable capacitance shunting the resistance reduces the impedance of the combination without limit as the frequency of the signal rises. In Fig. 162

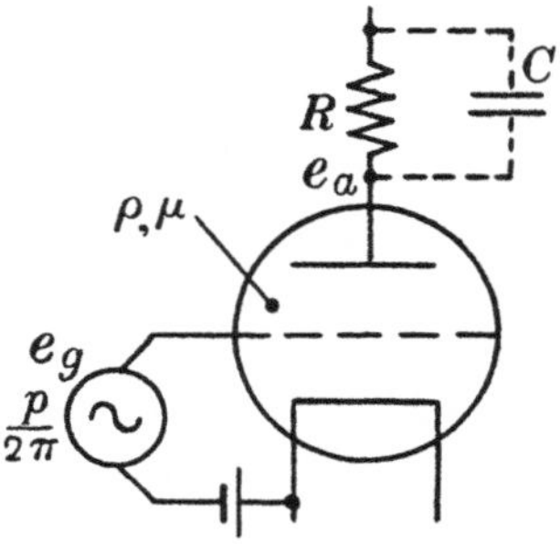

$$\frac{e_a}{e_g} = \mu \frac{\frac{1}{R}+jpC}{\rho+\dfrac{1}{\dfrac{1}{R}+jpC}} *,$$

$$= \mu \frac{R}{[\rho+R]+j[pC\rho R]}.$$

Fig. 162.

Hence the amplification is

$$m = \mu \frac{R}{\sqrt{(\rho+R)^2 + p^2 C^2 \rho^2 R^2}}.$$

In the example of p. 230, where $C = 0$, when an anode battery of 150 V was used with our reference triode and $R = 100$ kΩ, we found $\rho = 25$ kΩ, giving $\dfrac{m}{\mu} = 0\cdot 80$. But in reality C must have some finite value such as 10 $\mu\mu$F. The Table on p. 251 shows the value of $\dfrac{m}{\mu}$ calculated from the above formula for various frequencies $\dfrac{p}{2\pi}$ when $C = 10\ \mu\mu$F.

For the whole acoustic range of frequencies, and higher, resistance amplifiers can be constructed to give remarkably constant amplification; but with continued rise of frequency the amplification must

---

* This expression can be written down at sight by remembering the equivalent simple circuit of Fig. 130.

6. HIGH-FREQUENCY AMPLIFIERS

ultimately fall off. Fig. 163 shows a two-stage amplifier* with wire-wound resistance coils whose amplification over a wide range of

Frequency $\dfrac{p}{2\pi}$	Corresponding wavelength	$\dfrac{m}{\mu}$
0	—	0·80
10^5 c/s	3000 m	0·795
3×10^5 c/s	1000 m	0·75
10^6 c/s	300 m	0·50
3×10^6 c/s	100 m	0·20

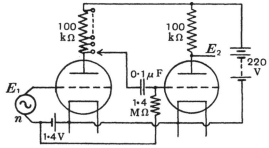

Fig. 163. An amplifier for acoustic measurements.

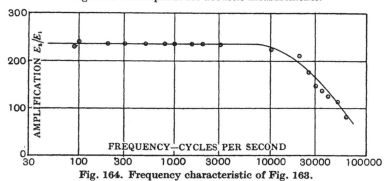

Fig. 164. Frequency characteristic of Fig. 163.

frequencies has been measured. Fig. 164 is the amplification-frequency curve for this amplifier; the points marked are the experimental readings. The triodes used were both Osram "D.E. 5B". According to the manufacturers' characteristics, $\mu = 20$ and the amplification

* Constructed for laboratory acoustic measurements. In the observations quoted below, the tap on the potential divider was at the anode end, as shown.

with 100 kΩ should be about $14 \cdot 5^2 = 210$. The output E_2 was maintained at 1·48 V (R.M.S.) throughout.

6 (b). *Rejector-circuit impedance*. While a capacitance shunt on a resistance can only reduce the impedance, on an inductance it may increase or decrease the impedance according to the frequency. It is customary in high-frequency amplifiers to turn the unavoidable stray capacitance to good account in this way.

In Fig. 165, C is a capacitance shunting a coil whose reactance pL is much larger than its resistance R.

We can write at sight*

$$\frac{e_a}{e_g} = \mu \frac{\dfrac{1}{\dfrac{1}{R+jpL}+jpC}}{\rho + \dfrac{1}{\dfrac{1}{R+jpL}+jpC}},$$

$$= \mu \frac{(R)+jpL}{[\rho(1-p^2LC)+(R)]+jpL\left[1+\dfrac{C}{L}\rho R\right]}.$$

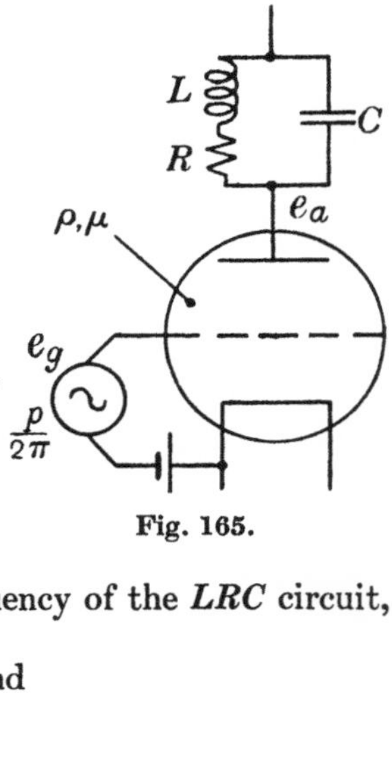

Since $R \ll pL$, the two terms written (R) are neglected. Writing

Fig. 165.

$p_0^2 LC = 1$, where $\dfrac{p_0}{2\pi}$ is the resonant frequency of the LRC circuit,

$$\delta = \frac{\pi R}{p_0 L}, \text{ where } \delta \text{ is its decrement, and}$$

$$x \equiv \frac{p}{p_0},$$

we obtain

$$\frac{e_a}{e_g} = \frac{j\mu p_0 L \cdot x}{\rho(1-x^2)+jp_0 L\left(1+\dfrac{\delta}{\pi}\cdot\dfrac{\rho}{p_0 L}\right)x}.$$

The amplification is therefore

$$m = \frac{\mu}{\sqrt{\left[\dfrac{\rho}{p_0 L}\left(\dfrac{1}{x}-x\right)\right]^2 + \left[1+\dfrac{\delta}{\pi}\cdot\dfrac{\rho}{p_0 L}\right]^2}}.$$

At resonance $x = 1$ and $m = \dfrac{\mu}{1+\dfrac{\delta}{\pi}\cdot\dfrac{\rho}{p_0 L}}$.

This approximates to the limiting value μ if $\dfrac{\delta}{\pi} \ll \dfrac{\rho}{p_0 L}$. Whereas

* See footnote on p. 250.

6. HIGH-FREQUENCY AMPLIFIERS

without the shunting capacitance C the amplification can approximate to μ only if $(p_0L)^2 \gg \rho^2$; in the presence of C, even though $p_0L < \rho$, the amplification approximates to μ *for signals of a critical frequency* if the decrement of the LRC circuit is sufficiently low.

If the condition for full amplification at the resonance frequency—viz. $\dfrac{\delta}{\pi} \ll \dfrac{\rho}{p_0L}$—*does not* obtain, the amplification must vary with the frequency. If the condition *does* obtain, we have

$$m \doteqdot \frac{\mu}{\sqrt{\left[\dfrac{\rho}{p_0L}\left(\dfrac{1}{x}-x\right)\right]^2 + 1}};$$

but again the amplification must fall off on each side of resonance unless $(p_0L)^2 \gg \rho^2$, i.e. unless the services of the shunting condenser were unnecessary in arriving at $m \doteqdot \mu$.

With high-frequency signals of short wavelength the designer cannot make $(p_0L)^2 \gg \rho^2$, since he cannot reduce C below the unavoidable stray capacitance. An amplifier for short waves must, therefore, be selective; and it can give good amplification even at the optimum frequency only if the decrement is sufficiently small. But for signals of lower frequency—long-wave or acoustic frequencies—the designer can choose dimensions to produce either a selective or a non-selective amplifier*. For high selectivity (sharp tuning) he must make $\dfrac{\rho}{p_0L}$ large and δ small; for low selectivity (blunt tuning) he must make $\dfrac{\rho}{p_0L}$ small, and he may disregard the value of δ.

The circuit of Fig. 166 is that of an amplifier intended to amplify signals over a range of frequencies extending below and above $\dfrac{p_0}{2\pi}$, where $p_0^2 LC = 1$. By making L large and C small, provided that the resistance R is kept sufficiently small, good amplification can be got at the optimum frequency $\dfrac{p_0}{2\pi}$, whether the anode tap is high or

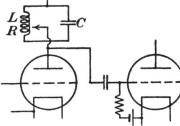

Fig. 166. Amplifier with adjustable selectivity.

---

* Selective acoustic amplifiers are anathema for telephony; but they are a valuable aid in avoiding interference in the reception of Morse signals by the heterodyne method.

low on the coil; but if high, the amplification will fall off rapidly with change of frequency on either side of $\frac{p_0}{2\pi}$; if low, not so rapidly*.

The foregoing conclusions as to the dependence of selectivity on the dimensions of the anode circuit are well illustrated in the following numerical example. The triode whose curves are given in Fig. 131 (p. 205: $\rho = 14$ kΩ, $\mu = 37\cdot5$) is to be employed with the circuit of Fig. 165 to amplify signals of frequency 50 kc/s ($\lambda = 6000$ m). A non-selective amplifier is sought by choosing the dimensions

$$L = 100{,}000 \ \mu\text{H}, \qquad C = 101\cdot5 \ \mu\mu\text{F};$$

and, alternatively, a selective amplifier by choosing the dimensions

$$L = 1{,}000 \ \mu\text{H}, \qquad C = 10{,}150 \ \mu\mu\text{F}.$$

In each design the decrement of the LRC circuit is assumed to be 0·03. The amplification-frequency curves in Fig. 167 have been calculated by inserting these numerical values in the formula for m on p. 252.

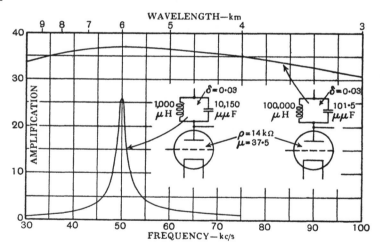

Fig. 167. Selective and unselective designs.

If the triode employed with these circuits were our reference triode ($\rho = 8$ kΩ, $\mu = 13$) both curves would be flatter, and the amplifications at the peaks would be changed from 37 and 25·5 to 13 and 8·5.

* In practice, the total equivalent resistance R would not be constant, since it contains a part due to absorption of power in the grid connections of the second triode. The magnitude of this component of R increases as the anode tap is lowered.

7. CAPACITANCE BETWEEN ANODE AND GRID

7 (a). *Effects of anode-grid capacitance.* Hitherto in this Chapter no consideration has been paid to the effect of capacitance between anode and grid (C in Fig. 168). Although this can be kept small by careful disposition of the electrodes and their connections*, it cannot in practice be reduced below about $2\ \mu\mu F$. If the source of the input alternating E.M.F. e—conventionally shown in the figures as an alternator—possessed no impedance, this stray capacitance between anode and grid would only have nearly the same effect as an equal capacitance between anode and filament. But the apparatus or circuit wherein e is impressed usually possesses considerable impedance Z_g. Any current caused to flow through C must flow also through this impedance, and therefore reacts on the grid potential e_g.

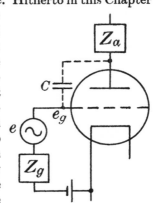

Fig. 168. Anode-grid capacitance.

The effects of C are not only to modify the amplification, but also to throw a load upon the source of the input signal e, since there is now a path which can be traversed by current through the source. The magnitude and sign of each of these effects depends not only on the magnitudes of C, Z_a and Z_g, but also on the character of the impedances. When e and Z_g in Fig. 168 are provided by a preceding triode as in Fig. 156 (p. 238), Z_g is practically the anode resistance of this triode; when as in Fig. 157, Z_g is approximately s^2 times this resistance, where s is the step-up ratio of the transformer; and when the transformer is leaky, or loose coupling is intentionally used as in high-frequency amplifiers, Z_g may possess also large reactance. Analysis involves tedious algebra even in special cases; it is therefore economical to proceed as far as practicable by a generalised treatment, as in Chapter X.

For the amplifier of Fig. 214 (p. 316), with the generalised impedances α, β, γ, the performance is given by equation (33) on p. 317; it is

$$\frac{e_a}{e} = \frac{-\alpha(g\gamma - 1)}{\alpha + \beta + (g+a)\alpha\beta + \gamma(1+a\alpha)},$$
$$= \frac{-\alpha(\mu\gamma - \rho)}{\rho(\alpha+\beta) + (\mu+1)\alpha\beta + (\rho+\alpha)\gamma} \quad \ldots\ldots\ldots(1).$$

* E.g., as in triode D, Plate XIII, where the grid and anode connections are led out at opposite points on the bulb.

We will use this formula (1) to exhibit the effect of anode-grid capacitance on the performance of amplifiers by applying it to two special cases, viz.:

where α and β are pure resistances R_1 and R_2 (Fig. 169), and

where α and β are inductive reactances jpL_1 and jpL_2 (Fig. 170).

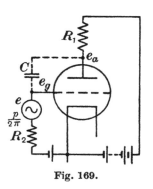

Fig. 169.

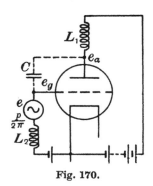

Fig. 170.

Writing $\gamma \equiv \dfrac{1}{jpC}$, the above formula (1) becomes:

for Fig. 169,

$$\frac{e_a}{e} = \frac{-(\mu - jpC\rho)}{\left[1 + \dfrac{\rho}{R_1}\right] + jpC\left[(\mu+1)R_2 + \rho\left(1 + \dfrac{R_2}{R_1}\right)\right]} \quad \ldots\ldots(2);$$

and for Fig. 170,

$$\frac{e_a}{e} = \frac{-(\mu - jpC\rho)}{[1 - (\mu+1)p^2L_2C] - j\dfrac{\rho}{pL_1}[1 - p^2(L_1 + L_2)C]} \quad \ldots(3).$$

For the resistance amplifier, formula (2) shows that as C is increased from zero the amplification $\left(\text{i.e. magnitude of } \dfrac{e_a}{e}\right)$ must be reduced from

$$\dfrac{\mu}{1 + \dfrac{\rho}{R_1}} \text{ when } C = 0 \text{ to } \dfrac{1}{(\mu+1)\dfrac{R_2}{\rho} + 1 + \dfrac{R_2}{R_1}} \text{ when } C = \infty.$$

The presence of C cannot greatly (if at all) increase the amplification, since it does not reduce either the real or the imaginary term of the denominator.

Plate I

Vertical antenna with flat top: Budapest broadcast transmitter (p. 28).
(Gesellschaft für drahtlose Telegraphie m. b. H.)

Inductance coils at Rugby (p. 64).
(Engineering Department, Post Office.)

PLATE III

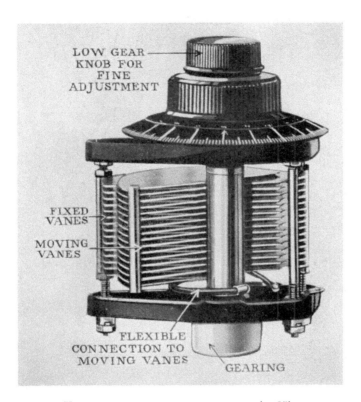

VARIABLE AIR-DIELECTRIC CONDENSER (p. 65).
(Pye Radio Ltd.)

PLATE IV

MICA CONDENSERS AT RUGBY (p. 65).
(Engineering Department, Post Office.)

PLATE V

FIXED SPARK GAP (p. 111).
(Marconi's Wireless Telegraph Co., Ltd.)

PLATE VI

MULTIPLE QUENCHED-SPARK GAP (p. 115).
(Gesellschaft für drahtlose Telegraphie m. b. H.)

PLATE VII

SPARK TRANSMITTER: $1\frac{1}{2}$ KW (p. 116).
(Radio Communication Co., Ltd.)

PLATE VIII

ELWELL-POULSEN ARCS AT LEAFIELD: 240 KW (p. 131).
(Engineering Department, Post Office.)

PLATE IX

FEDERAL-POULSEN ARC: 500 kW (p. 132).
(Federal Telegraph Co.)

PLATE X

ALEXANDERSON ALTERNATOR : 200 kW ; 22,000 c/s (p. 136).
(Institute of Radio Engineers.)

PLATE XI

BETHENOD-LATOUR ALTERNATORS AT SAINTE ASSISE (p. 187).
(Compagnie Générale de Télégraphie sans Fil.)

PLATE XII

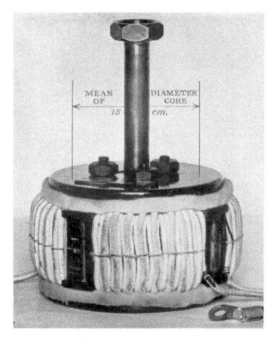

FREQUENCY MULTIPLIER WITH HIGH SATURATION : 10 kW (p. 143).
(C. Lorenz Aktiengesellschaft.)

PLATE XIII

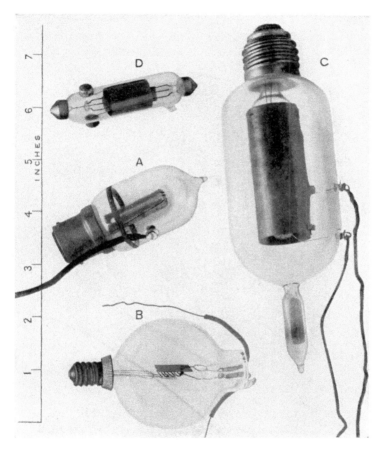

EARLY THERMIONIC TUBES.
A. FLEMING VALVE (p. 172).   C. MARCONI "N" TRIODE (p. 185).
B. DE FOREST "AUDION" (p. 172).   D. MARCONI "Q" TRIODE.

PLATE XIV

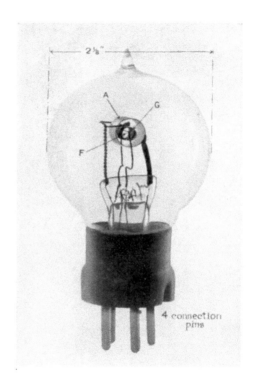

FRENCH OR ARMY "R" TRIODE (pp. 173 and 200).

PLATE XV

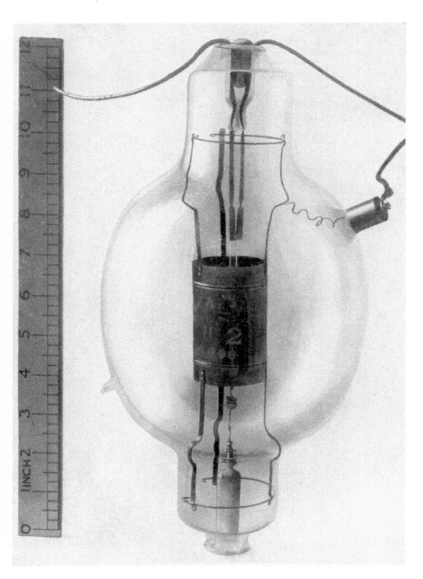

VALVE (DIODE) FOR RECTIFYING H.T. ALTERNATING CURRENT (p. 187).

Plate XVI

Typical modern receiving triode : actual size (p. 208).
(General Electric Co., Ltd.)

PLATE XVII & PLATE XVIII

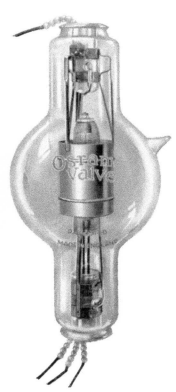

GLASS TRANSMITTING TRIODE: ANODE DISSIPATION 200 W; BULB DIAM. 12 CM (p. 210).
(General Electric Co., Ltd.)

WATER-COOLED TRIODE: ANODE DISSIPATION 7·5 KW; OVERALL LENGTH 54 CM (p. 210).
(General Electric Co., Ltd.)

PLATE XIX

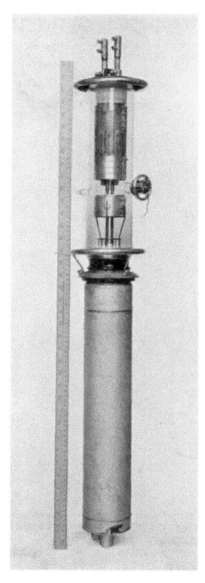

HIGH-POWER WATER-COOLED TRIODE: ANODE
DISSIPATION 50 kW; OVERALL LENGTH 108 CM (p. 212).
(Marconi's Wireless Telegraph Co., Ltd.)

PLATE XX and PLATE XXI

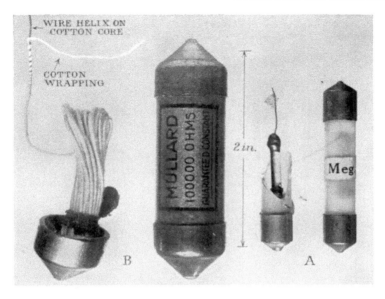

A. EDISWAN GRID LEAK (pp. 233 and 263).
B. MULLARD WIRE-WOUND ANODE RESISTOR (p. 234).

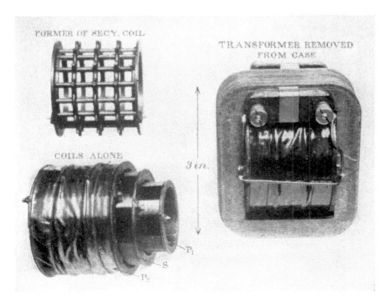

FERRANTI TRIODE TRANSFORMER (pp. 242, 246).

PLATE XXII

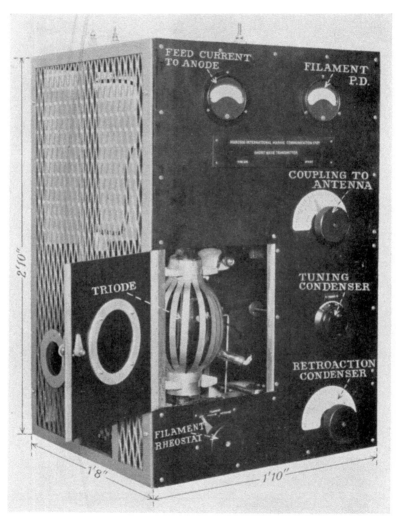

SHORT-WAVE TELEGRAPH TRANSMITTER (p. 294).
(Marconi International Marine Communication Co., Ltd.)

PLATE XXIII

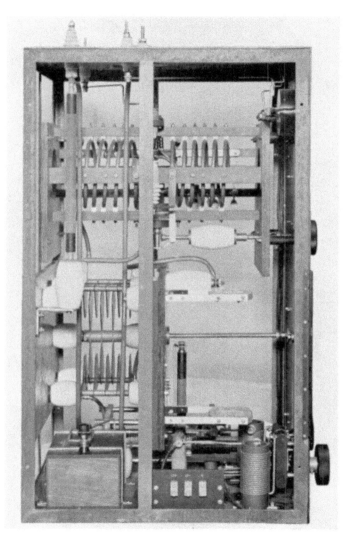

INTERIOR OF PLATE XXII (p. 294).

PLATE XXIV and PLATE XXV

RECEIVER FOR WAVELENGTHS 15 M–20,000 M (p. 370).
(Marconi International Marine Communication Co., Ltd.)

INTERIOR OF PLATE XXIV (p. 370).

PLATE XXVI

WIRELESS CABIN OF MODERN LINER (p. 371).
(Marconi International Marine Communication Co., Ltd.)

PLATE XXVII

BROADCAST TRANSMITTER, BROOKMAN'S PARK (p. 425).
(British Broadcasting Corporation.)

PLATE XXVIII

SHORT-WAVE TELEPHONE RECEIVER, BALDOCK (p. 429).
(Engineering Department, Post Office.)

PLATE XXIX

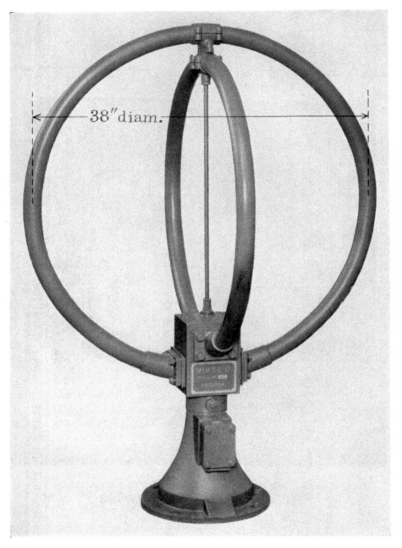

MARINE D.F. ANTENNAS, BELLINI-TOSI SYSTEM (p. 457).
(Marconi International Marine Communication Co., Ltd.)

PLATE XXX

SHORT-WAVE ANTENNA ARRAYS AT GRIMSBY (p. 475).
(Engineering Department, Post Office.)

Plate XXXI

Junction box of Plate XXX (p. 475).

7. CAPACITANCE BETWEEN ANODE AND GRID

For the reactance amplifier, formula (3) indicates the possibility of the more complicated effects associated with resonance phenomena. As C is increased from zero, both the real and the imaginary terms of the denominator are reduced. A value C_1 of C is reached at which one or other of these terms passes through zero; further increase of C causes this term to increase (negatively), the other term still decreasing until, when $C = C_2$, it passes through zero.

If both real and imaginary terms pass through zero together—i.e. if $C_1 = C_2$—the amplification becomes infinite. Expressed otherwise, a finite alternating anode potential e_a is produced with an infinitesimal input E.M.F. e; the amplifier is unstable, and any momentary disturbance will cause it to oscillate spontaneously. The condition for $C_1 = C_2$ is obviously

$$(\mu + 1) p^2 L_2 C = 1 = p^2 (L_1 + L_2) C,$$

i.e. $L_1 = \mu L_2$ ....(4).

Instability of amplifiers due to the anode-grid capacitance C is a well-known and very troublesome phenomenon. Equation (4) is the condition which must be fulfilled before instability can occur in the amplifier of Fig. 170; and at first glance it might be thought a simple matter to ensure that equation (4) shall not be fulfilled. It would be a simple matter if L_1 and L_2 were independent of frequency, or if the frequency of the self-oscillation were under control. But the amplifier will generate oscillation of any frequency whatever which makes equation (4) true; and in reality L_1 and L_2 must vary with the frequency, since each of them is not a pure inductance, but is the equivalent inductance value $\dfrac{p\mathscr{L} - \dfrac{1}{p\mathscr{C}}}{p}$ of some inductance $\mathscr{L}$ shunted by some stray capacitance $\mathscr{C}$. In the presence of stray capacitance, therefore, equation (4) may hold for *some* frequency*.

7 (b). *Numerical example.* Fig. 171 shows a short-wave amplifier ($\lambda = 30$ m) in which the anode and grid stray capacitances are taken as 5 $\mu\mu$F each, and the anode-grid capacitance is K.

---

* *May* hold, not must hold. As shown in X–4 (b), the reactances of grid and anode circuits must both be positive for self-oscillation to occur. This rules out frequencies above the lower of the resonance frequencies of the grid and anode circuits, and gives the designer an opportunity of selecting circuit dimensions with which self-oscillation does not have to be repressed by the introduction of resistance. The condition for self-oscillation in the presence of resistances is given by equation (14a) on p. 313.

VIII. THE TRIODE AS AMPLIFIER

The impedances α and β are found from equation (5) on p. 102 (with $L'' = 0 = M$) to be

$$\alpha = (9 + 2100j)\ \Omega, \quad \text{and} \quad \beta = (0.6 + 780j)\ \Omega;$$

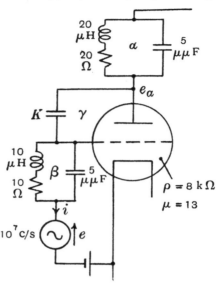

Fig. 171. A short-wave amplifier.

and the impedance γ is $\dfrac{-16000j}{K}$ if K is in $\mu\mu\text{F}$. The Table shows the performance $\dfrac{e_a}{e}$ and the input impedance $-\dfrac{e}{i}$, calculated for our

Anode-grid capac. $K\,\mu\mu\text{F}$	Performance $\dfrac{e_a}{e}$	Input impedance $-\dfrac{e}{i}\ \Omega$
0	$-0.8 - 3.2j$	$\infty\,(-1 - 0.6j)$
0.5	$-0.7 - 3.6j$	$-7100 - 3500j$
1	$-0.6 - 4.1j$	$-3300 - 1300j$
2	$3.5 - 5.1j$	$-1400 - 170j$
3	$2.6 - 5.8j$	$-730 + 200j$
4	$4.9 - 4.2j$	$-420 + 380j$
5	$5.2 - 1.9j$	$-220 + 490j$
8	$3.2 + 0.2j$	$60 + 660j$
12	$1.7 + 0.2j$	$210 + 740j$

7. CAPACITANCE BETWEEN ANODE AND GRID

reference triode by use of equations (33) and (34) on p. 317, for values of K between 0 and 12 $\mu\mu$F.

The magnitudes of $\dfrac{e_a}{e}$ and of $-\dfrac{e}{i}$, viz. the amplification and the magnitude of the input impedance, are plotted in Fig 172.

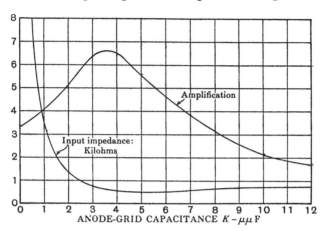

Fig. 172. Effects of anode-grid capacitance in Fig. 171.

7 (c). *Methods of preventing anode-grid capacitance current.* The effects of stray capacitance are, of course, most in evidence at high frequencies. The designer of a high-frequency amplifier usually strives either (i) to avoid all retroaction from anode to grid, or (ii) to ensure that the self-oscillation condition approached shall be of the signal frequency, and that the degree of retroaction—the closeness of approach—shall be under control. Aim (ii) presents no difficulty if (i) has previously been achieved.

There are two distinct methods for removing the retroaction caused by anode-grid capacitance. One is that of the screen-grid tetrode, in which the capacitance between anode and grid is reduced to an extremely small value by the interposition of a constant-potential screening grid as described in VII–9 (d). In the other method the capacitance is not removed, but is balanced by some form of bridge circuit. The usual disposition is shown in principle in Fig. 173, and often goes by the name of "neutrodyne." The mid point (instead of one end) of the impedance Z_a is joined to the anode battery; one end (as usual) is connected to the anode; and the other end is connected through a small variable capacitance NC to the grid. The alternating component of the anode-grid P.D. sends a current I across the stray capacitance SC to the grid; but the equal anti-phased P.D. between the other end of Z_a and the grid sends a current $-I$ across the neutralising capacitance NC when this is adjusted to equality with SC. Consequently no current flows through Z_g and the grid potential is unaffected by anode changes.

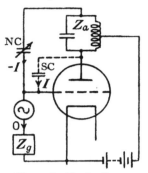

Fig. 173. Neutrodyne balance.

In practice each of these methods is very effective; though the second requires more care in carrying out, and is prone to need re-adjustment of NC when the wavelength of the signals is changed. One or other must be resorted to in constructing any selective amplifier giving high amplification—and therefore in any amplifier giving high amplification for short-wave signals—especially if several stages of amplification are employed.

8. Strictly aperiodic amplifiers giving high amplification

Multi-stage amplifiers of the type of Fig. 155 (p. 238), with the impedances $Z_1, Z_2, \ldots$ pure resistances, are capable of amplifying changes of potential e_{g1} of any form, including the long-maintained changes such as may occur in a long telegraphic Morse dash and in various physical and physiological processes where an electrical indication is required. But with high amplifications, such dispositions are very inconvenient to operate.

There are two alternative methods of obtaining high aperiodic amplification. In one, an aperiodic (resistance) form of retroaction is employed. This can be effected only by the use of at least two triodes; but it yields results at all frequencies analogous to those obtained at one particular frequency from a single triode with retroactive oscillatory circuits, as discussed in Chapter IX. The two-triode retroactive aperiodic amplifier shown in Fig. 174* was found capable of giving amplification $\dfrac{E}{e_1}$ as high as 2000. The

* See L. B. Turner, "The Kallirotron, an aperiodic negative-resistance triode combination," *Radio Review*, Vol. I (1920), p. 317.

8. APERIODIC AMPLIFIERS GIVING HIGH AMPLIFICATION

amplification in either triode alone was only about 3; so that the amplification, for signals of any form, was enhanced by retroaction some 220 times.

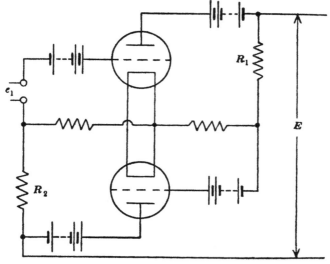

Fig. 174. Aperiodic retroaction.

In the other method also, negative differential (or slope) conductance is utilised; but it is obtained not as the mathematical expression of the result of circuit retroaction, but as the direct physical effect of the secondary emission of electrons from a cold electrode in a single thermionic tube*.

An instance and explanation of negative conductance so produced were given in VII-9(e) (Fig. 143); and the screen-grid tetrode, although not normally used in this condition, is a convenient form of tube for obtaining the desired effect. Let the slope $\dfrac{\partial i_a}{\partial e_a}$ of the characteristics in the neighbourhood of 30 V in Fig. 143 (p. 221) be the negative anode conductance $-\alpha$, where α is a positive quantity (viz. about 0·0167 mA/V in Fig. 143); and let $g \equiv \dfrac{\partial i_a}{\partial e_g}$ be the mutual

---

* This was first utilised by A. W. Hull, in his "Dynatron" and "Pliodynatron." See A. W. Hull, "The Dynatron, a vacuum tube possessing negative resistance," *Proc. I.R.E.* Vol. 6 (1918), p. 5.

conductance in that region (viz. about 0·45 mA/V in Fig. 143). Then (Fig. 175)

$$\delta i_a = g \cdot \delta e_g + (-\alpha) \delta e_a,$$
$$\delta e_a = -R \cdot \delta i_a;$$
$$\therefore \quad -\frac{\delta e_a}{\delta e_g} = \frac{g}{\alpha} \cdot \frac{\alpha R}{1 - \alpha R}.$$

This tends to ∞ as R is increased towards the critical value $\frac{1}{\alpha}$, the magnitude of the (negative) slope resistance of the anode. If R exceeds $\frac{1}{\alpha}$, there is instability, and the representative point cannot be maintained in that region of the characteristic*.

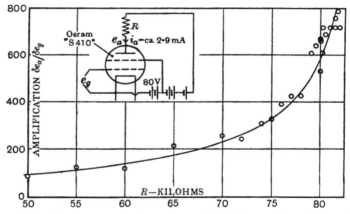

Fig. 175. Performance of a resistance amplifier with negative conductance.

The curve in Fig. 175 is plotted from the observed values of δe_g producing an anode change δe_a of about 2 V, with different values of R, using a screen-grid tetrode very like that of Fig. 143. Controllable amplification up to nearly 800 was obtained; and although in these tests, for the sake of experimental simplicity, A.C. signals (of 90 c/s) were employed, the same amplification would have been found with signals of any form and duration.

* This is readily seen on imagining the application to the terminals of R of an infinitesimal externally generated P.D., for example by the occurrence of a potential gradient in the atmosphere. The current change in R thereby induced would *reinforce* the applied P.D.

8. APERIODIC AMPLIFIERS GIVING HIGH AMPLIFICATION

Both these methods depend on effecting an adjustment of circuit conditions, whereby the amplification tends towards infinity as the critical condition of unstable equilibrium is approached. The amplification obtainable in practice is limited by imperfections in steadiness of the circuit conditions, forbidding too close an approach towards instability. An amplifier of the type of Fig. 156 is always to be preferred for signals whose duration is sufficiently brief. If $C_2 = 2\ \mu\text{F}$ and $r_2 = 2\ \text{M}\Omega$, a signal of the character of a Morse dash lasting as long as 1 second would be fairly faithfully reproduced. The need for strictly aperiodic amplifiers does not arise in the ordinary processes of telegraphy and telephony.

Where a resistance amplifier of any type is to be used over a very wide range of frequencies, the impurity of the resistor may be important. Measurements* at various frequencies between 0 and 390 kc/s showed that the 100 kΩ resistor (Mullard) of p. 234 (B, Plate XX) had an impedance of only 53 kΩ at 52 kc/s, falling continuously to 20 kΩ at 390 kc/s. The capacitance shunting responsible for this fall doubtless is due to the close bunching of the cable within the protecting tube, and is avoidable by using the cable not bunched. The impedance then would probably be very slightly (negligibly) inductive. Another form of wire resistor (Ferranti), made up with the same external form but wound with fine wire in a slotted former, showed a more constant impedance. Its D.C. resistance was 75 kΩ, and the impedance rose with frequency only to 77·2 kΩ at 53 kc/s. Over the whole range from 0 to 390 kc/s the measured impedance remained very nearly that of 75 kΩ in series with 44 mH, the whole shunted by 3·14 $\mu\mu\text{F}$.

* Kindly made for the author by Mr L. A. Meacham.

CHAPTER IX

THE TRIODE AS OSCILLATOR

1. General principle of retroaction in triodes

THE mechanism of a clock is a mechanical oscillator to which a counter is attached. The pendulum continues to swing, despite the damping due to air resistance and other energy losses, because a source of energy (the mainspring or the driving weight) is allowed (by a mechanism called the escapement) to apply a driving impulse every cycle. The essence of the escapement is that it causes the motion of the thing which is to be driven itself to turn on and off the source of energy in such a way that force is applied in the direction of motion at the instant. An LC circuit is the electrical analogue of the pendulum, and the triode makes it easy to devise analogues of the escapement.

A clock is not required to serve as an engine, but as a time-keeper; the design of the escapement is therefore mainly conditioned by the desire to obtain the most perfect possible isochronism of the swinging pendulum *. To this end, the damping of the pendulum is reduced as far as can be contrived; the driving power is only what is required to make up the unavoidable loss, and is provided by a force acting on the pendulum during only a small fraction of the period. On the other hand, a triode oscillator is required (usually) to act as an alternator delivering power to external apparatus; the driving E.M.F. is applied to the oscillator more or less continuously throughout the cycle; and the chief function of the circuital analogue of the escapement mechanism is to ensure that this E.M.F. shall always (or in the main) be directed *with* the oscillatory current.

An appreciation of how oscillation must arise in the circuits of Fig. 176 when certain conditions are met is a short cut to an understanding of the retroactive principle underlying the triode oscillator. LRC is a lightly damped oscillatory circuit inserted between grid and filament, and is coupled by a mutual inductance M to a small inductance inserted between anode and filament. We will suppose

---

* A good ship chronometer may hold its rate constant to within about 1 in 87,000 (a second a day).

1. GENERAL PRINCIPLE OF RETROACTION IN TRIODES

the batteries are chosen to place the representative point at P on the anode characteristic, where it is approximately straight and where the grid current is zero.

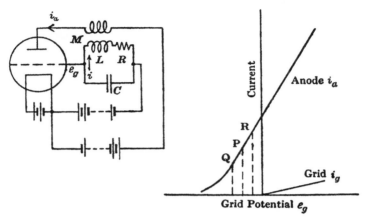

Fig. 176. Self-oscillation by retroaction.

Let the *LRC* circuit be disturbed by any means and left oscillating with its natural frequency given by

$$p \equiv 2\pi n \leftrightharpoons \frac{1}{\sqrt{LC}},$$

so that P oscillates between Q and R; and let $\mathscr{I}$ be the R.M.S., and i the instantaneous, value of current in *LRC*. Under the action of the ohmic E.M.F. $-Ri$, energy is being removed from the circuit at the mean rate $R\mathscr{I}^2$, and unless some E.M.F. is impressed from without, compensating for the ohmic drop in R, the amplitude of the oscillation will die down according to the damping factor $\epsilon^{-\frac{R}{2L}t}$. Owing to the reactance drop in the grid coil L, the grid potential fluctuates, lagging 90° on the current i and with an R.M.S. value $pL\mathscr{I}$. Assuming for simplicity that the anode resistance $\rho \equiv \frac{\partial e_a}{\partial i_a}$ is very large, the anode current i_a approximately follows, in phase, the variations of grid potential, so that in the anode coil there is an alternating current $g \cdot pL\mathscr{I}^*$ lagging 90° on the current in L. This

---

* $g \equiv \dfrac{\partial i_a}{\partial e_g}$, $a \equiv \dfrac{\partial i_a}{\partial e_a}$, as heretofore.

anode-coil current induces in the grid coil an E.M.F. $pM \cdot g \cdot pL\mathscr{I}$ syn-phased or anti-phased with respect to the current i according as M is positive or negative. If the sense of M is such that a rising anode current induces an E.M.F. tending to raise the grid potential (as is the case with the connections and directions of winding shown in Fig. 176 if the two coils are coaxial), there is introduced into the *LRC* circuit an E.M.F.

$$p^2gML\mathscr{I}$$

in phase with the current, and therefore contributing energy to the circuit at the rate

$$p^2gML\mathscr{I}^2.$$

The net loss of energy from the *LRC* circuit is therefore at the rate

$$R\mathscr{I}^2 - p^2gML\mathscr{I}^2$$
$$= \mathscr{I}^2(R - p^2gML).$$

The rate of decay of the oscillation is thus reduced; and if the second term is numerically greater than the first, the rate of decay is negative, and the amplitude of the initial oscillation instead of dying down increases. This means that as soon as such a circuit is brought into being, as by lighting up the filament or coupling the coils together, an oscillation will start and build itself up to some finite amplitude. The final value we do not here investigate.

The condition for an oscillation to maintain itself is clearly

$$R - p^2gML = 0,$$

i.e. $$R = \frac{gM}{C}, \text{ since } p^2 = LC,$$

i.e. $$M = \frac{RC}{g}.$$

Whenever, by manipulation of M or R or C or g, the L.H.S. of this equation is made greater than the R.H.S., there is instability; and on the smallest disturbance occurring, the system sets itself into oscillation.

An accurate derivation of the conditions of self-oscillation in this circuit is given in X–3 (d) (equations (26) and (27)); but the foregoing approximate examination is likely to throw more light on the physical aspect of the retroactive principle.

2. Analysis of a usual circuit from first principles

2 (a). *Inception of oscillation*. There are many ways of connecting up a triode and one or more oscillatory circuits so as to produce sustained oscillation. The principle in all is, of course, to let the oscillatory current so stimulate the grid that a supply of power is introduced into the oscillatory circuit from the source in the anode circuit, viz. the high-tension battery. This is what we have called the retroactive principle. We will study here from first principles and in some detail one simple arrangement, very commonly used and shown in Fig. 177, in which the oscillatory circuit is in series with the anode (instead of the grid as in Fig. 176). The resistance R of the oscillatory circuit covers, of course, not merely the ordinary Ohm's-law resistance of the coil, but also all losses of power from the circuit, such as condenser losses and losses due to coupling to an external circuit where work is required to be done. The actual location or locations of R in the oscillatory circuit is not a matter of any importance; it is shown as a part of the inductance coil for simplicity, and because it is usually from the inductance coil that power is drawn to the external circuit.

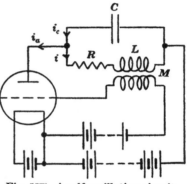

Fig. 177. A self-oscillating circuit.

Let i_c, i be the instantaneous values of current *change* in C and L, and let i_a, e_g, e_a be the instantaneous values of anode current and grid and anode potential *changes**, from the non-oscillating condition. Then
$$i_c - i = i_a = a \cdot e_a + g \cdot e_g.$$
Since grid current is zero,
$$e_g = -M \frac{di}{dt} \dagger.$$
Also
$$e_a = Ri + L \frac{di}{dt},$$
and
$$i_c = -C \frac{de_a}{dt} = -CR \frac{di}{dt} - CL \frac{d^2 i}{dt^2}.$$

---

* We are concerned with the steady (non-oscillating) values only indirectly, as determining the regions of the characteristics over which the triode is operating.

† The coils are supposed to be oppositely wound as shown, and coupled as if coaxial.

Making these substitutions in the first equation,

$$CR\frac{di}{dt} + CL\frac{d^2i}{dt^2} + i = gM\frac{di}{dt} - aRi - aL\frac{di}{dt};$$

i.e. $$CL\frac{d^2i}{dt^2} + (CR + aL - gM)\frac{di}{dt} + (1 + aR)i = 0.$$

If $$(CR + aL - gM)^2 < 4CL(1 + aR),$$

the roots of the auxiliary equation are unreal, and the current i is oscillatory, viz.

$$i = Ie^{bt}\sin(pt + \theta),$$

where $$b = -\frac{CR + aL - gM}{2CL},$$

$$p = \sqrt{\frac{1 + aR}{CL} - \frac{(CR + aL - gM)^2}{4C^2L^2}},$$

and I and θ are determined by initial conditions.

Owing to its association with the triode, the damping exponent b has been changed from $-\frac{1}{2L}R$ to

$$-\frac{1}{2L}\left(R + \frac{aL}{C} - \frac{gM}{C}\right).$$

The first term within the brackets is the resistance proper to the circuit; the second term is the resistance added owing to the anode-filament resistance $\rho \equiv \frac{1}{a}$ shunting the condenser; and the third term is a negative resistance, introduced by anode-grid retroaction. According as b is negative, zero, or positive, i.e. as

$$gM \lesseqgtr (CR + aL),$$

the amplitude of any initial oscillation dies down, remains constant, or increases with time (Fig. 178*).

The first condition, $gM < (CR + aL)$ (Fig. 178, C), is used, as we shall see in XII-2, in retroactive amplifiers, where retroaction between anode and

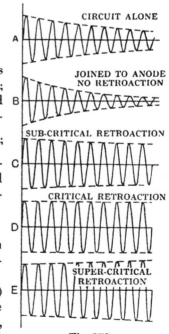

Fig. 178.

* The oscillations in Fig. 178 are drawn approximately to scale, with the same initial amplitude in each curve, the decrements being 0·1, 0·2, 0·02, 0 and −0·02.

2. ANALYSIS OF A CIRCUIT FROM FIRST PRINCIPLES

grid circuits is provided in order to decrease the damping of an oscillatory circuit, but not enough to produce self-oscillation. The third condition,

$$gM > (CR + aL) \quad \text{(Fig. 178, E)},$$

is the condition for the oscillation to increase in amplitude, and may be referred to as the *condition for growth*. It must lead to an amplitude at which a and g can no longer be taken as constants, and/or the grid current as zero. The second or intermediate condition,

$$gM = (CR + aL) \quad \text{(Fig. 178, D)},$$

is the *condition for maintenance* of the oscillation unchanged in amplitude. Then $p = \sqrt{\dfrac{1 + aR}{CL}}$. The frequency is therefore greater than the resonant frequency of the oscillatory circuit when isolated; but since the term aR is usually extremely small compared with unity, the frequency is usually very nearly the resonant frequency.

2 (b). *Amplitude of steady oscillation.* For the inception of oscillation the foregoing analysis is accurate, provided that there is negative grid bias to justify the assumption of zero grid current. It is not necessary that the triode characteristics should be rectilinear, i.e. that a and g should be truly constants, since for an indefinitely small amplitude of oscillation the characteristics are indistinguishable from their tangents. As the amplitude grows, however, the representative point on (say) the lumped characteristic extends its travel; and unless the region traversed is rectilinear, a and g can no longer be regarded as constants. Moreover, if the amplitude grows sufficiently to bring the grid potential to a positive value during a portion of the cycle, grid current will flow and the grid potential will no longer be exactly $M \dfrac{di}{dt}$. In either event—viz. a and g ceasing to be constant, or grid current ceasing to be zero—it is obvious that harmonics of the fundamental frequency must appear in the oscillation*.

Nevertheless, in examining an oscillation of finite amplitude (and not merely the conditions of its inception), it is by no means necessarily unpractical to take a and g as constants and grid current as zero. Except when the oscillator is required to show a high efficiency

---

* Experience shows that an effect of large grid current is also to lower the fundamental frequency.

as a converter of electrical power from the D.C. to the A.C. form, it is entirely feasible to choose the static anode and grid potentials so that an extensive region of sensibly rectilinear characteristics may be worked over, and so that no grid current may flow. For example, with the characteristics of Fig. 132 (p. 206), if the static potentials of anode and grid were 200 V and -6 V, the anode current fluctuation i_a might reach some 5 mA without departure from approximate rectilinearity, and the grid potential fluctuation e_g might reach some 6 V without sensible grid current occurring. Oscillators required to produce alternating current of especially pure sine form are, in fact, arranged to operate under such conditions. Moreover, even when grid current does flow during part of the cycle, it is often so much smaller than the anode current, and still more the power consumed at the grid is so much smaller than the power in the anode circuit, that it is permissible to ignore the effect of the grid current except when the purity and/or the exact frequency of the oscillation are of prime importance. This we do in the following examination of what happens in the arrangement of Fig. 177 after oscillation has grown up.

The condition for growth has been found to be

$$gM > (CR + aL).$$

This may be rewritten as

$$\mu M > L + CR\rho.$$

If therefore the mutual inductance M between the two coils (Fig. 177) is gradually increased until it just exceeds the critical value, an oscillation, once started, increases in amplitude until the condition no longer obtains. It ceases to obtain chiefly because $\rho \equiv \dfrac{\partial e_a}{\partial i_a}$ no longer remains constant throughout the cycle. Thus in Fig. 132, ρ remains nearly constant at 8000 Ω until the anode current falls to (say) 5 mA; thereafter it rises, reaching about 13,000 Ω when the anode current is (say) 3 mA. Further growth of the amplitude of fluctuation of anode current, so that it penetrates into regions of increasing ρ, can occur only if M is increased to compensate.

2 (c). *Output and efficiency in the linear régime.* In order to investigate the amplitude to which the oscillation can grow without departing from sensibly sinoidal conditions, let us assume that

2. ANALYSIS OF A CIRCUIT FROM FIRST PRINCIPLES

$M = \dfrac{CR + aL}{g}$, and that the circuit is in the steady state with current $i = I \sin pt$ flowing in the inductance L, where

$$p = \sqrt{\dfrac{1+aR}{CL}} \doteqdot \sqrt{\dfrac{1}{CL}}.$$

Then from the equations on p. 267,

$e_g = -pMI \cos pt;$

$e_a = RI \sin pt + pLI \cos pt,$

$\quad = I\sqrt{p^2L^2 + R^2} \cos(pt - \alpha)$, where $\tan \alpha \equiv \dfrac{R}{pL}$;

$i_a = I[aR \sin pt + (aL - gM)p \cos pt]$

$\quad = IR[a \sin pt - pC \cos pt]$

$\quad = -IR\sqrt{a^2 + p^2C^2} \cos(pt + \beta)$, where $\tan \beta \equiv \dfrac{a}{pC} \doteqdot \dfrac{pL}{\rho}$.

In practical triode oscillators, $R^2 \ll p^2L^2$ and $a^2 \ll p^2C^2$, so that these relations between the several alternating potentials and currents are as follows:

If the current in the inductance L is

$$i = I \sin pt,$$

then $e_a \doteqdot pLI \cos(pt - \alpha)$, where α is the small angle $\tan^{-1} \dfrac{R}{pL}$;

$e_g = -pMI \cos pt;$

$i_a \doteqdot -pCRI \cos(pt + \beta),$

$\quad \doteqdot -\dfrac{R}{pL} I \cos(pt + \beta)$, where β is the small angle $\tan^{-1} \dfrac{pL}{\rho}$.

The physical significance of these formulas is more clearly exposed on writing

$\dfrac{R}{pL} = f$, where f is the power factor* of the oscillatory circuit *per se*;

$$\alpha = \tan^{-1} \dfrac{R}{pL} \doteqdot f;$$

$$\beta = \tan^{-1} \dfrac{pL}{\rho} \doteqdot \dfrac{pL}{\rho};$$

and $\qquad pM = \dfrac{pCR + paL}{g} \doteqdot \dfrac{1}{\mu}(f\rho + pL).$

* See IV–2(f).

IX. THE TRIODE AS OSCILLATOR

The formulas then become:

If $i = I \sin pt$,

then $e_a \doteqdot pLI \cos(pt - f) = E_a \cos(pt - f)$, say;

$e_g \doteqdot -\dfrac{1}{\mu}(f\rho + pL)I \cos pt = -E_g \cos pt$, say;

$e_l = e_a + \mu e_g = pLI \cos(pt - f) - (f\rho + pL)I \cos pt,$

$= -f\rho I \cos\left(pt + \dfrac{pL}{\rho}\right);$

$i_a \doteqdot -fI \cos\left(pt + \dfrac{pL}{\rho}\right) = -I_a \cos\left(pt + \dfrac{pL}{\rho}\right)$, say.

For picturing the physical conditions during oscillation, it is more helpful to think of the lumped characteristic* than the i_a, e_a or i_a, e_g characteristics. In a set of characteristics such as Fig. 132 the representative point does not run up and down any one characteristic; it glides across the characteristics as the anode and grid potentials simultaneously change. On eliminating e_a and t from the above formulae for e_a, e_g and i_a, an equation between i_a and e_g is obtained showing that the path of the representative point on an i_a, e_g diagram (e.g. Fig. 132) is an ellipse with major axis inclined less steeply than the slopes of the static characteristics.

We may read from the above formulas the following conclusions:

Alternating anode current i_a. Amplitude I_a equals amplitude I of oscillatory current multiplied by power factor f of oscillatory circuit. It is nearly in quadrature with oscillatory current i, and syn-phased to grid potential e_g.

Alternating anode potential e_a is in quadrature with oscillatory current i except for small angle f, and is anti-phased to anode current i_a except for small angle $\left(f + \dfrac{pL}{\rho}\right)$.

Effective impedance between anode and filament is nearly the pure resistance $\dfrac{pL}{f}$ of a tuned rejector circuit. (See IV-8(c).)

Alternating grid potential e_g is nearly syn-phased to anode current i_a and anti-phased to anode potential e_a.

Ratio $\dfrac{E_a}{E_g}$ between amplitudes of anode and grid potentials is nearly

$$\mu \dfrac{pL}{f\rho + pL} = \dfrac{\mu}{1 + f\dfrac{\rho}{pL}}.$$

* See p. 197.

2. ANALYSIS OF A CIRCUIT FROM FIRST PRINCIPLES

We must now consider total values of current and potentials, instead of the alternating components only. We will write J_0, B_a, $-B_g$ for the mean, or non-oscillating, values of anode current, anode potential and grid potential respectively. The total values when oscillating are then:

Total anode current $= J_0 - I_a \cos\left(pt + \dfrac{pL}{\rho}\right)$; $I_a = fI$.

Total anode potential $= B_a + E_a \cos(pt - f)$; $E_a = pLI$.

Total grid potential $= -B_g - E_g \cos pt$; $E_g = \dfrac{1}{\mu}(f\rho + pL)I$.

Total lumped potential $= B_a - \mu B_g - E_l \cos\left(pt + \dfrac{pL}{\rho}\right)$; $E_l = f\rho I$.

These are shown in Fig. 186 (p. 282) (for an oscillation in which the total grid potential just rises to zero*).

An investigation of amplitudes and efficiencies is facilitated by substituting for the actual circuits of Fig. 177 the equivalent amplifier disposition of Fig. 179. The resistance R_r in the anode circuit is the rejector-circuit resistance $\dfrac{pL}{f}$ of the actual oscillatory circuit LCR, and exists only for the alternating component of the anode current; and the grid E.M.F. e_g is actually the retroaction E.M.F. impressed from the oscillatory circuit.

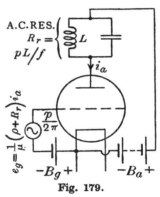

Fig. 179.

The condition for not departing from the sinoidal régime (linear region of triode characteristics) is that the total anode current shall not fall below a certain minimum value $J_{\min.}$, below which the triode characteristics become curved and above which they are sensibly straight†. This is the point A on the lumped characteristic of Fig. 180‡. $P_{\min.}$ is the corresponding minimum total lumped potential. C is the static representative point P_0, J_0. An oscillation

---

* The small phase angles are neglected. The curves are drawn to scale for the conditions of the 4th line ($R_r = 16$ kΩ) of the Table on p. 277.

† Also, of course, that the total anode current shall not rise *above* the value where curvature begins owing to incipient saturation. We assume adequate total emission.

‡ This curve is actually the lumped characteristic of the "DEL 410" triode of Fig. 132, plotted from the 200 V characteristic by taking

(lumped potential) = (anode potential) + 13 (grid potential).

Scale: 1 cm = 20 V, 1 cm = 2 mA.

IX. THE TRIODE AS OSCILLATOR

with the just adequate condition of maintenance will assume a steady amplitude causing the representative point to travel back and forth over the range ACB of the lumped characteristic. The mean current

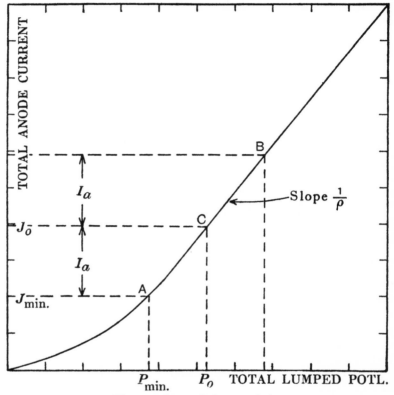

Fig. 180. Lumped characteristic.

from the battery B_a remains J_0 whether oscillation is occurring or not. The Table gives the power balance sheet.

	Not oscillating	Oscillating
$W_o \equiv$ power out (A.C.) to oscillatory circuit	0	$\tfrac{1}{2} R_r I_a{}^2$
$W_i \equiv$ power in (D.C.) from anode battery	$B_a J_0$	$B_a J_0$
$W_i - W_o =$ power heating anode	$B_a J_0$	$B_a J_0 - \tfrac{1}{2} R_r I_a{}^2$

2. ANALYSIS OF A CIRCUIT FROM FIRST PRINCIPLES

The amplitude of anode current is limited to $I_a = J_0 - J_{\min}$. But this amplitude can be reached only if the lumped characteristic *is* followed; and this implies certain restrictions on the value of the alternating part e_a of the anode potential. For example, if during the cycle the total anode potential became negative, the total anode current would remain zero however large a positive value the total grid potential might have. Even though the total anode potential remained positive throughout the cycle, if the total grid potential ever exceeded the total anode potential the phenomenon of secondary emission* at the anode would reduce the anode current from the assumed lumped-characteristic value. To obtain the travel ACB (Fig. 180), therefore, not only must the static total lumped potential $(B_a - \mu B_g)$ be P_0, but B_a and B_g must separately be large enough to make the minimum total anode potential $(B_a - R_r I_a)$ sufficiently exceed the maximum total grid potential $\left(-B_g + \dfrac{1}{\mu}\,\overline{\rho + R_r I_a}\right)$; for the anode minimum and the grid maximum occur nearly simultaneously†.

When grid current is debarred we are relieved of the difficulty of assessing how much this excess of anode over grid potential must be. For then we have the condition

$$-B_g + \frac{1}{\mu}(\rho + R_r) I_a \not> 0,$$

i.e. $\quad B_g \not< \dfrac{1}{\mu}(\rho + R_r) I_a.$

* See VII–9 (e).
† For some purposes the rough approximation that the lumped characteristic is a straight line through the origin (i.e. $J_{\min}=0$, $P_{\min}=0$, in Fig. 180) is useful. If this characteristic were followed without restriction as to the values of anode and grid potentials separately, we should have (from Fig. 180 and Table on p. 274):

> Maximum output when $I_a = J_0$, $R_r = \rho$;
> D.C. input $= B_a J_0$;
> Anode heating $= B_a J_0 - \tfrac{1}{2} R_r I_a^2 = \tfrac{1}{2} B_a J_0$;
> Efficiency at maximum output $= 50\,°/_\circ$.

Many writers have derived this result as the maximum output and the efficiency which could be obtained in the sinoidal régime if the triode characteristics remained straight down to zero anode current. (See H. J. van der Bijl, *The thermionic vacuum tube* (1920), p. 300; L. B. Turner, *Outline of wireless* (1921), p. 135; C. Gutton, *La lampe à trois électrodes* (1923), p. 71; J. Zenneck and H. Rukop, *Lehrbuch der drahtlosen Telegraphie* (1925), p. 570; E. Mallett, *Telegraphy and Telephony* (1929), p. 335.) But the result is reached only on the tacit and grossly unreal supposition that the lumped characteristic is followed even when the grid potential exceeds the anode potential.

If B_g is large enough to meet this condition, the A.C. output of the triode, $W_o = \tfrac{1}{2} I_a^2 R_r$, does not depend on the values of B_a and B_g individually, but only on the lumped potential, $P_0 = B_a - \mu B_g$. But the D.C. input, $W_i = B_a J_0 = B_a (J_{\min.} + I_a)$, is proportional to B_a. Hence the efficiency, $\eta = \dfrac{W_o}{W_i}$, rises as B_a and B_g are reduced towards the minimum values which allow the lumped characteristic to be followed, viz. towards

$$B_g = \frac{1}{\mu}(\rho + R_r) I_a ;$$

$$\begin{aligned} B_a &= P_0 + \mu B_g, \\ &= P_{\min.} + \rho I_a + \mu B_g, \\ &= P_{\min.} + (2\rho + R_r) I_a. \end{aligned}$$

Then

$$I_a = \frac{B_a - P_{\min.}}{2\rho + R_r},$$

$$B_g = \frac{1}{\mu}(B_a - P_{\min.}) \frac{\rho + R_r}{2\rho + R_r},$$

$$W_o = \frac{(B_a - P_{\min.})^2 R_r}{2(2\rho + R_r)^2},$$

$$W_i = B_a \left(J_{\min.} + \frac{B_a - P_{\min.}}{2\rho + R_r} \right),$$

$$\eta = \frac{W_o}{W_i} = \frac{(B_a - P_{\min.})^2}{2 B_a} \cdot \frac{R_r}{(2\rho + R_r)(2\rho + R_r . J_{\min.} + B_a - P_{\min.})}.$$

For choice of R_r, the output is a maximum when $R_r = 2\rho$, and is then

$$(W_o)_{\max.} = \frac{(B_a - P_{\min.})^2}{16\rho}.$$

The efficiency is then

$$(\eta)_{W_o=\max.} = \frac{(B_a - P_{\min.})^2}{4 B_a} \cdot \frac{1}{4 J_{\min.} \rho + B_a - P_{\min.}}.$$

Both output and efficiency increase with
    increase of anode battery B_a;
    decrease of minimum permissible anode potential $P_{\min.}$
        or current $J_{\min.}$;
    decrease of anode resistance ρ.

2. ANALYSIS OF A CIRCUIT FROM FIRST PRINCIPLES

If the lumped characteristic were a straight line through the origin—as, for a rough approximation, it may sometimes be taken to be—$P_{min.} = 0$, $J_{min.} = 0$, giving

$$(W_o)_{max.} = \frac{B_a^2}{16\rho},$$

$$(\eta)_{W_o=max.} = 25\,\%.$$

With this ideal characteristic, the output falls and the efficiency rises with increase of R_r towards the limiting values

$$W_o = 0 \text{ and } \eta = 50\,\% \text{ when } \frac{R_r}{\rho} = \infty\,*.$$

2 (d). *Numerical examples of linear régime.* These results for maximum sinoidal output and efficiency may be illustrated with reference to the triode of Fig. 132 (p. 206)†. We take $\rho = 8\,\text{k}\Omega$, $\mu = 13$, $J_{min.} = 4\,\text{mA}$, $P_{min.} = 76\,\text{V}$, $B_a = 200\,\text{V}$. The Table shows, for a range of values of the output resistance R_r, values given by the formulae established in the last Subsection (p. 276).

Output resistance R_r	Grid bias B_g	Output power W_o	Input power W_i	Efficiency η	Inst. max. of anode current
4 kΩ	5·7 V	0.077 W	2·04 W	3·8 %	16·4 mA
*8	6·4	0·107	1·83	5·8	14·4
12	6·8	0·118	1·68	7·0	12·8
†16	7·2	0·120	1·57	7·6	11·7
24	7·6	0·116	1·47	7·9	10·2
32	7·95	0·107	1·32	8·1	9·2
40 kΩ	8·2 V	0·098 W	1·24 W	7·9 %	8·4 mA

* $R_r = \rho$.
† W_o = maximum. These conditions are plotted to scale in Fig. 186.

The smallness of the output obtainable (whether as generator or as output triode of an amplifier) in this example, viz. 0·120 W, is very striking. When the output is the maximum, of the total power supplied over 92 % is expended in heating the anode. This particular triode is, of course, designed primarily to serve as an

---

* Cf. the erroneous result, W_o = max. and $\eta = 50\,\%$ when $\frac{R_r}{\rho} = 1$. See footnote on p. 275.

† They apply equally to the calculation of the "maximum undistorted power" obtainable from the triode of an amplifier with a non-reactive load. For an experimental check of the validity of the analysis in the last Subsection, see L. B. Turner and L. A. Meacham, "The triode oscillation generator and amplifier," *Proc. Camb. Phil. Soc.* XXVI (1930), p. 507.

278   IX. THE TRIODE AS OSCILLATOR

amplifier of small potential changes, not to deliver power. It is instructive to make a similar calculation for a triode designed specifically to deliver large power without serious departure from sinoidal conditions. We take the M.O. Valve Co's. "C.A.M. 1"*, $\rho = 2500\ \Omega$, $\mu = 7$, $J_{\min.} = 0.1$ A (say), $P_{\min.} = 700$ V, $B_a = 8000$ V. When $R_r = 2\rho = 5000\ \Omega$, we find:

$B_g = 780$ V, $W_o = 1330$ W, $W_i = 6600$ W, $\eta = 20\ \%$,

instantaneous maximum of anode current = 1·56 A.

3. OTHER RETROACTIVE OSCILLATOR CIRCUITS

In the last Section the circuit of Fig. 177, oscillating in the linear régime, has been thoroughly investigated. Other retroactive circuits may be analysed in like manner. Although exact solutions are often troublesome to obtain, the insight into the physical conditions acquired during the investigation of this circuit will often enable us to ascertain the approximate oscillatory conditions in other circuits.

In Fig. 177, and in the equivalent amplifier disposition of Fig. 179, the load in the anode circuit was a resistance $R_r = \dfrac{pL}{f}$, where f is the power factor $\dfrac{R}{pL}$ of the *LRC* circuit.

In Subsections 2 (c) and 2 (d), the influence of the value of R_r was considered. Any desired value of R_r may, of course, be obtained by the selection of appropriate values of L and C, while maintaining the product LC at the value giving the desired frequency. But a more convenient course is to provide an inductance L larger, and a capacitance C smaller, than these particular values, and to obtain the desired R_r by the use of an *anode tap* on the coil, as shown in Fig. 181. The farther the tapping point is moved to the right, the smaller is the value R_r of the rejector-circuit resistance interposed between anode and filament†. An exact analysis would be complicated by

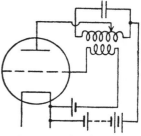

Fig. 181. Anode tap.

---

* See W. J. Picken, *loc. cit.* p. 210, Fig. 11. This triode is employed under conditions resembling those in the B.B.C. transmitters "5 GB" (Daventry) and "2 LO" (Brookman's Park).

† See IV-8 (c).

3. OTHER RETROACTIVE OSCILLATOR CIRCUITS

the existence of two currents in the two parts of the inductance coil to left and right of the tapping point; but since in any practical conditions these currents are only slightly unequal, it is usually unnecessary to distinguish between the dispositions of Fig. 177 and Fig. 181.

Figs. 182 and 183 show two other circuits which are clearly capable of producing self-oscillation if suitable magnitudes are taken. The anode and grid potential fluctuations are approximately anti-phased; the load in the anode circuit is adjustable by the tap T_a or the mutual inductance M_a; the magnitude of the grid potential fluctuation is adjustable by T_g or M_g; so the conditions in the oscillator of Fig. 177 can be simulated, as regards both the inception of self-oscillation and the steady amplitude reached.

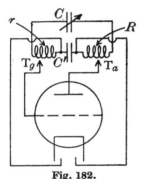
Fig. 182.
Another self-oscillating circuit.

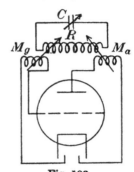
Fig. 183.
Another self-oscillating circuit.

Fig. 182, when the taps T_a, T_g are placed at the ends of the coil*, is a particularly useful oscillator disposition for very high frequencies, when the anode-grid stray capacitance K (not shown in Fig. 182) becomes much in evidence. C' is an indefinitely large capacitance serving merely to separate the D.C. components of anode and grid potential. Since C and K are bunched, the wavelength is proportional to $\sqrt{C+K}$, and C may be made as small as desired—even zero (Fig. 184)—when very short wavelength is required.

The approximate conditions of oscillation in the circuit of Fig. 183 may be quickly found as follows. Let the existence be

* The circuit is then sometimes called the Hartley circuit. Its exact analysis is included in X–2.

assumed of a steady oscillation of frequency $\frac{p}{2\pi}$, in which I, I_a, E_a, E_g are the vector quantities representing the alternating components of current and potential shown in Fig. 185*.

Then
$$E_a = jpM_a I - jpL_a I_a;$$
$$E_g = -jpM_g I;$$
$$I_a = \frac{1}{\rho} E_a + \frac{\mu}{\rho} E_g,$$
$$= \frac{jp}{\rho}(M_a - \mu M_g) I - \frac{jp}{\rho} L_a I_a;$$
$$\therefore I_a = \frac{jp(M_a - \mu M_g)}{\rho + jpL_a} I.$$

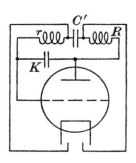
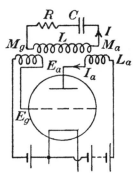

Fig. 184. Omission of C in Fig. 182.   Fig. 185. Analysis of Fig. 183.

But for the **LRC** circuit
$$jpM_a I_a = \left(R + jpL + \frac{1}{jpC}\right) I;$$
$$\therefore \frac{jp(M_a - \mu M_g)}{\rho + jpL_a} - \frac{R + j\left(pL - \frac{1}{pC}\right)}{jpM_a} = 0.$$

Equating the real and imaginary parts separately to zero, we find
$$p = \frac{1}{\sqrt{C\left(L + \frac{R}{\rho} L_a\right)}} \doteqdot \frac{1}{\sqrt{CL}};$$
$$p^2 M_a(\mu M_g - M_a) = \rho R + \frac{L_a}{C}(1 - p^2 LC),$$
$$\doteqdot \rho R.$$

* Coils supposed all wound in same sense and coaxial.

4. OSCILLATORS OF HIGH EFFICIENCY

Hence self-oscillation cannot occur unless M_g exceeds $\frac{1}{\mu}M_a$. The smallest possible grid potential swing for any given oscillatory current I is obtained with the smallest possible M_g. This occurs when $pM_a \rightleftharpoons \sqrt{\rho R}$; then $pM_g \rightleftharpoons \frac{2}{\mu}\sqrt{\rho R}$.

Any circuit may be investigated for self-oscillation by assuming the existence of a steady sinoidal régime in this manner.

4. OSCILLATORS OF HIGH EFFICIENCY

4 (a). *The high-efficiency cycle.* It was shown in IX-2 (c) that the efficiency of an oscillator functioning strictly sinoidally, and with the circuits adjusted for maximum output, must be less than $25\,°/_o$. The difference between input power and output power is the kinetic-energy power of the electrons as they impinge on the anode with a velocity proportional to the square root of the anode potential at the instant. It appears as heat at the anode, and the instantaneous rate of heating is the product of anode potential and anode current. In order to obtain better efficiencies, the cycle must be so modified that the anode current remains small (even zero) throughout that part of the cycle during which the anode potential is not small. This is effected by an increase in the negative grid bias; with, of course, an appropriate increase in the retroaction giving, for the same current in the oscillatory circuit, a larger alternating component of grid potential.

This will be understood on studying the conditions in the numerical case of a high-efficiency cycle depicted in Fig. 187, which should be compared with the sinoidal conditions in Fig. 186*. Whereas in Fig. 186 the dimensions of the circuits were so chosen that the representative point on the lumped characteristic (Fig. 180) never left the straight portion, in Fig. 187 the fluctuations of anode and grid potential, and the anode and grid batteries provided to deal with them, are such as to bring the lumped potential not only below the rectilinear region, but even below zero; so that the anode current remains zero for a considerable fraction (over a quarter) of the cycle. Although the anode current is now far from sinoidal, the

---

* These Figures refer to the same triode, that of Fig. 132 and the lumped characteristic in Fig. 180. The scales in Figs. 186 and 187 are the same, viz.
$1\,\text{cm} = 5\,\text{mA}, \quad 1\,\text{cm} = 100\,\text{V}, \quad 1\,\text{cm} = 1\,\text{W}.$

282 IX. THE TRIODE AS OSCILLATOR

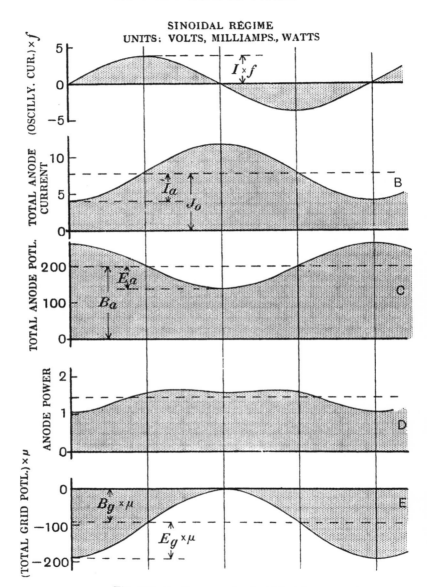

Fig. 186. Cyclic curves in sinoidal régime.

4. OSCILLATORS OF HIGH EFFICIENCY 283

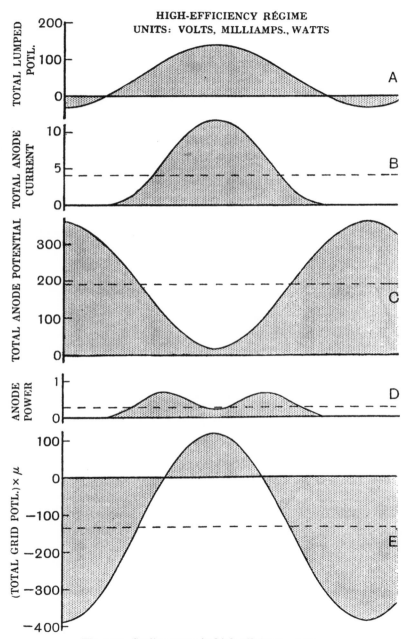

Fig. 187. Cyclic curves in high-efficiency régime.

IX. THE TRIODE AS OSCILLATOR

oscillatory current and the anode potential still fluctuate nearly sinoidally, owing to the balance-wheel action of the low-decrement oscillatory circuit *.

Some of the dimensions of the oscillator of Fig. 187 are, of course, arbitrarily chosen; the choice has been influenced by the desire to keep the high-efficiency oscillator fairly closely comparable with the sinoidal oscillator of Fig. 186. For the high-efficiency cycle we elect to have the same maximum total anode current as before, viz. 11·7 mA. This requires a maximum total lumped potential of 136 V. In order to obtain a fair efficiency, we make the total lumped potential fall below zero, to -30 V; the total anode current is then (we shall see) less than 2 mA for half the cycle, and is actually zero for over a quarter of the cycle. The total lumped potential fluctuates between -30 V and 136 V. Disregarding the small harmonic terms which it must in fact contain, the total lumped potential is

$$P_0 - E_l \cos pt,$$

where $P_0 = \dfrac{136 - 30}{2} = 53$ V, and $E_l = \dfrac{136 + 30}{2} = 83$ V. Curve A in Fig. 187 shows this total lumped potential throughout one cycle.

The total anode current, curve B, is obtained by collating curve A with the lumped characteristic of the triode. On measuring the area below curve B, the mean value of the total anode current is found to be 4·1 mA. The mean input power is therefore $0·0041 B_a$.

The P.D. across the output circuit is $e_a \doteqdot E_a \cos pt$. The output power W_o is therefore the mean over a period of the product of $(-e_a)$ and the total anode current. Examination of curve B shows that the mean value of the product of the ordinate and $\cos pt$ is $-2·92$ mA.

Hence $W_o = 0·00292 E_a$; $E_a = 343 W_o$.

Since $\mu E_g - E_a = E_l$,

$$E_g = \frac{E_a + 83}{13} = 26·4 W_o + 6·4.$$

The output power increases without limit as E_a and E_g are increased. We elect to have an output $W_o = 0·5$ W, requiring

$$E_a = 172 \text{ V}, \qquad E_g = 19·6 \text{ V}.$$

---

* The balance wheel or the pendulum of a clock oscillates very nearly sinoidally although the driving force is not applied sinoidally but in brief spasms at relatively long intervals. See the next Subsection.

The input power W_i being $0{\cdot}0041B_a$, B_a is $244 W_i$. Since
$$B_a - \mu B_g = P_0,$$
$$B_g = \frac{B_a - 53}{13} = 18{\cdot}7 W_i - 4{\cdot}1.$$
The input power is the less the smaller are B_a and B_g; but these must be made large enough, with the values of E_a and E_g obtaining, to carry the anode current through its assumed cycle as given by the lumped characteristic. The total anode current is at its maximum (11·7 mA) when the total anode potential is at its minimum $B_a - E_a = 244 W_i - 172$, and when the total grid potential is at its maximum
$$-B_g + E_g = -18{\cdot}7 W_i + 4{\cdot}1 + 19{\cdot}6 = -18{\cdot}7 W_i + 23{\cdot}7.$$
We will assume that the lumped characteristic is followed provided that the anode potential is positive and exceeds twice the grid potential*. This limits W_i to the value given by writing
$$244 W_i - 172 = 2(-18{\cdot}7 W_i + 23{\cdot}7);$$
$$\text{i.e. } W_i = 0{\cdot}78 \text{ W}.$$
With this value of W_i,
$$B_a = 244 W_i = 190 \text{ V};$$
$$B_g = 18{\cdot}7 W_i - 4{\cdot}1 = 10{\cdot}4 \text{ V};$$
$$\eta = \frac{W_o}{W_i} = 64\,\%.$$
Curve C in Fig. 187 shows the total anode potential
$$B_a + E_a \cos pt = 190 + 172 \cos pt;$$
and curve D shows the product of curve B and curve C, viz. the power expended in heating the anode. Finally, curve E shows the total grid potential
$$-B_g - E_g \cos pt = -10{\cdot}4 - 19{\cdot}6 \cos pt.$$
In practice, high-power triode oscillators and amplifiers are usually adjusted to work at efficiencies of the order of $70\,\%$, except in special circumstances where peculiar freedom from harmonics is required. High efficiency is desired not only to avoid wasteful expense of power lost as heat at the anodes, but also because of the difficulty of removing that heat. In tubes with glass or silica envelopes†, this difficulty is specially obtrusive; the heat leaves the

---

* It will be seen that very little difference would result from taking a coefficient 1 or 4 in place of 2.

† See p. 210.

anode almost entirely by radiation, and the permissible load is determined by the permissible rise of temperature of the anode, i.e. by the permissible *anode dissipation*. If P_i, P_o, P_a are the input, output and anode dissipation powers respectively, and if η is the efficiency,

$$P_o = \eta P_i;$$
$$P_a = P_i - P_o = (1 - \eta) P_i;$$
$$\therefore P_o = \frac{\eta}{1-\eta} P_a.$$

Hence if $\eta = 25\,°/_°$, $P_o = \tfrac{1}{3} P_a$; whereas if $\eta = 75\,°/_°$, $P_o = 3 P_a$. A triode loaded up to its rated anode dissipation, therefore, can deliver 9 times as much power if worked at 75 °/° as at 25 °/° efficiency. The Table on p. 293 of working conditions in the powerful telegraph transmitter at the Rugby wireless station shows an efficiency of 72·5 °/°.

The required negative grid bias may be provided by a battery or a dynamo; but it is often obtained, either wholly or in part, by inserting a resistance R_g, shunted by a condenser, between grid and filament, as in Fig. 188. When oscillation occurs, grid current

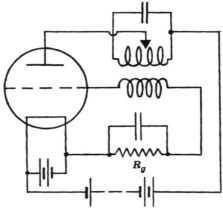

Fig. 188. Automatic grid bias.

flows during that part of the cycle during which the grid potential is positive. The grid-filament path is a rectifier, and the rectified current I_r (the mean grid current) flows through R_g, while the large capacitance of the condenser shunt constitutes a bye-pass of negligible impedance for the alternating components of the grid current. The mean grid potential is lowered by an amount $R_g I_r$, exactly as

4. OSCILLATORS OF HIGH EFFICIENCY

though a grid bias battery of P.D. $R_g I_r$ were inserted in place of R_g. Since the value of the alternating component of grid potential does not depend upon R_g, any desired grid bias can be obtained by a suitable choice of R_g.

4 (b). *Harmonics produced*. Harmonic analysis of curve B in Fig. 187 shows that the total anode current is represented by the Fourier series:

Total anode current in milliamperes

$$= 4{\cdot}1 - 5{\cdot}8 \cos pt + 2{\cdot}0 \cos 2pt + 0{\cdot}0 \cos 3pt - 0{\cdot}1 \cos 4pt + \ldots.$$

It thus contains a prominent second harmonic whose amplitude is about 35 % of the amplitude of the fundamental. Referring to Fig. 177 (p. 267), the fluctuating anode current i_a contains a component I_1 of frequency $\frac{p}{2\pi}$, and a component I_2 of frequency $2\frac{p}{2\pi}$; and in our numerical example $I_2 = 0{\cdot}35 I_1$. We have to enquire what are the relative strengths of the fundamental and harmonic components of current in L and C corresponding to the fundamental and harmonic components in the anode current. We will keep the enquiry general by taking the fundamental and the hth harmonic, where h may be any integer.

Since in all practical cases $R \ll pL$, the impedances* of the circuit inserted between anode and filament, for the fundamental and hth harmonic frequencies, are approximately:

for the fundamental frequency $\frac{p}{2\pi}$,	for the hth harmonic frequency $h\frac{p}{2\pi}$,
$pL \cdot \frac{\pi}{\delta}$;	$\dfrac{1}{\dfrac{1}{jhpL} + jhpC} = -\dfrac{jhpL}{h^2 - 1}$;
i.e. a resistance of magnitude $pL \cdot \frac{\pi}{\delta}$, where δ is the decrement of the LRC circuit.	i.e. a capacitative reactance of magnitude $pL \cdot \dfrac{h}{h^2 - 1}$.

The components of P.D. across L and C therefore have the magnitudes

$$E_1 = I_1 \cdot pL \cdot \frac{\pi}{\delta}; \qquad E_h = I_h \cdot pL \cdot \frac{h}{h^2 - 1};$$

---

* See IV–8 (c).

and the corresponding current magnitudes are:

in the L branch,
$$\frac{E_1}{pL} = I_1 \cdot \frac{\pi}{\delta};$$
in the C branch,
$$E_1 \cdot pC = I_1 \cdot \frac{\pi}{\delta}.$$

in the L branch,
$$\frac{E_h}{hpL} = I_h \cdot \frac{1}{h^2 - 1};$$
in the C branch,
$$E_h \cdot hpC = I_h \cdot \frac{h^2}{h^2 - 1}.$$

The ratio $\frac{h\text{th harmonic current}}{\text{fundamental current}}$ is therefore:

in the L branch,
$$\frac{I_h}{I_1} \cdot \frac{\delta}{\pi} \cdot \frac{1}{h^2 - 1};$$

in the C branch,
$$\frac{I_h}{I_1} \cdot \frac{\delta}{\pi} \cdot \frac{h^2}{h^2 - 1}.$$

In both branches of the oscillatory circuit this ratio is therefore much less than the ratio $\frac{I_h}{I_1}$ of the anode current components; and in the L branch the harmonic is only $\frac{1}{h^2}$ of its value in the C branch.

In our numerical example, $\frac{I_h}{I_1} = 0.35$, $h = 2$; so that if the decrement δ were 0·1, the ratio $\frac{\text{harmonic current}}{\text{fundamental current}}$ would be 0·37 °/₀ in the L branch and 1·48 °/₀ in the C branch.

In a transmitting station it is very undesirable to radiate harmonics even so relatively weak as these percentages indicate. It is

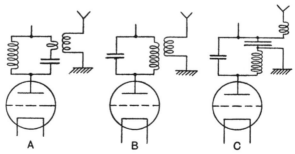

Fig. 189. Harmonics in the antenna.

therefore important to keep the decrement of the LRC circuit small, and to couple the load circuit (e.g. the antenna) to the L branch

5. THE TRIODE TELEGRAPH TRANSMITTER

rather than to the C branch. Further, for eliminating the harmonics, a coupling between the two circuits of the type of Fig. 56 B (p. 94) is h^2-times better than a coupling of the type of Fig. 56 A or Fig. 55. In Fig. 189, dispositions A, B and C are in ascending order of excellence as regards freedom from harmonics in the antenna; B is h^2-times better than A, and C is h^2-times better than B. A offers no attractions, but B is often employed in preference to C on account of the greater ease in adjusting the coefficient of coupling (for the fundamental frequency) by a change of mutual inductance than by a change of capacitance.

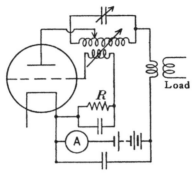

Fig. 190. Cultivation of harmonics.

Cases arise, especially in laboratory practice, where the presence of a set of well-pronounced harmonics is desired. The load circuit is then preferably coupled to the oscillator by an inductive relation with the anode current, as shown in Fig. 190. If the grid circuit resistance R and the milliammeter A are included, the sudden drop in the reading of A when oscillation starts is a convenient indication showing whether the circuit is oscillating or not.

5. THE TRIODE TELEGRAPH TRANSMITTER

5 (a). *Connection of oscillator to antenna.* When a triode oscillator is used to excite an antenna, the latter may be a separate circuit coupled to the oscillatory circuit with concentrated L and C as shown in previous diagrams. This arrangement is illustrated in Fig. 191. There are then two separate oscillatory circuits to be adjusted for wavelength, two mutual inductances, and the anode tap; and power is wasted in heating two oscillatory circuits instead of one. When,

IX. THE TRIODE AS OSCILLATOR

therefore, the extra oscillatory circuit is not necessary for filtering out harmonics, the closed oscillatory circuit is often replaced by the antenna itself, as at A in Fig. 192.

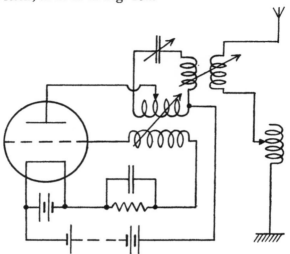

Fig. 191. Antenna separate from oscillator.

A practical inconvenience of the arrangement shown at A in Fig. 192 is that the potential of the filament battery differs from earth by the high tension B_a of the anode supply. Alternatively, if a condenser is inserted in the earth lead and the filament battery is earthed, the mean potential of the whole antenna is B_a above earth, which is also objectionable. This is avoided in the modification shown at B, where Ch is a choke coil to prevent the high-frequency component of anode current from being diverted from the antenna coil, to which this current passes through the condenser SC. The reactance of Ch should be so great that no appreciable high-frequency current traverses it. Circuit C is a variation in which the separate grid coil is dispensed with, but at the cost of requiring high-frequency isolation of the filament from earth. In unusual cases where it is inconvenient to have as large an inductance inserted in the antenna as would be needed to give the proper anode tap in the circuits A, B and C, the anode coil may be a separate larger inductance closely coupled to the antenna coil, as shown at D.

In very low-power transmitters where only two or three hundred volts is required, the source of high-tension supply is usually a

5. THE TRIODE TELEGRAPH TRANSMITTER

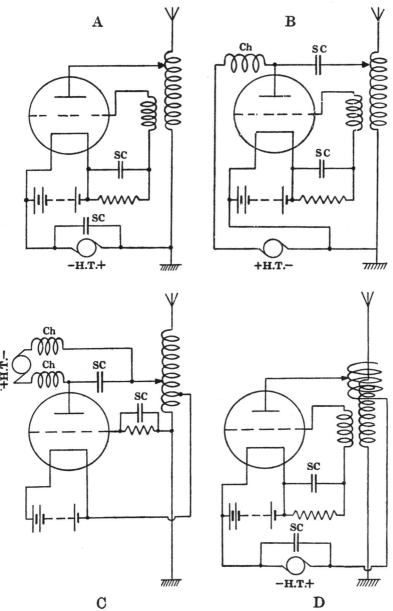

Fig. 192. Circuits where antenna is part of oscillator.

battery of dry cells or accumulators, or a D.C. lighting circuit. For larger powers a special high-tension dynamo is sometimes used. Recent installations at the Rugby Station and at the Broadcasting Station at Brookman's Park have shown that machines supplying hundreds of kilowatts at upwards of ten thousand kilovolts can give excellent service. It is more usual, however, to obtain high-voltage D.C. supplies by the rectification of alternating current. (See p. 187, and Fig. 258, p. 382.)

5 (b). *Master-oscillator system.* So far we have assumed that the oscillator triode, or group of triodes in parallel, generated the full-power to be utilised in the antenna. But this need not be. A feeble oscillator, generating alternating current of the radiation frequency, may be used to excite the input terminals of a high-frequency amplifier of one or several stages, the output circuit of the last stage being the antenna or a circuit coupled therewith. In this arrangement, the frequency of the radiation can be made quite independent of the natural frequency of the antenna, so that, for example, the swaying of the antenna in the wind produces no wandering of the radiation frequency. Further, it is then easier to effect the signalling modulation—Morse or telephonic—since, however great the power radiated, modulation can be imposed at a stage where the power is small. This is an advantage particularly marked in telephony transmitters, for there modulation must be effected over a very wide range of frequencies and with the highest degree of faithfulness. The oscillation generator, as distinct from the amplifier which follows it, is often called the *master oscillator*.

Since the master oscillator is of small power*, its efficiency as a converter from D.C. to A.C. power is of no significance; interest in power losses rests no longer with the oscillation generator, but with the amplifier, especially the output end if there are several stages. Nevertheless, the investigation of efficiency and the production of harmonics, carried out in Sections 2 and 4 of this Chapter with reference to self-oscillators, is applicable also to the master-oscillator type of triode transmitter. For the self-oscillator is nothing but an output amplifier in which the grid stimulus is provided from a particular source, which happens to be the anode circuit of the

---

* In an extreme case, at the Rugby telegraph transmitter, its output is of the order of 10^{-11} times the antenna power. See Subsection 6 (b) of this Chapter.

same triode. In the self-oscillator the power, if any, consumed by the grid is taken from the oscillatory circuit, and therefore contributes something—in practice very little—to the resistance of the oscillatory circuit. In our treatment of the lost power in the self-oscillating triode, we ignored any grid loss; so that the results of the analysis in Sections 2 and 4 of this Chapter are (if anything) truer of the power amplifier than of the oscillation generator itself.

As an example of a modern triode amplifier with large power output, certain particulars of the final stage in the Rugby telegraph transmitter are here quoted*. The output stage of the amplifier consisted of a bank of 54 Western Electric Co.'s "VT 26" triodes, with water-cooled anodes, connected in parallel. Each triode had an anode resistance of about 7 kΩ, an amplification factor of about 40, and a filament emission of several amperes. When used in the conditions tabulated below, in round numbers for each triode the grid bias was -500 V, the amplitude of the alternating component of grid potential 1600 V, the mean grid current 0·11 A, and the power dissipated at the grid owing to the occurrence of grid current 0·14 kW. This grid loss is only 1·6 % of the power supplied to the anode circuit. The Table shows the performance of the complete bank.

Antenna current	740 A
Antenna power (if resistance 0·61 Ω)	335 kW
Total D.C. input to anodes	6·3 kV × 73 A = 460 kW
Total filament power	48 kW
Efficiency of transmitter:	
Excluding filament power	72·5 %
Including ,, ,,	66 %
Peak P.D. between antenna and earth	235 kV

5 (c). *Keying*. Whereas the keying of a high-power transmitter is a serious problem when the generator is of the spark, arc or alternator type, it presents no difficulty in triode transmitters. There are obviously many ways in which the oscillation can be stopped and started by means of a Morse key. With small powers, simple interruption of the high-tension supply is often adopted; but with high powers it is preferable to place the keying contact in

* Mainly from Hansford and Faulkner, *loc. cit.* p. 64.

a grid circuit where relatively very small power is passing. For example, in any of the circuits shown in Fig. 192, the key might be made to throw in and out extra grid bias, either by introducing the P.D. of a battery or dynamo, or by increasing the grid circuit resistance (R_g in Fig. 188). With master-oscillator transmitters when keying is performed at a low-power stage, in the subsequent stages, auxiliary keying contacts are provided serving either to reduce the

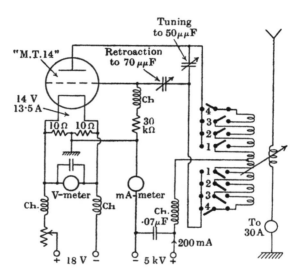

Fig. 193. A short-wave telegraph transmitter.

anode potential or to increase the negative grid bias. This is necessary to prevent an actual increase in the development of heat at the anodes during a Morse space, more especially if the grid bias is wholly or partly provided by the oscillation itself through the agency of a grid resistance (R_g, Fig. 188, p. 286).

5 (d). *Example of short-wave telegraph transmitter.* A short-wave transmitter comprising a single oscillating triode, with the anode circuit coupled directly to the antenna, is shown in Plates XXII and XXIII.* This transmitter is intended for long-range marine

* "Short wave transmitter, type 376" of the Marconi International Marine Communication Co., to whom the author is indebted also for the further information in Fig. 193. In Plate XXIII the triode has been removed.

telegraph service on wavelengths adjustable between 15 m and 40 m. Fig. 193 shows the circuit, with approximate electrical dimensions. The inductance coil is in sections, controlled by switches which are behind the coil in Plate XXIII; for $\lambda = 40$ m the whole of the coil is in circuit, and the inductance is then about 9 μH. Of the four choke coils marked Ch, those leading to the anode and the grid have a natural wavelength of about 13 m.

6. OTHER TYPES OF TRIODE OSCILLATOR

6 (a). *Negative conductance without retroaction.* In Fig. 175 we had an amplifier, comprising a pure resistance R in association with

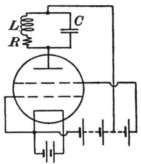

Fig. 194. Self-oscillation with screen-grid tetrode.

a tube exhibiting negative conductance, in which the amplification tended to infinity when a certain critical relation between the electrical dimensions was approached. If R is replaced by an oscillatory rejector circuit LRC, we revert to the amplifier disposition of Fig. 165 with only the modification (from the mathematical standpoint) that the positive slope resistance ρ of Fig. 165 (p. 252) is replaced by the negative slope resistance $-\dfrac{1}{\alpha}$. The analysis for Fig. 165 remains applicable on substituting $-\dfrac{1}{\alpha}$ for ρ. The amplification becomes infinite—i.e. finite anode potential fluctuation occurs with infinitesimal grid potential fluctuation—if in the expression for $\dfrac{e_a}{e_g}$ on p. 252 both terms in the denominator simultaneously vanish, i.e. if

$$jpL\left(1 + \frac{C}{L}\rho R\right) = 0 \quad \text{and} \quad \rho(1 - p^2 LC) + R = 0.$$

Putting $-\dfrac{1}{\alpha}$ for ρ, this condition becomes

$$\frac{C}{L}R = \alpha \quad \text{and} \quad p^2 LC = 1 - R\alpha.$$

Expressed in words, self-oscillation, at approximately the resonant frequency of the oscillatory circuit interposed between anode and filament, occurs if the rejector-circuit resistance exceeds the magnitude of the negative anode slope resistance of the tube. Fig. 194,

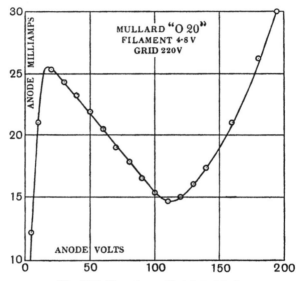

Fig. 195. Dynatron effect in a triode.

when this condition is met, is a self-oscillating circuit, generating alternating current of frequency nearly equal to

$$\frac{1}{2\pi\sqrt{LC}}.$$

This is sometimes a convenient way of causing an acoustic-frequency circuit, or a high-frequency circuit of low decrement such as a wavemeter, to generate oscillation; for it avoids the necessity of tampering with the circuit in order to provide the mutual inductance, or the tapping point on the coil, required to cause self-oscillation in the usual manner by triode retroaction.

6. OTHER TYPES OF TRIODE OSCILLATOR

When the tetrode is employed in this way merely to throw an oscillatory circuit into self-oscillation by the introduction of negative conductance, the control grid is not utilised as such. Larger values of negative slope conductance are easily obtainable in a triode if its grid is maintained at a suitable potential above that of the anode, and the filament emission is limited. Fig. 195 shows a curve found with a small transmitting triode*. With the grid at 220 V there

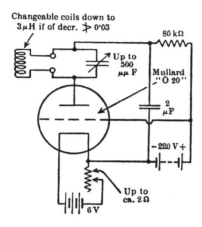

Fig. 196. A dynatron oscillator.

is a nearly constant negative conductance $\frac{1}{7\cdot 2\,\text{k}\Omega}$ at all anode potentials between 25 V and 100 V. A convenient disposition with this triode is shown in Fig. 196. Oscillations of frequencies ranging from hundreds to millions of cycles per second can thus be got by merely changing the inductance coils, and without the necessity for very low resistances.

* Mullard's "O 20"; rating: $\rho = 6\cdot 5\,\text{k}\Omega$, $\mu = 5$; pure tungsten filament, 5·4 V, 1·8 A; permissible anode battery 400 V, dissipation 20 W. For use in this way as a dynatron (see p. 222), a triode with pure tungsten filament is desirable. Even with the best vacuum it seems that any coated filament so far devised undergoes rapid deterioration if submitted to a strong inward electric field, unless the emitting surface is shielded by an excess of emitted electrons.

298　　　　IX. THE TRIODE AS OSCILLATOR

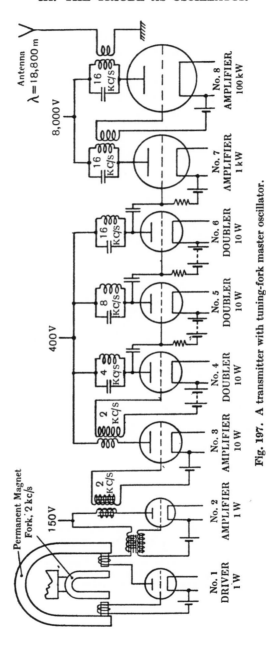

Fig. 197. A transmitter with tuning-fork master oscillator.

6 (b). *Tuning-fork master oscillator.* The precision with which the frequency of a tuning-fork can be measured, and its great constancy, have suggested its use as a basis for the calibration of wavemeters. Since satisfactory forks cannot be made with frequencies higher than about 2 kc/s, comparison between the wireless frequency and the fork is made indirectly, by the method of harmonics. It has been shown, in Subsection 4 (b) of this Chapter, that unless special precautions are taken, triode oscillators and amplifiers are prone to generate oscillation containing well-marked harmonics of the fundamental; and it is easy to intensify this action, so that harmonics of a high order such as 10 or 20 are easily perceptible. If the fundamental frequency of the triode oscillator is made an acoustic frequency (say 1 kc/s), which is easily measurable in terms of the frequency of a fork, we have in (say) the 10th harmonic (selected and amplified as may be necessary) an oscillation known in terms of the fork frequency to be 10 kc/s. If now a second triode oscillator is adjusted*, to have this same frequency as its fundamental frequency, its 10th harmonic (for example) furnishes an oscillation of 100 kc/s.

Procedure of this kind, with various complications and refinements, provides a technique of striking beauty and convenience for measuring high (radiation) frequencies as known integral (harmonic) multiples of a low (acoustic) frequency. If the first triode oscillator could be not only adjusted—as by turning a condenser handle—to have a frequency equal to that of a tuning-fork, but also be caused to maintain that equality automatically, the radiation frequency—the 100 kc/s of our example—would be as constant as the fork frequency itself. A fork arranged to maintain itself in steady vibration can indeed be made to constrain to its own frequency a self-oscillating triode system whose independent frequency is not far removed from that of the fork; but it is probably more satisfactory to omit the self-oscillating property from the triode system altogether, leaving the fork as the sole generator.

A system of this kind is shown diagrammatically in Fig. 197. Numerical values are suggested, but only in order to provide a concrete reasonable example†. The fork is maintained in steady vibration

* By the method of beats, using the heterodyne reception method. See IV-7.
† It is obvious the designer has great freedom of choice as to the order of the harmonics selected, the patterns of triode, the number of stages, and the manner of connecting the stages. A specified output to the antenna could be provided satisfactorily from a specified fork in a variety of ways.

by means of triode No. 1 *, a small receiving tube capable of dissipating perhaps 1 W. A feeble alternating current of the fork frequency flows in the anode circuit, and through a transformer stimulates the grid of a small amplifier triode, No. 2. The 2 kc/s oscillation is further amplified in No. 3, a larger triode capable of dissipating perhaps 10 W. The grid of triode No. 4 has a large negative bias, so that the anode current contains a strong second harmonic, producing a nearly pure 4 kc/s oscillation in the anode circuit †. This is passed on to triode No. 5, also adapted for frequency doubling; and the frequency is again doubled in triode No. 6—and in as many further stages (not shown) as may be necessary. The 16 kc/s output from No. 6 is amplified by triode No. 7 (say 1 kW dissipation), and again in the 100 kW triode (or group of triodes in parallel) No. 8. The output from this, the last amplifying stage, is passed to the antenna.

If the fork is a good one and is not over driven, its frequency is almost independent of the electrical conditions in the driving circuit. Maintained at a uniform temperature, it can be relied upon to vibrate indefinitely with extreme constancy of frequency ‡. Moderate changes in any of the subsequent stages can, it is true, affect the power delivered, but can have no influence on the frequency, which must remain precisely eight times the fork frequency.

Contrasting examples of the uses of a fork as a master oscillator are (i) the wavemeter at the National Physical Laboratory§; (ii) the Post Office telegraph transmitters at the Northolt and Rugby stations; and (iii) some of the British Broadcasting Corporation's transmitters. At Northolt‖ the 22nd, and at Rugby¶ the 9th, harmonic of the feeble oscillation of fork frequency is selected at an early stage, and is thereafter amplified to a high power. In the B.B.C. transmitters the frequency is repeatedly doubled in some ten successive stages.

6 (c). *Quartz-crystal master oscillator.* The steel tuning-fork is a mechanical oscillatory system of low decrement which is easily associated with electrical systems by the use of magnetic pole-pieces

---

* This excellent manner of driving a tuning-fork was devised by W. H. Eccles and F. W. Jordan in 1918. The mode of action will be obvious.
† See Subsection 4 (b) of this Chapter.
‡ To within ± 2 in 100,000, during observations extending over 18 months. See D. W. Dye, "A self-contained harmonic wavemeter," *Phil. Trans.* A, Vol. 224 (1924), p. 284.   § See Dye, *loc. cit.* above.
‖ See A. G. Lee, "Tuning-fork generator," *Electrician*, Vol. xciv (1925), p. 510.   ¶ See E. H. Shaughnessy, *loc. cit.* p. 82.

6. OTHER TYPES OF TRIODE OSCILLATOR

and coils after the fashion of stage 1 in Fig. 197. It can be made to confer on the electrical system the definiteness and constancy of frequency with which the mechanical system, the fork, is endowed. But the fork is an oscillator of acoustic frequency, and when the electrical system is to be the seat of radio-frequency current, a far-fetched harmonic multiplication of some kind is necessary for relating the two frequencies. An example of such procedure has been described in the last Subsection. A mechanical oscillator of vastly higher frequency, but resembling the fork in low damping and ease of linkage with electrical systems, is available in the quartz crystal. In common with certain other crystalline substances quartz possesses the property, termed *piezo-electric*, of developing electric potential gradient under the action of mechanical strain, and vice versa. This property was investigated, and first applied to the measurement and control of wireless frequencies, by W. G. Cady. His results were published in 1922*; and since this time quartz crystals have been employed for frequency measurement and control in a great variety of ways†.

Fig. 198 shows, in fine dotted lines, portion of a perfect quartz crystal, the optical axis of which is in the Z direction. In preparing a piece of quartz for use as a piezo-electric oscillator or resonator, the cut generally accepted as the most favourable is shown by the rectangular slab drawn in heavier lines with axes X, Y, Z. An electric stress in the Y direction is accompanied by mechanical strains in the X and Y directions (but not in the Z direction). If metal plates are laid against the two XZ faces of the slab—so forming a condenser with quartz dielectric instead of the usual mica, oil or air—this condenser exhibits peculiar

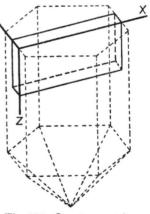

Fig. 198. Quartz crystal.

---

* "The piezo-electric resonator," *Proc. I.R.E.* Vol. 10 (1922), p. 83.

† For a short general summary see G. W. N. Cobbold and A. E. Underdown, "Some practical applications of quartz resonators," *Journ. I.E.E.* Vol. 66 (1928), p. 855: and for recent investigations of the constancy of quartz oscillators, see J. E. P. Vigoureux, "The valve-maintained quartz oscillator," *Journ .I.E.E.* Vol. 68 (1930), p. 265; and H. J. Lucas, "Some developments of the piezo-electric crystal as a frequency standard," same vol., p. 855.

properties when subjected to an alternating P.D. It has capacitance, as with dielectrics which do not possess the piezo-electric property; but in addition there occurs a conversion of electric energy into the energy of mechanical oscillation (and thence, owing to elastic hysteresis, into heat). Since the mechanical oscillatory system formed by the quartz slab has an exceedingly low decrement, its absorption of energy is exceedingly small except at frequencies very near a natural frequency of mechanical oscillation, when, for a given amplitude of P.D. between the plates, the amplitude of the mechanical oscillation becomes relatively very great*.

The effect of the quartz condenser on the electrical circuit in which it is inserted is therefore that of a capacitance K (Fig. 199) shunted by a path LRC, where $\dfrac{1}{2\pi\sqrt{LC}}$ is the natural frequency,

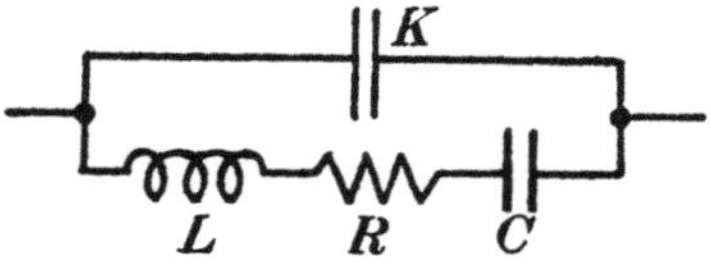

Fig. 199. Equivalent all-electric system.

and $\pi R\sqrt{\dfrac{C}{L}}$ is the decrement, of the mechanical oscillation.

The density and elastic constants of quartz are such that for the two principal modes of oscillation, viz. those of alternating stretch in the X direction and in the Y direction, the natural frequencies are approximately

$$\dfrac{270}{l_x} \text{ kc/s} \quad \text{and} \quad \dfrac{285}{l_y} \text{ kc/s},$$

where l_x cm and l_y cm are the length and thickness of the slab respectively. These frequencies correspond with radiation wavelengths

$$1110\, l_x \text{ metres} \quad \text{and} \quad 1050\, l_y \text{ metres}$$

respectively. A third, intermediate, well-marked natural frequency also occurs, whose value depends upon the ratio between l_x and l_y; and often other resonance frequencies also are found. A good deal of anomalous behaviour is sometimes experienced, and two slabs similarly cut and dimensioned may show piezo-electric differences which cannot be foretold by optical examination.

Quartz slabs of various shapes and sizes have been used. In selecting the form of slab, reasonable guiding considerations seem to be as follows. The slab should be thin (i.e. l_y small) to facilitate the

* It is necessary to guard against destruction of the slab by mechanical shattering due to excessive excitation.

6. OTHER TYPES OF TRIODE OSCILLATOR

application of sufficient electric stress. It should preferably be of bar form (i.e. long and narrow), and used in longitudinal mechanical oscillation, to make the mode of resonant oscillation as definite as possible; but its volume must be large enough to make its energy-absorbing property sufficiently evident. Consequently, for the longer waves, the bar form in length-oscillation is the more suitable; and for very short waves, the plate form in thickness-oscillation. In the latter case, the XY faces are preferably circular instead of rectangular, and are ground as nearly plane and parallel as possible. Typical examples are

2 cm long × 0·5 cm wide × 0·15 cm thick for 135 kc/s ($\lambda = 2220$ m);

and 1 to 2 cm diam. × 0·1 cm thick for 2850 kc/s ($\lambda = 105$ m).

A typical bar crystal resonating at 89·9 kc/s was found* to have the following equivalent electrical dimensions (Fig. 199):

$$K = 3\cdot 57 \,\mu\mu\text{F}, \quad L = 140 \text{ H}, \quad R = 16 \text{ k}\Omega, \quad C = 0\cdot 023 \,\mu\mu\text{F}.$$

Its decrement was therefore 0·0006— far lower than that of an electrical oscillatory system. Slightly off resonance, this crystal simulates a nearly pure capacitative reactance of about 500 kΩ; while at its resonant frequency a shunt resistance of 16 kΩ appears (Fig. 200).

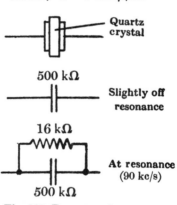

Fig. 200. Resonance in a quartz crystal.

To produce an oscillation generator it is necessary to provide the crystal (as the tuning-fork) with a source of power, i.e. to maintain it in oscillation by some retroactive triode connection. On forgetting what the crystal looks like, and thinking of it as the electrical circuit of Fig. 199, it is seen that no new problem is presented. It is obvious that many familiar triode oscillator dispositions will serve to keep the circuit of Fig. 199 in oscillation; and because it has such a remarkably low decrement, the condition for maintenance is particularly easily met.

---

* By K. S. Van Dyke; see *Physical Review*, Vol. 25 (1925), p. 895. This crystal was 3·07 cm × 0·41 cm × 0·14 cm, and had been used in Cady's investigations, *loc. cit.* p. 301.

A simple disposition is that of Fig. 201. We will calculate the conditions for oscillation with the crystal whose equivalent electrical dimensions were quoted on p. 303, and with the triode whose characteristics are given in Fig. 131 ("DEH 410," $\rho = 14\,\text{k}\Omega$, $\mu = 38$). We will take the two inductances as of equal value L' and just large enough to produce self-oscillation; and in order to simplify the calculation we will neglect the capacitance K in Fig. 199, assuming that the inductances will turn out to be small enough to make its influence very slight*.

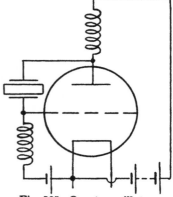

Fig. 201. Quartz oscillator.

Using the generalised treatment of triode circuits in Chapter X, the conditions of self-oscillation are given by equating separately to zero the real and unreal parts of the denominator of (33) on p. 317. In this expression α and β are each jpL'; and γ is the impedance of the crystal, viz. $16{,}000 + jX$, where X is the reactance of the crystal at the frequency $\dfrac{p}{2\pi}$ of the oscillation (which is approximately 89·9 kc/s, but whose exact value remains to be determined). In this way we find

$$L' = 4100\,\mu\text{H} \quad \text{and} \quad X = -7200\,\Omega.$$

Since the equivalent inductance L of the crystal (Fig. 199) is 140 H, $pL \doteqdot 7\cdot 9 \times 10^7\,\Omega$, and the exact frequency of the oscillation exceeds the natural frequency of the crystal by $\dfrac{\frac{1}{2} \times 7200\,\Omega}{7\cdot 9 \times 10^7\,\Omega}$, i.e. by 1 in 22,000. It is, of course, the smallness of this departure from the natural frequency of the crystal *per se* which makes the crystal oscillator so constant. Even a large change in the triode and its circuits would have very slight effect on the frequency generated by the crystal-triode combination.

Although we have found the circuit of Fig. 201 to be easily set into self-oscillation, it is usually preferable to associate the crystal with a circuit which is predisposed to oscillate at about the frequency of the desired mode of mechanical oscillation of the crystal; for

* It will be seen below on evaluating L' that this assumption is justified.

6. OTHER TYPES OF TRIODE OSCILLATOR

there are several possible modes, each with a different, though quite definite, frequency. Fig. 201 may for this reason be modified advantageously to Fig. 202; or even to Fig. 203, where the adjustable retroaction serves to encourage self-oscillation but is inadequate

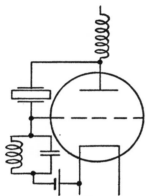
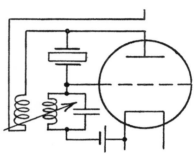

Fig. 202. Modification of Fig. 201.

Fig. 203. Quartz oscillator with inductive retroaction.

to produce it in the absence of the crystal. In circuits such as Figs. 202 and 203, the crystal is competent to maintain the oscillation almost exactly at its own resonant mechanical frequency, although the resonant frequency of the associated electrical circuit may differ fairly widely therefrom.

Every quartz crystal oscillator circuit, like any other triode oscillator, must include some form of anode-grid retroaction. In diagrams which do not show retroaction, such as Fig. 204, it is in fact provided by the anode-grid stray capacitance.

Quartz crystal oscillators, suitably mounted and kept at a constant temperature, are capable of maintaining the same almost perfect constancy of frequency as are tuning-fork oscillators. And whereas the latter are available for frequencies up to about 2 kc/s, the former can be made to function satisfactorily up to 3000 or 4000 kc/s.

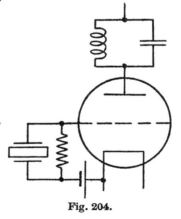

Fig. 204.

A lower limit to the frequencies of quartz resonators is imposed by the difficulty of obtaining suitably cut homogeneous quartz bars of sufficient length. For frequencies below about 50 kc/s composite steel-quartz resonators may be used. A steel bar is given a length such that its fundamental natural frequency in longitudinal vibration, with its centre a node of displacement, is the desired resonant frequency; and the requisite electro-mechanical linkage is provided piezo-electrically by much shorter quartz plates cemented to the steel. Cady describes* such a resonator, represented about full size in Fig. 205, the approximate dimensions in millimetres being:

Steel Bar, $95 \times 9 \times 3$;
each Quartz Plate, $9 \times 10 \times 1$.

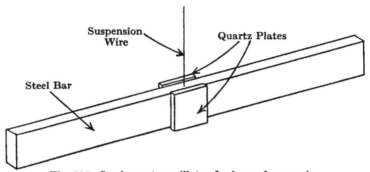

Fig. 205. Steel-quartz oscillator for lower frequencies.

The natural frequency is 30 kc/s, corresponding with a radiation wavelength of 10,000 m. Tinfoil coatings on the outer faces of the quartz plates form one electrical terminal, and the steel bar itself forms the other.

The piezo-electric quartz oscillator, like the electromagnetic tuning-fork oscillator, is necessarily very limited in power, and is used as the master oscillator driving a chain of amplifiers.

7. Constancy of Frequency

In considering how much the frequency of a transmitter may be permitted to wander, two distinct effects of a change of frequency must be weighed. Firstly, there is the drop of response at the receiver consequent upon its imperfect resonance with the received

* *Loc. cit.* p. 301.

7. CONSTANCY OF FREQUENCY

E.M.F.; this depends on the percentage change of frequency, and on the selectivity of the receiving circuits. Secondly, there is the change of pitch of the note in a receiver of heterodyne type; this is equal to the absolute change of frequency. In order to make a numerical computation, we will take a wavelength of 15,000 m (20 kc/s) and a selectivity equal to that of a single circuit whose decrement is 0·005. A change of frequency of 20 c/s, i.e. 0·1%, then reduces the response to about 0·62 of its tuned value (see Fig. 36, p. 72), and changes the heterodyne note by 20 c/s.

Now if the operator receiving the heterodyned signal by ear* were not subject to interfering signals, these changes in strength and pitch would not be serious. He might, for example, set his heterodyne oscillator to yield a note of 1000 c/s when the transmitter is on its correct frequency; he would then have to read a Morse signal wandering in strength between 0·62 and 1, and in frequency between 980 and 1020 c/s. But wireless signalling congestion is such that the frequency intervals between long-wave telegraph transmitters are forced down to intervals like 200 c/s. Let us therefore consider two transmitters, the desired D and the undesired U, working on wavelengths near 15,000 m, differing in frequency nominally by 200 c/s but each liable to wander by ± 20 c/s (i.e. by ± 0·1%). We will suppose that, at our receiving station, U is intrinsically 10 times as strong as D—a condition with which the operator should be able to cope by adjustments of his tuning dials and heterodyne oscillator.

The second column of the Table gives the frequencies of the transmitters in their nominal and in their worst wandered conditions; the other columns show the conditions at the receiver, with alter-

Transmitter	Frequency of transmitter	Field strength	Phone amplitude	If heterodyne 21,000 c/s, pitch of note	If heterodyne 20,200 c/s, pitch of note
D	20,000 c/s	1	1	1000 c/s	200 c/s
U	20,200 c/s	10	0·8	800 c/s	0
D	19,980 c/s	1	0·6	1020 c/s	220 c/s
U	20,220 c/s	10	0·7	780 c/s	20 c/s
D	20,020 c/s	1	0·6	980 c/s	180 c/s
U	20,180 c/s	10	0·9	820 c/s	20 c/s

* Or by relay with selective circuits tuned to the acoustic frequency.

native adjustments of the heterodyne frequency (which is, of course, entirely under the operator's control). The strength of D when at its nominal frequency is taken as unity. With the heterodyne set at 20,200 c/s, although D is sometimes weaker than U, and although its frequency wanders between 220 and 119 c/s, U's frequency is always so low that interference is avoided. D is not *jammed* by U.

But suppose there is a second undesired transmitter U' of frequency 19,800 c/s ± 20 c/s, also of field strength 10 at our receiver. The transmitter frequency situation then is indicated in Fig. 206. The heterodyne setting which avoided interference from U brings in a note from U' sometimes stronger than the 180–220 c/s note

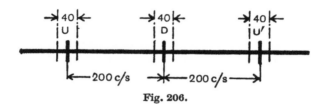

Fig. 206.

from D, and wandering in frequency between 380 and 420 c/s*. Our operator must now try to read the warbling note from D through the warbling note from U', although the latter is sometimes weaker and sometimes stronger than the former and is always of a pitch to which his ear is more sensitive. He cannot be expected to succeed.

It is clear from this simple numerical illustration that it is important that transmitters even of such great wavelength should not wander in frequency by anything like ± 20 c/s (i.e. ± 0·1°/$_o$).

In a report† of some measurements of the frequency wandering of many of the high-power transmitters of the world, its authors regard ± 10 to 12·5 c/s as the largest permissible frequency wandering in a satisfactorily steady transmitter, whatever the wavelength. This corresponds with ± 0·05 % at 15,000 m and is to be compared with the conclusion arrived at in our numerical example, that ± 0·1 % is much too high. They tabulate for 39 transmitters the mean observed wandering, during oscillographic records

---

* There are also weak interfering signals wandering between 360 and 440 c/s due to beats between U and U'; but these are here ignored.

† Schuchmann and Schreihage, *Telefunken-Zeitung*, Oct. 1925.

7. CONSTANCY OF FREQUENCY

covering 5 to 10 seconds, as a percentage of the permissible wandering. Most fall well below 100 %, the only offending transmitters being:

Amongst alternators	Saint Assise	150 °/₀
	Grodno	208 ,,
	Peking	200 ,,
Amongst triode generators	Moscow	175 ,,
Amongst arc generators	Bukarest	180 ,,

The steadiest transmissions tabulated are Kootwijk, 22 % and Ongar, 28 %. About half are under 50 %.

These figures witness that adequate constancy of frequency, even for brief epochs such as 10 seconds, is not easily attained without recourse to the master-oscillator method. Fluctuations in the power supply, changes of temperature, swaying of the antenna in the wind, the effects of keying, all have to be reckoned with. The master-oscillator system, more especially if of fork or crystal type, provides a complete solution.

With shorter waves, when reception is by the heterodyne method, the demands for constancy are proportionally more severe. At the very short wavelength of 30 m the frequency is 10,000 kc/s, and the heterodyne note changes through 100 c/s for a change of 1 in 100,000 in the transmitter frequency. The highest available constancy, such as that of a tuning-fork master oscillator*, is therefore needed. Even when, as in telephony, heterodyne reception is not practised, the rapidly growing world congestion of short-wave transmitters calls for the closest possible spacing of their frequencies. Acoustic considerations† limit the interval to some 10 kc/s, i.e. at $\lambda = 30$ m to about 1 in 1000. Hence the transmitters should remain fixed in frequency, from second to second and from day to day, to within (say) 1 in 10,000.

The short-wave (15–50 m) telephony transmitters of the British Post Office‡ have quartz crystal master oscillators, with frequency doublers and amplifiers bringing the antenna power to about 10 kW.

* See footnote on p. 300.
† See XIII–4 (a).
‡ At Rugby. At present (1930) there are six, and it is anticipated that by 1933 there will be twelve. See A. G. Lee, "The radio communication services of the British Post Office," *Proc. I.R.E.* Vol. 18 (1930), p. 1690.

CHAPTER X

GENERALISED TREATMENT OF COMPLEX TRIODE CIRCUIT IN STEADY-STATE RECTILINEAR OPERATION

NOTE. *Without exception the formulas derived in this Chapter are free from approximations. In numerical applications it will often be found that many of the terms are negligible.*

1. MOST GENERAL CIRCUIT CONSIDERED

BY a complex triode circuit is meant a triode with a set of A.C. paths of any character between anode and filament, a set between grid and filament, a set between anode and grid, and with magnetic coupling (mutual inductance) between one path and another. Clearly no analysis could cover all dispositions lying within that description. But most dispositions met in practice consist of the whole or part of the disposition shown in Fig 207. Here $\alpha_1, \alpha_2, \beta_1, \beta_2, \beta_3, \gamma$ are the impedances of paths through which flow the currents marked; and δ is the quasi-impedance jpM associated with a mutual inductance M* between self-inductances which are comprised in α_1 and β_1 and which carry the currents I_1 and i_1. When the disposition is used as an amplifier the input is the impressed E.M.F. E in the β_3 path, or alternatively E' in the β_1 path; when it is a generator E and E' are zero.

No simpler general network seems possible if it is to comprehend certain common dispositions. Thus, if the α_1

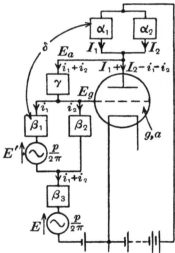

Fig. 207. Generalised triode circuit.

path is a coil and the α_2 path a condenser, they cannot be represented by a single impedance α unless $\delta = 0$. Similarly with the paths β_1 and β_2. Moreover, even when $\delta = 0$, if an E.M.F. E' is impressed in one of the two paths β_1 and

---

* M is taken as positive when it has the sense obtaining if the two coils are coaxial and wound in the same direction.

1. MOST GENERAL CIRCUIT CONSIDERED

β_2, these cannot be represented by a single impedance β. When the γ path exists, any investigation of amplification which ignores the impedance β_3 does not sufficiently represent practical conditions*.

Accordingly, the network of Fig. 207 is the basis of our present investigation. If we could find expressions for the four currents in Fig. 207 in terms of the impedances $\alpha - \delta$, and the triode conductances g and a, and the impressed E.M.F. (if any) E or E', we should be equipped with formulas whence for most triode arrangements the amplification and the input impedance of the amplifier, or the condition for self-oscillation and the frequency of the oscillator, could be quickly written down; any irrelevant terms in the general formula being omitted in the special case. Unfortunately, the expressions tend to become so unwieldy that it is not profitable to attempt the general algebraic solution for Fig. 207. After deriving the four equations for the four currents—which can be solved in numerical terms—we will proceed to algebraic solutions of certain special cases (i.e. special circuits and qualities of impedances).

Putting E_a and E_g for anode and grid potentials, we can write down the following six equations for the complex circuit in Fig. 207:

via α_2: $\quad E_a = -\alpha_2 I_2$ ....(1);

via α_1 and α_2: $\quad -\alpha_2 I_2 + \alpha_1 I_1 + \delta i_1 = 0$ ....(2);

via γ: $\quad E_a - E_g = \gamma(i_1 + i_2)$ ....(3);

via β_2 and β_3: $\quad E_g = E + \beta_3(i_1 + i_2) + \beta_2 i_2$ ....(4);

via β_1 and β_2: $\quad E' + \beta_1 i_1 + \delta I_1 - \beta_2 i_2 = 0$ ....(5);

triode: $\quad I_1 + I_2 - i_1 - i_2 = aE_a + gE_g$ ....(6).

Eliminating E_a and E_g from these, we have the following four equations for the four currents:

(2) is $\quad \alpha_1 I_1 - \alpha_2 I_2 \quad + \delta i_1 \qquad\qquad\qquad = 0$ ......(7);
(5) is $\quad \delta I_1 \qquad\qquad + \beta_1 i_1 \quad - \beta_2 i_2 \qquad\qquad = -E'$...(8);
from (1), (3) and (4) $\quad -\alpha_2 I_2 \quad -(\beta_3 + \gamma)i_1 \ -(\beta_2 + \beta_3 + \gamma)i_2 = E$ ......(9);
from (1), (4) and (6) $\quad I_1 + (1 + a\alpha_2)I_2 - (1 + g\beta_3)i_1 - (1 + g\beta_2 + g\beta_3)i_2 = gE$ ...(10).

With numerical values for the impedances—or some of them—equations (7)–(10) can be solved without very much trouble. With

---

* Several writers have analysed more or less generalised triode circuits, but have taken the grid potential E_g, and not the impressed E.M.F. E, as the starting point. See F. M. Colebrook, "A generalised analysis of the triode valve equivalent network," *Journ. I.E.E.* Vol. 67 (1929), p. 157. The circuit there investigated is, in effect, that of Fig. 207 with α_1, β_1, β_3 and δ omitted, and the γ path a condenser.

simpler networks than that of Fig. 207, the equations are simplified by the omission of the irrelevant terms. Thus, if there is no path between anode and grid, $\gamma = \infty$, $i_1 = -i_2$, and β_3 has no influence; if there is no mutual inductance, $\delta = 0$; if there is no path α_2, $\alpha_2 = \infty$, $I_2 = 0$, and $\alpha_2 I_2 = \alpha_1 I_1 + \delta i_1$; if there is no path β_2, $\beta_2 = \infty$, $i_2 = 0$, $\beta_2 i_2 = \beta_1 i_1 + \delta I_1 + E'$; and β_3 and E' may be omitted from the network—i.e. each put equal to zero in the equations—without loss of generality.

In practice, the impedance γ is often that of a capacitance—either the stray capacitance alone or with an added condenser.

When used as an amplifier, the performance is given as the ratio $\dfrac{E_a}{E}\left(\text{or }\dfrac{E_a}{E'}\right)$. If this ratio becomes infinite, i.e. if the denominator vanishes, a finite oscillation persists even when E (or E') is zero. This is the critical condition between that of an amplifier, responsive to E.M.F. impressed from without, and that of a generator, maintaining itself in spontaneous oscillation at a frequency determined from within the system. By equating separately to zero the real and imaginary parts of the denominator of expressions such as (13), two equations are obtained which express the condition for self-oscillation and the frequency (or frequencies) of self-oscillation.

It should be noticed that with $\gamma = \infty$ the amplifier input impedance to the impressed E.M.F. E is always infinite*.

2. Nature of impedances specified; anode and grid circuits both single; Fig. 208

The impedances of Fig. 207 are here:

$$\alpha_2 = \infty;\quad \beta_2 = \infty;$$
$$\alpha_1 \equiv \alpha = R + jpL;$$
$$\beta_1 \equiv \beta = r + jpl;$$
$$\gamma = \frac{1}{jpK};$$
$$\delta = jpM.$$

Equations (9) and (10) then become

$$(\alpha + \delta) I + (\beta + \gamma + \delta) i = -E \quad \ldots (11);$$
$$(1 + a\alpha - g\delta) I - (1 + g\beta - a\delta) i = gE \quad (12).$$

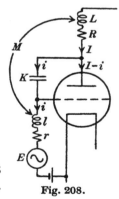

Fig. 208.

---

* This assumes, of course, that there is no grid thermionic current—as indicated in the Figures by the negative grid bias battery.

3. CASE OF $\gamma = \infty$. (NO ANODE-GRID PATH)

Eliminating i from (11) and (12) we find

$$\frac{I}{E} = -\frac{1 - g\gamma - (g+a)\delta}{\alpha + \beta + \gamma + 2\delta + (g+a)(\alpha\beta - \delta^2) + \gamma(a\alpha - g\delta)} \quad \ldots(13).$$

Substituting the specified impedances in (13), and equating the real and imaginary parts of the denominator separately to zero, we find for self-oscillation the two equations

$$\frac{1 + aR}{K[L + l + 2M + (g+a)(Rl + rL)]} = p^2$$

$$= \frac{K[R + r + (g+a)Rr] + aL - gM}{K(g+a)(Ll - M^2)} \quad \ldots(14).$$

Fig. 208 is sometimes called the Hartley circuit. It is referred to on p. 279.

It is interesting to examine the condition of oscillation for Fig. 208 in the special case of $M = 0$. Eliminating p^2 from equations (14) and putting $M = 0$, we find

$$\frac{K}{L}[R + r + (g+a)Rr] = \frac{gl - aL[1 + (g+a)r]}{L + l + (g+a)(Rl + rL)} \quad \ldots(14\mathrm{a}).$$

Since every symbol in (14 a) stands for a real positive quantity, if $\mu l < L$, self-oscillation cannot occur, whatever the value of K and however small the resistances R and r may be.

3. CASE OF $\gamma = \infty$. (NO ANODE-GRID PATH)

3 (a). *Nature of impedances not specified; Fig.* 209. Putting $i_1 = -i_2 \equiv i$, we have:

from (7) $\quad \alpha_1 I_1 \quad -\alpha_2 I_2 + \delta i \quad = 0 \ldots\ldots\ldots\ldots(15);$

from (8) $\quad \delta I_1 \quad +(\beta_1 + \beta_2)i = -E' \ldots\ldots(16);$

from (10) $\quad I_1 + (1 + a\alpha_2)I_2 + g\beta_2 i \quad = gE \ldots\ldots\ldots(17).$

Eliminating I_1 and i between (15), (16) and (17), we find

$E_a = -\alpha_2 I_2,$

$$= \frac{g\alpha_2(\alpha_1\beta_1 + \alpha_1\beta_2 - \delta^2)E \quad \text{or} \quad \alpha_2(g\alpha_1\beta_2 - \delta)E'}{(\alpha_1 + a\alpha_1\alpha_2 + \alpha_2)(g\beta_2\delta - \beta_1 - \beta_2) - \delta(1 + a\alpha_2)(g\alpha_1\beta_2 - \delta)} \quad (18).$$

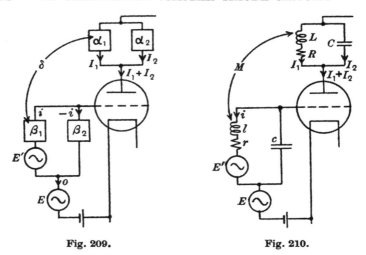

Fig. 209. Fig. 210.

3 (b). *Nature of impedances specified; anode and grid circuits both double; Fig.* 210. The impedances are specified as:

$$a_1 = R + jpL; \qquad \beta_1 = r + jpl;$$
$$a_2 = \frac{1}{jpC}; \qquad \beta_2 = \frac{1}{jpc};$$
$$\delta = jpM.$$

Substituting in the denominator of (18), and equating to zero the imaginary and real terms, we find respectively

$$gM - [aL + CR + cr(1 + aR)] + p^2 c[a(Ll - M^2) + C(Lr + lR)] = 0 \quad \ldots\ldots(19);$$

and

$$p^4 Cc(Ll - M^2) - p^2 [LC + lc + ac(Lr + lR) + CcRr] + (1 + aR) = 0 \quad \ldots\ldots(20).$$

These may be treated as simultaneous equations for M and p^2, the solution of which gives the values of M producing self-oscillation and the possible frequencies of self-oscillation.

3 (c). *Nature of impedances specified; anode circuit double; grid circuit single; Fig.* 211. Putting $\beta_2 = \infty$, (18) becomes

$$E_a = \frac{-ga_1 a_2 E}{(a_1 + aa_1 a_2 + a_2)(1 - g\delta) + ga_1 \delta(1 + aa_2)} \quad \ldots\ldots(21).$$

3. CASE OF $\gamma = \infty$. (NO ANODE-GRID PATH)

Substituting in (21) for a_1, a_2 and δ as in Subsection 3(b), we find

$$\frac{E_a}{E} = \frac{-g(R+jpL)}{[1 - p^2 LC + aR] - jp[aL + CR - gM]} \quad \ldots (22).$$

The condition for self-oscillation is therefore

$$aL + CR - gM = 0 \quad \ldots (23);$$

and the frequency of self-oscillation is

$$\frac{p}{2\pi} = \frac{1}{2\pi}\sqrt{\frac{1+aR}{LC}} \quad \ldots (24).$$

The equations (23) and (24) for self-oscillation have been obtained already by a different method in IX–2(a). They are also found on putting $c = 0$ in (19) and (20) (Fig. 210).

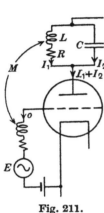

Fig. 211.

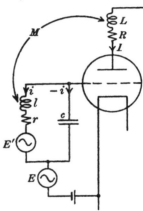

Fig. 212.

3(d). *Nature of impedances specified; anode circuit single; grid circuit double; Fig. 212.* Putting $a_2 = \infty$, (18) becomes

$$E_a = \frac{g(a_1 . \overline{\beta_1 + \beta_2} - \delta^2) E \text{ or } (ga_1\beta_2 - \delta) E'}{-(\beta_1 + \beta_2)(1 + aa_1) + g\beta_2\delta + a\delta^2} \quad \ldots (25).$$

The performance of Fig. 212 as an amplifier is given by (25) on substituting for β_1, β_2 and δ.

Self-oscillation is investigated most easily by putting $C = 0$ in (19) and (20). We thus find

$$a(1 + aR) M^2 - g[l + a(Lr + lR)] M$$
$$+ r[a^2 L^2 + cl + arcL + acR(2l + a . \overline{Lr + lR})] = 0 \ldots (26);$$

316   X. GENERALISED COMPLEX TRIODE CIRCUITS

and
$$p^2 = \frac{1+aR}{c\,[l+a\,(Lr+lR)]} \quad \ldots\ldots\ldots\ldots(27).$$

In Fig. 211 there was a single frequency and a single critical value of M; here there is also a single frequency, but two critical values of M (both of which are positive).

4. Case of $\delta = 0$. (No anode-grid mutual inductance)

4(a). *Nature of impedances not specified; Figs. 213 and 214.* With the removal of mutual inductance it is permissible to amalgamate the α_1 and α_2 paths. Putting $\alpha_2 = \infty$, $I_2 = 0$, $\alpha_2 I_2 = \alpha_1 I_1$, and then omitting the subscripts from α_1 and I_1, the general equations (8), (9) and (10) become:

from (8) $\qquad \beta_1 i_1 \qquad\qquad - \beta_2 i_2 \qquad\qquad = -E'$ (28);

from (9) $\qquad -\alpha I - (\beta_3 + \gamma) i_1 - (\beta_2 + \beta_3 + \gamma) i_2 = E$ ...(29);

from (10) $(1+a\alpha) I - (1 + g\beta_3) i_1 - (1 + g\beta_2 + g\beta_3) i_2 = gE$ (30).

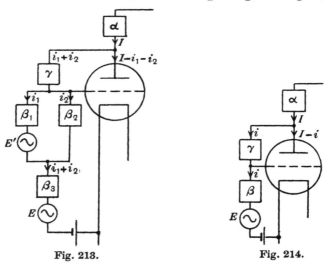

Fig. 213.   Fig. 214.

Much simplification ensues when E' can be abandoned, for then β_1 and β_2 may be amalgamated in the same way as α_1 and α_2, and β_3 may be omitted. Fig. 213 then simplifies to Fig. 214 and we have:

from (29) $\qquad -\alpha I - (\beta + \gamma) i = E \quad \ldots\ldots\ldots\ldots(31);$

from (30) $\qquad (1 + a\alpha) I - (1 + g\beta) i = gE \quad \ldots\ldots\ldots\ldots(32).$

4. CASE OF $\delta = 0$. (NO ANODE-GRID MUTUAL INDUCTANCE)

Solving (31) and (32) we find

$$\frac{E_a}{E} = \frac{-\alpha I}{E} = \frac{-\alpha(g\gamma - 1)}{a + \beta + \overline{g + a} \cdot \alpha\beta + \gamma(1 + a\alpha)} \quad \ldots\ldots(33);$$

$$\text{input impedance} = -\frac{E}{i} = \frac{\text{denominator of (33)}}{1 + \overline{g + a} \cdot \alpha} \quad \ldots\ldots(34).$$

It should be noticed that the advantage (in amplifiers) of infinite input impedance for the impressed E.M.F. E possessed by the dispositions of Section 3 necessarily disappears with the introduction of the γ path. Equations (33) and (34) show that the input impedance tends to zero as the retroaction is increased to make the amplification tend to infinity*.

One or both of the α and β paths in Fig. 214 may consist of an inductance and capacitance in parallel. The algebraic analysis is then very cumbersome. But if each is sensibly a single path, tolerably simple dispositions are obtained. One such disposition is that of Section 2 (Fig. 208) with M put equal to zero.

A numerical application of equations (33) and (34) is given in VIII–7 (b).

4 (b). *Nature of γ path specified; Fig. 215.*

When $\gamma = \frac{1}{jpK}$, the condition for self-oscillation—viz. that the denominator of (33) vanishes—becomes

$$\alpha + \beta + \overline{g + a} \cdot \alpha\beta - j\frac{1 + a\alpha}{pK} = 0 \ldots\ldots(35).$$

Putting $R + jX$ for α, and $r + jx$ for β, in (35), and separating the real and imaginary parts, we get

$$R + r + \frac{aX}{pK} = (g + a)(Xx - Rr) \quad \ldots(36),$$

and

$$X + x + (g + a)(rX + Rx) = \frac{1 + aR}{pK} \ldots\ldots\ldots\ldots(37).$$

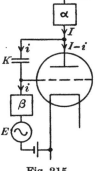

Fig. 215.

Now in equations (36) and (37), every symbol stands for a real

---

* It seems to be a sound general principle of amplifier circuit design that γ should be kept sensibly infinite where possible, i.e. that tuning or retroaction should not be furnished by a condenser connected between anode and grid. But in high-frequency amplifiers for short waves the anode-grid stray capacitance alone effectively prevents the γ path from being negligible.

quantity, and all except X and x are necessarily positive. On eliminating x between (36) and (37), an equation for X is obtained which can be satisfied only if X is positive. Since (36) can not be satisfied with X positive and x negative, it follows that both X and x must be positive. Expressed in physical terms, self-oscillation in Fig. 215 is possible only if α and β are both inductive impedances.

4 (c). *Nature of impedances specified; anode and grid circuits both single; Fig. 216.* In Fig. 216—sometimes called the Colpitts circuit —the condenser coil and grid leak shown dotted are for affecting the D.C. conditions only; they are to be supposed large enough to have no effect on the A.C. conditions. The γ path is now a coil of inductance S and resistance W*.

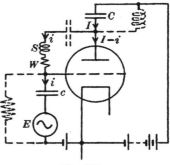

Fig. 216.

Putting $\alpha = \dfrac{1}{jpC}$,

$\beta = \dfrac{1}{jpc}$,

$\gamma = W + jpS$,

we obtain from equation (33) (Fig. 214)

$$\frac{E_a}{E} = \frac{-pc\left[(1-gW)-jgpS\right]}{p\left[C+c(1+aW-p^2CS)\right]-j\left[g+a-p^2c(aS+CW)\right]}.$$
$$\dotfill(38).$$

On equating the denominator of (38) to zero, we find for self-oscillation

$$\frac{C+c(1+aW)}{CcS} = p^2 = \frac{g+a}{c(aS+CW)} \quad \dotfill(39).$$

---

* Memorised as S for self-inductance; W for *Widerstand* (German for resistance).

CHAPTER XI

THE TRIODE AS RECTIFIER

1. THE TRIODE IN RELATION TO OTHER RECTIFIERS

IN Chapter VI the rectification effected by non-ohmic conductors in general was investigated. The rectifier might have been anything inside a closed box with two terminals, provided only that the relation between the P.D. v and the current i between those terminals was a curvilinear relation. In one particular form of rectifier there considered, the curved characteristic is exhibited by the contact

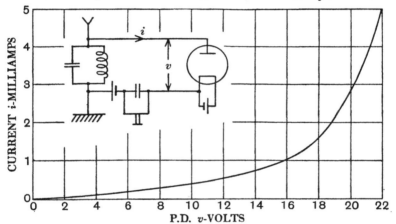

Fig. 217. Characteristic of a Fleming Valve.

between two dissimilar conductors. (Crystal rectifier, Figs. 98 and 101.) In another class of rectifying conductors the curvature of characteristic is the outcome of the space-charge between two electrodes in a vacuous envelope. (Thermionic diode, Fig. 122.) Thermionic currents were, indeed, first utilised in wireless telegraphy in this way—by J. A. Fleming in 1904. Fig. 217 shows the characteristic* of a *Fleming Valve*, and the circuit with which it might be employed as a detector.

Such a box with two terminals may be called a primary rectifier, because v and i are the P.D. and current between a single pair of

---

* Replotted from Fig. 38 of J. A. Fleming—*The thermionic valve and its development in radiotelegraphy and telephony* (1919). In these early diodes, remanent gas had a large influence on the characteristics.

XI. THE TRIODE AS RECTIFIER

terminals. But if the box were provided with two pairs of terminals, and i_2 were the current between one pair accompanying a P.D. v_1 between the other pair, curvilinearity of the graph plotted from pairs of valves v_1, i_2 would still imply a rectification in the sense that an alternating change in v_1 whose mean value is zero produces a change in i_2 whose mean value is not zero. Such a box with two pairs of terminals may be called a secondary rectifier. The fact that it can be used as a secondary rectifier (v_1, i_2) does not, of course, preclude its simultaneous or alternative partial use as a primary rectifier (v_1, i_1 or v_2, i_2).

It is obvious that a triode and its batteries may occupy our box with two pairs of terminals (Fig. 218), and may be fully used as a secondary rectifier (v_1, i_2) or be used as a diode (primary rectifier)

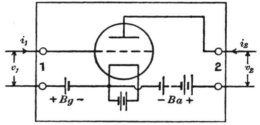

Fig. 218.

leaving either anode or grid out of service. If, for example, the triode were the "DEH 410" of Fig. 131 (p. 205), it might be used as a rectifier in the dispositions shown in the Table. Corresponding

Dispo-sition	Between termls. 1	Between termls. 2	B_g	B_a	Remarks
A	Input alternating P.D.	Output rectified current	−1·4 V	200 V	Secondary rectifier Zero input current
B	Input alternating P.D., and output rectified current	—	+1·4 V	—	Primary rectifier Anode not utilised in rectification
C	Short circuit	Input alternating P.D., and output rectified current	+3·8 V	0	Primary rectifier Grid utilised only to affect shape of anode-filament characteristic

2. THE ANODE RECTIFIER

dispositions are indicated in Fig. 219, where the input P.D. is shown as the P.D. across part of a high-frequency circuit, and the output is the P.D. produced by the rectified current flowing through the resistance R_a or R_g. Disposition A, usually termed a (triode) *anode rectifier*, is in very common use; it is dealt with in the next Section. Disposition B, with the anode unused as shown, would offer no advantage over the simple diode rectifier or Fleming Valve (Fig. 217); but elaborated so that the anode is brought into service, it is in very common use under the name of *grid rectifier*; it is discussed in Section 3 of this Chapter. Disposition C is not in common use, but possesses certain advantageous features; it is discussed in Section 6 of this Chapter.

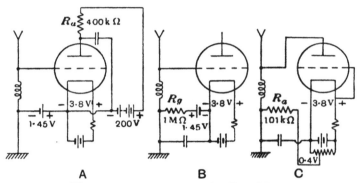

Fig. 219. Three triode rectifier dispositions.

When the rectifier is required to deal with a telephonic or telegraphic signal, the input high-frequency P.D. is modulated at an acoustic or Morse-key frequency. The resistance R_a or R_g may then be replaced by a reactive impedance for the modulation frequency, with the advantages and disadvantages already examined in VIII-2 with respect to amplifiers.

2. THE ANODE RECTIFIER

In Chapter VI, v was the P.D. across the rectifier and i the current through it. The argument in VI-2 (b) is directly applicable to the anode rectifier (Fig. 220) if we put $\dfrac{\partial^2 i_a}{\partial e_g^2}$ in place of $\dfrac{d^2 i}{dv^2}$, and $\dfrac{\partial i_a}{\partial e_g}$ in place of $\dfrac{di}{dv}$. Accordingly, in Fig. 220 a weak input signal $\mathscr{E}$

produces an output signal (fall of e_a due to rise of mean i_a) of magnitude

$$E_s = \frac{\mathscr{E}^2}{2} \cdot \frac{\partial^2 i_a}{\partial e_g^2} \cdot \frac{R_a}{1 + R_a \dfrac{\partial i_a}{\partial e_a}} \quad \ldots\ldots\ldots\ldots\ldots\ldots(1).$$

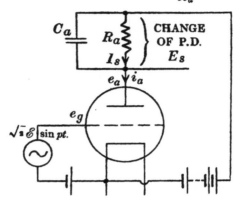

Fig. 220. Anode rectifier.

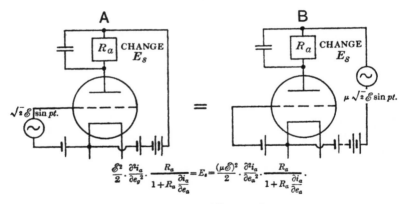

$$\frac{\mathscr{E}^2}{2} \cdot \frac{\partial^2 i_a}{\partial e_g^2} \cdot \frac{R_a}{1 + R_a \dfrac{\partial i_a}{\partial e_a}} = E_s = \frac{(\mu\mathscr{E})^2}{2} \cdot \frac{\partial^2 i_a}{\partial e_a^2} \cdot \frac{R_a}{1 + R_a \dfrac{\partial i_a}{\partial e_a}}.$$

Fig. 221. Anode rectifier as H.F. amplifier plus rectifier.

In so far as the curved feet of the triode characteristics conform to the basic triode equation

$$i_a = f(e_a + \mu e_g),$$

the equivalence exhibited in Fig. 130 (p. 204) must hold good; and, as regards the relation between $\mathscr{E}$ and E_s, the secondary rectifier at A in Fig. 221 is equivalent to the primary rectifier at

2. THE ANODE RECTIFIER

B in the same Figure fed with μ-times the input signal. Baldly stated, in the anode rectifier the grid may be said to be used for amplifying the high-frequency E.M.F. before it is rectified at the anode.

The *anode* rectifier is so called, not because the input signal is directly applied to the anode, but because rectification depends upon anode-current curvature. By applying the signal in the grid circuit (A, Fig. 221) instead of the anode circuit (B, Fig. 221), provided there is adequate negative grid bias, absorption of power from the input circuit is avoided.

As found in VI–3 (a) for primary rectifiers, so with the (secondary) anode rectifier, in order to obtain the highest $\frac{\text{volts out}}{\text{volts in}}$ coefficient of performance $C_{V/V}$ with weak signals, R_a should be made large in comparison with the anode slope resistance $\frac{\partial e_a}{\partial i_a}$. The output is then determined by the shape of the anode characteristic according to the relation

$$E_s \propto \frac{\partial^2 i_a}{\partial e_a^2} \bigg/ \frac{\partial i_a}{\partial e_a}.$$

The appropriate part of the characteristic over which to work is therefore a part where the ratio $\frac{\text{rate of change of slope}}{\text{slope}}$ is large, unless this leads to so large a value of the slope resistance $\frac{\partial e_a}{\partial i_a}$ as to make it impracticable to keep R_a much larger.

The curved foot of an e_a, i_a characteristic of the Osram "DEH 410" triode (Fig. 131, p. 205) at the fixed grid potential $-1\cdot45$ V is plotted to a large scale in Fig. 222, together with the derived curves for $\frac{\partial i_a}{\partial e_a}$ and $\frac{\partial^2 i_a}{\partial e_a^2}$. From these curves, and by direct observation of the amplification factor μ at the anode currents stated, the following data are collected:

i_a	e_a	$\frac{\partial i_a}{\partial e_a}$	$\frac{\partial^2 i_a}{\partial e_a^2}$	μ
50 μA	77 V	3·0 μA/V	0·120 μA/V/V	34
300 μA	117 V	9·8 μA/V	0·235 μA/V/V	36·5

324   XI. THE TRIODE AS RECTIFIER

Inserting these values in the formula in Fig. 221, viz.

$$E_s = \mathscr{E}^2 \cdot \frac{\mu^2}{2} \cdot \frac{\partial^2 i_a}{\partial e_a^2} \cdot \frac{R_a}{1 + R_a \dfrac{\partial i_a}{\partial e_a}} \quad \ldots\ldots\ldots\ldots(2),$$

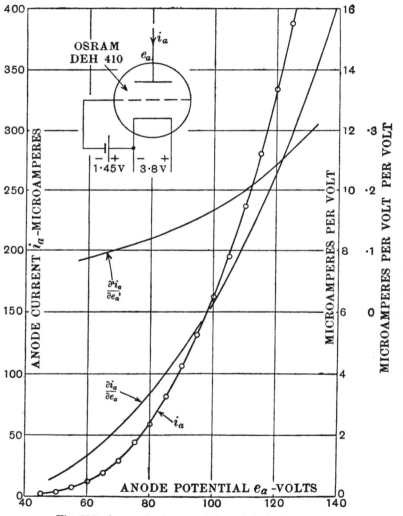

Fig. 222. An anode characteristic and derived curves.

2. THE ANODE RECTIFIER

when $R_a = 400\,\kappa\Omega$, we find the following theoretically predicted relations between input $\mathscr{E}$ and output E_s:

when $i_a = 50\,\mu A$, $\quad E_s = \mathscr{E}^2 \times 12\cdot 6$;
when $i_a = 300\,\mu A$, $\quad E_s = \mathscr{E}^2 \times 12\cdot 7$*.

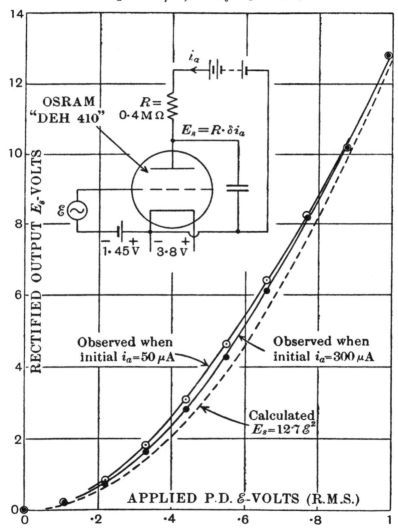

Fig. 223. Performance of anode rectifier, Fig. 222.

* It is, of course, fortuitous that these figures are almost equal. With another value of R_a it would not be so.

$E_s = 12{\cdot}7\;\mathscr{E}^2$ is plotted as a dotted line in Fig. 223; and in the same Figure are plotted the measured values of E_s obtained under the same conditions with the same no-signal state-points ($i_a = 50\;\mu\text{A}$ and $300\;\mu\text{A}$), using input signals $\mathscr{E}$ of frequency 90 c/s. In view of the difficulty of determining with much precision the second differential of the observed characteristic, the agreement between the observed output of the rectifier, and the output calculated entirely from the static characteristics, is probably as close as could be expected.

3. The grid rectifier

The disposition shown at B, Fig. 219, operates precisely as described in VI–2 (b) with reference to Fig. 94. The curvilinear v, i characteristic is now the grid-current characteristic of the triode. The output P.D. of the rectifier is the change of P.D. across R_g; and with a weak input signal consisting of an alternating P.D. of R.M.S. value $\mathscr{E}$, this output P.D. is

$$\frac{\mathscr{E}^2}{2} \cdot \frac{\partial^2 i_g}{\partial e_g{}^2} \cdot \frac{R_g}{1 + R_g \dfrac{\partial i_g}{\partial e_g}} \quad\ldots\ldots\ldots\ldots\ldots\ldots(3).$$

The rectified P.D. would naturally be applied to the grid of an amplifier; and since in this grid rectifier it is already applied between the grid and filament of the rectifying tube itself, and the anode of this tube is unemployed, it is obviously convenient to utilise the same tube as amplifier. A bald description, analogous to that of the anode rectifier given on p. 323, is that in the (complete) grid rectifier the grid may be said to be used for rectifying the high-frequency signal, the rectified output from the grid being amplified at the anode.

Since the grid-filament relation is that of a primary rectifier, the grid bias must be such as to allow an appreciable grid current to flow. The grid rectifier, therefore, in distinction from the anode rectifier, must absorb power from the input circuit.

The complete grid rectifier circuit is accordingly that of Fig. 224. Ignoring for the moment the capacitance C_a, the change of anode potential consequent on a signal $\sqrt{2}\;\mathscr{E}\sin pt$ is a copy of the changes

3. THE GRID RECTIFIER

of grid potential magnified $-m$ times, where

$$m = \mu \frac{R_a}{R_a + \rho};$$

i.e. is $\quad m \left(\sqrt{2}\, \mathscr{E} \sin pt - \frac{\mathscr{E}^2}{2} \cdot \frac{\partial^2 i_g}{\partial e_g^2} \cdot \frac{R_g}{1 + R_g \frac{\partial i_g}{\partial e_g}} \right) \quad \ldots\ldots\ldots(4).$

If the high-frequency component of anode potential is unwanted —and it would usually be a nuisance in the low-frequency circuits following the rectifier—it is reduced to a sufficiently negligible amount by the shunting condenser C_a, whose impedance may be supposed to be very small (in comparison with ρ) for the high

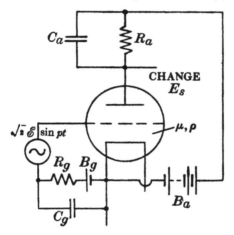

Fig. 224. Grid rectifier.

frequency $\frac{p}{2\pi}$, and very large in comparison with R_a for the low frequency (telephonic, or Morse-key, signalling frequency). The triode is then a high-frequency rectifier and a low-frequency amplifier combined, and the over-all output is the rise of anode potential

$$E_s = \left(\frac{\mathscr{E}^2}{2} \cdot \frac{\partial^2 i_g}{\partial e_g^2} \cdot \frac{R_g}{1 + R_g \frac{\partial i_g}{\partial e_g}} \right) \times \left(\frac{\mu R_a}{R_a + \rho} \right) \ldots\ldots\ldots(5).$$

The selection of suitable grid and anode batteries B_g and B_a, and resistances R_g and R_a, presents no difficulty. Firstly, since the function of the anode is that of an amplifier, not a rectifier, the no-signal anode current should be large enough to prevent excursions under the action of the signal into the curvilinear regions of the anode-current characteristics. If the concave feet of the anode characteristics are reached, anode rectification will occur; and since anode rectification implies fall of mean anode potential, whereas grid rectification implies rise of mean anode potential, the rectified output is reduced. A no-signal anode current of (say) not less than 4 or 5 mA is therefore desirable (see Fig. 131, p. 205).

Secondly, as found in VI-3(a), for good $\dfrac{\text{volts out}}{\text{volts in}}$ coefficient of performance, $C_{V/V}$, of the rectifier proper (i.e. grid-filament), the

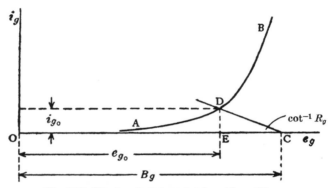

Fig. 225. No-signal state-point in grid rectifier.

grid potential should be such that in the absence of a signal grid current flows with a large value of $\dfrac{\partial^2 i_g}{\partial e_g^2} \Big/ \dfrac{\partial i_g}{\partial e_g}$, and R_g should be large compared with $\dfrac{\partial e_g}{\partial i_g}$. With triodes of the type of Figs. 131–133, this implies that e_g should be slightly positive (see Fig. 136, p. 213*); and R_g is conveniently of the order of a megohm.

Thirdly, for good $\dfrac{\text{volts out}}{\text{watts in}}$ coefficient of performance $C_{V/W}$ also, $\dfrac{\partial i_g}{\partial e_g}$ should be small. Since in grid current characteristics $\dfrac{\partial^2 i_g}{\partial e_g^2} \Big/ \dfrac{\partial i_g}{\partial e_g}$ does not vary widely† even down to quite small values of i_g, it is desirable to bring the no-signal state-point to a region of small

---

* With pure tungsten filaments, slightly negative.
† See Sub-section 4(b) of this Chapter.

3. THE GRID RECTIFIER

$\frac{\partial i_g}{\partial e_g}$, i.e. of large slope resistance such as an appreciable fraction of a megohm. With a given grid characteristic AB (Fig. 225), the no-signal state-point is determined by the intersection of AB with the "grid leak line" CD. CD is drawn from the point C, given by the grid bias battery B_g, at an inclination given by the grid leak R_g. The no-signal grid current $i_{g_0} = $ DE is the ordinate of the characteristic at the no-signal grid potential $e_{g_0} = $ OE, and is also the current flowing through a resistance R_g across which the P.D. is (OC − OE). Hence D is the no-signal state-point. The filament P.D. itself is sometimes chosen as the value for B_g, since this allows R_g to be

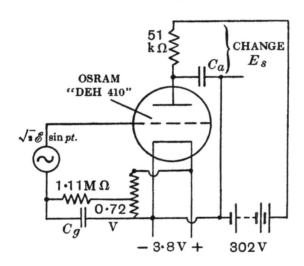

Fig. 226. A grid rectifier.

simply connected to the positive end of the filament. The same state-point can, of course, be arrived at with any number of pairs of values of B_g and R_g. Thus if AB in Fig. 225 were the grid characteristic drawn to scale in Fig. 227, the point D making $i_g = 0.16\,\mu$A would be given by $B_g = 0.72$ V, $R_g = 1.11$ MΩ; or by $B_g = 3.8$ V, $R_g = 20$ MΩ. A potential divider connected across the filament conveniently provides any value of B_g which may be desired. Such a large value for R_g as 20 MΩ would be objectionable for two reasons: it is very difficult to construct durable grid leaks of such large resistance; and the effects of the shunts provided by unavoidable stray capacitances and imperfect insulation would be too great.

As a numerical example of a complete grid rectifier the disposition of Fig. 226 is taken, in which the triode is that of Fig. 131 (p. 205). The grid characteristic is given in Fig. 227, together with the derived curves for $\dfrac{\partial i_g}{\partial e_g}$ and $\dfrac{\partial^2 i_g}{\partial e_g^2}$. (The i_g characteristics at $e_a = 110$ V and $e_a = 190$ V are to be ignored for the present.) It is

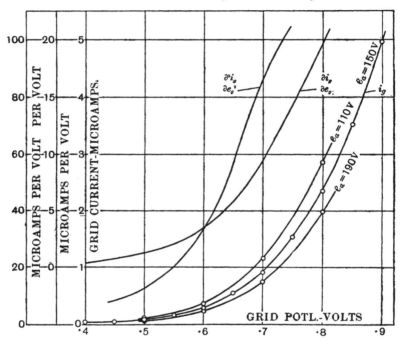

Fig. 227. A grid characteristic and derivatives.

found from these curves, and from observations of anode current, that at the no-signal state-point in Fig. 226,

$$e_g = 0{\cdot}55 \text{ V}, \qquad e_a = 150 \text{ V},$$
$$i_g = 0{\cdot}16 \,\mu\text{A}, \qquad i_a = 2{\cdot}97 \text{ mA},$$
$$\frac{\partial i_g}{\partial e_g} = 2{\cdot}0 \,\mu\text{A/V}, \qquad \rho = 26{\cdot}3 \text{ k}\Omega,$$
$$\frac{\partial^2 i_g}{\partial e_g^2} = 20 \,\mu\text{A/V}, \qquad \mu = 38.$$

3. THE GRID RECTIFIER

Accordingly formula (5) for weak signals becomes, in volt units:

$$\frac{E_s}{\mathscr{E}^2} = \left(\tfrac{1}{2} \times 20 \times 10^{-6} \times \frac{1\cdot 11 \times 10^6}{1 + 1\cdot 11 \times 2\cdot 0}\right) \times \left(\frac{38 \times 51}{51 + 26\cdot 3}\right),$$

$$= (3\cdot 45) \times (25) = 86.$$

In Fig. 228 observed values of E_s* are plotted against $\mathscr{E}$, together

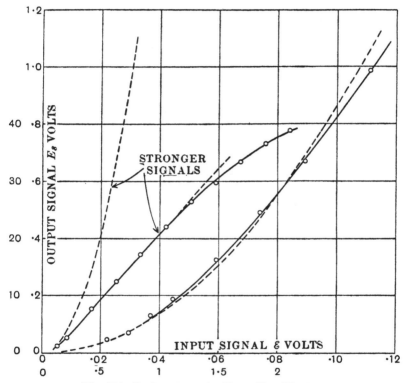

Fig. 228. Performance of grid rectifier, Fig. 226.

with the dotted curve $E_s = 86\mathscr{E}^2$. It will be seen that the agreement is very close up to about $\mathscr{E} = 0\cdot 10$ V, $E_s = 0\cdot 83$ V. The slide-back of mean grid potential is $\dfrac{E_s}{m}$, where m is the amplification $\mu \dfrac{R_a}{R_a + \rho} = 25$; so that the grid slide-back is here only $\dfrac{0\cdot 83 \text{ V}}{25} = 0\cdot 033$ V.

---

* The measurements were made at 90 c/s; $C_g = 1\,\mu\text{F}$, $C_a = 5\,\mu\text{F}$.

The continuation of the curves at stronger signals shows that above $\mathscr{E} = 0\cdot1$ V the discrepancy gradually becomes very large. It is instructive to endeavour to see over what range our formula giving the calculated performance $E_s = 86\mathscr{E}^2$ may be expected to hold good, and to enquire whether, in this somewhat complicated device, there may be factors we have left out of account.

4. Closer Analysis of the Grid Rectifier

4 (a). *The higher differential coefficients.* The fundamental rectification formula hitherto used is trustworthy only in so far as the approximations on p. 149 are good; and for this, either the higher differential coefficients must tend to vanish or the signal must be

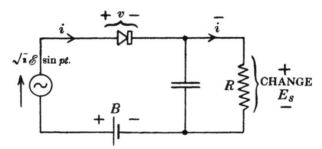

Fig. 229. Closer analysis of rectification.

very weak. It will be seen from Fig. 227 that the second differential of the grid characteristic is far from constant; that $\dfrac{\partial^3 i}{\partial e^3}$, and probably higher differentials, are much more prominent than in the anode characteristic of Fig. 222. We proceed to elaborate on the simple rectification formula of VI-2 (b), seeking an indication of the range over which it is applicable.

In Fig. 229,

if $\qquad i = f(v),$

$$\bar{i} = \frac{1}{\tau}\int_0^\tau i\,.\,dt = \text{mean value of } i, \text{ and}$$

$v_0 =$ value of v when $\mathscr{E} = 0$;

4. CLOSER ANALYSIS OF THE GRID RECTIFIER

then
$$i = f(v_0 - E_s + \sqrt{2}\mathscr{E} \sin pt),$$
$$= f(v_0 - E_s) + \sqrt{2}\mathscr{E} \sin pt \cdot f'(v_0 - E_s)$$
$$+ \frac{(\sqrt{2}\mathscr{E})^2}{\lfloor 2} \sin^2 pt \cdot f''(v_0 - E_s) + \dots;$$

$$\therefore \bar{i} = f(v_0 - E_s) + \frac{\mathscr{E}^2}{2} f''(v_0 - E_s) + \frac{\mathscr{E}^4}{16} f''''(v_0 - E_s) + \dots \quad (6).$$

On again expanding by Taylor's theorem, and putting
$$\bar{i} - f(v_0) = \frac{E_s}{R},$$
we obtain

$$E_s = \frac{\mathscr{E}^2}{2} \frac{\left[f''(v_0) - E_s f'''(v_0) + \frac{E_s^2}{\lfloor 2} f''''(v_0) - \dots \right] + \frac{\mathscr{E}^2}{8} [f''''(v_0) - E_s f^{\text{v}}(v_0) + \dots]}{\frac{1}{R} + f'(v_0) - \frac{E_s}{\lfloor 2} f''(v_0) + \frac{E_s^2}{\lfloor 3} f'''(v_0) - \dots}$$
$$\dots\dots(7).$$

Retaining as many terms as may be necessary, this equation (7) can theoretically be used to calculate the numerical relation between E_s and $\mathscr{E}$ in terms of the differential coefficients of the characteristic at the no-signal state-point v_0. If only the first term of the numerator and the first and second terms of the denominator are retained, the general equation (7) degenerates into the approximate formula for the simple rectifier derived in VI-2 (b), yielding (1), (3) and (5) in the foregoing Sections of this Chapter. In practice equation (7) is not of much use for numerical calculations unless the series are so rapidly convergent that the terms previously omitted are small enough to be treated as small corrections to the less accurate formula.

Applied to the complete grid rectifier, equation (7) changes the approximate formula (5) on p. 327 into

$$E_s = [\text{R.H.S. of (7)}] \times \frac{\mu R_a}{R_a + \rho} \quad \dots\dots\dots\dots(8).$$

The grid slide-back is $\dfrac{E_s}{\mu} \cdot \dfrac{R_a + \rho}{R_a}$.

In the rectifier whose characteristic and response curves were given in Figs. 227 and 228, $\dfrac{\partial^3 i_g}{\partial e_g^3}$ may be seen from Fig. 227 to be about 190 μA/V^3. Including all terms up to f'''', when $E_s \rightleftharpoons 1$ V, and therefore grid slide-back $\rightleftharpoons 0\cdot 040$ V, equation (8) in numerical terms is

$$E_s = \frac{\mathscr{E}^2}{2} \times \frac{20 - 7\cdot 6 + \ldots}{0\cdot 90 + 2\cdot 0 - 0\cdot 40 + 0\cdot 05 - \ldots} \times 25 = 61\mathscr{E}^2,$$

i.e. $\mathscr{E} = 0\cdot 128$ V when $E_s = 1$ V.

Fig. 228 shows that when $E_s = 1$ V the observed value of $\mathscr{E}$ was $0\cdot 112$ V, while the value calculated from the previous formula $E_s = 86\mathscr{E}^2$ was $0\cdot 108$ V. In view of the largeness of the correction terms now introduced, it cannot be expected that either the corrected or the uncorrected formula will give accurate values with signal outputs as large as 1 V, even if the f'''' terms are unimportant. Indeed, the range over which there is good agreement in Fig. 228 between the observed curve and the curve calculated from equation (5) seems to be fortuitously long.

4 (b). *Exponential characteristics.* The thermionic current crossing the vacuum between the hot cathode and a cold electrode only slightly above it in potential—e.g. the grid current in a grid rectifier—tends to be an exponential function of the P.D., in which, therefore, the differential coefficients of high order do not become negligible*.

Referring again to Fig. 229, if the characteristic is

$$i = \alpha \epsilon^{\beta v}$$

---

* This is due to the interference with the Langmuir $\tfrac{3}{2}$-power law of the finite emission velocities. The normal component of the most probable velocity of emission (see VIII-2 (c)) from a tungsten filament at ordinary temperature is that of about $0\cdot 1$ V, and 1 °/$_\circ$ of the emitted electrons have velocities exceeding that of $0\cdot 5$ V. These emission voltages are proportional to the absolute temperature of the emitter, and would therefore be about halved with the modern very dull filaments.

In the course of an extensive investigation of rectification, F. M. Colebrook has very fully analysed the operation of a rectifier with truly exponential characteristic. See "The rectification of small radio frequency potential differences by means of triode valves," *Experimental Wireless*, Vol. 2 (1925), p. 865. Colebrook examined a number of triodes with pure tungsten filaments, and one with a thoriated filament. He found the characteristics very nearly exponential in shape, and the calculated performances in close agreement with the observed. Unfortunately the modern filaments do not give characteristics so nearly exponential. See the next Sub-section.

4. CLOSER ANALYSIS OF THE GRID RECTIFIER

and the P.D. $R\bar{i}$ across the load resistance is written E,

$$i = \alpha \epsilon^{\beta(B - \bar{E} + \sqrt{2}\mathscr{E} \sin pt)};$$

$$\therefore \beta\bar{E} = \beta R \cdot \bar{i} = \beta R \cdot \alpha \epsilon^{\beta B} \cdot \epsilon^{-\beta \bar{E}} \cdot \frac{p}{2\pi} \int_0^{\frac{2\pi}{p}} \epsilon^{\sqrt{2}\beta\mathscr{E} \sin pt} dt;$$

i.e. $\quad (\beta\bar{E}) \epsilon^{(\beta\bar{E})} = R\alpha\beta\epsilon^{\beta B} \frac{p}{2\pi} \int_0^{\frac{2\pi}{p}} \epsilon^{\sqrt{2}\beta\mathscr{E} \sin pt} dt \quad \ldots\ldots\ldots\ldots(9).$

On expanding $\epsilon^{\sqrt{2}\beta\mathscr{E} \sin pt}$ and integrating the series term by term, we find

$$\frac{p}{2\pi} \int_0^{\frac{2\pi}{p}} \epsilon^{\sqrt{2}\beta\mathscr{E} \sin pt} dt = 1 + \frac{\left(\frac{\beta\mathscr{E}}{\sqrt{2}}\right)^2}{(\lfloor 1)^2} + \frac{\left(\frac{\beta\mathscr{E}}{\sqrt{2}}\right)^4}{(\lfloor 2)^2} + \frac{\left(\frac{\beta\mathscr{E}}{\sqrt{2}}\right)^6}{(\lfloor 3)^2} + \ldots,$$

$$= J_0(\sqrt{-1} \cdot \sqrt{2}\beta\mathscr{E}),$$

where J_0 is the Zero-order Bessel Function of the First Kind, values of which may be read from mathematical tables*, or for small values of $\beta\mathscr{E}$ may easily be computed from the series. For any assumed $\beta\mathscr{E}$, by the aid of tables of ϵ^x versus x, equation (9) may be solved in numerical terms for $\beta\bar{E}$. If $\bar{E}_0$ is the no-signal value of the P.D. across R (Fig. 229), equation (9) may be written

$$\frac{(\beta\bar{E}) \epsilon^{(\beta\bar{E})}}{(\beta\bar{E}_0) \epsilon^{(\beta\bar{E}_0)}} = J_0(\sqrt{-1} \cdot \sqrt{2}\beta\mathscr{E}) \quad \ldots\ldots\ldots\ldots(10).$$

The output signal is found as

$$E_s = \frac{\beta\bar{E} - \beta\bar{E}_0}{\beta}.$$

It is readily deduced† from the general equation (9) that for weak signals ($\beta\bar{E} \ll 1$), if $\bar{E}_0$ is the no-signal value of the P.D. across R (Fig. 229),

$$E_s \doteqdot \tfrac{1}{2}\mathscr{E}^2 \frac{\beta^2}{\beta + \dfrac{1}{E_0}} \quad \ldots\ldots\ldots\ldots\ldots\ldots(11);$$

---

* E.g. Jahnke und Emde, *Funktionentafeln* (1928), p. 130.

Above $x = 11$, $J_0(\sqrt{-1}x) = \dfrac{\epsilon^x}{\sqrt{2\pi x}}$ to within 1 %.

† As shown by Groenveld, v.d. Pol and Posthumus—"Gittergleichrichtung," *Zeitschr. f. Hochfrequenztechnik*, Vol. 29 (1927), p. 139. Figs. 230 and 231 are prepared from Figs. 11 and 10 in that paper.

and for strong signals ($\sqrt{2}\beta\mathscr{E} >$ (say) 5 or 10)

$$E_s \doteqdot \overline{E}_0 \left(\frac{\epsilon^{\beta(\sqrt{2}\mathscr{E} - E_s)}}{\sqrt{2\pi\sqrt{2}\beta\mathscr{E}}} - 1 \right). \quad\dots\dots\dots(12).$$

Equation (11) shows that for rectifying weak signals it is favourable to use a triode in which β is large, and that R (Fig. 229) should be large enough to make E_0 large compared with β. Equation (12) shows that in rectifying stronger signals, E_s must approach, but can never exceed, the peak value $\sqrt{2}\mathscr{E}$, whatever the values of β and

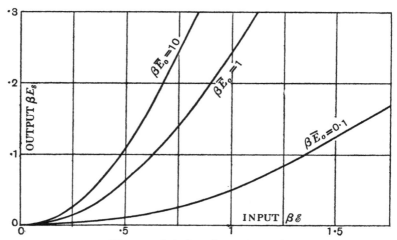

Fig. 230. Solution of equation (10).

E_0. The importance for weak signals, and the relative unimportance for strong signals, of β and E_0 are exhibited quantitatively by the curves in Figs. 230 and 231, which are calculated from equation (10) for several constant values of $\beta\overline{E}_0$.

$\frac{1}{\beta}$ is the rise of grid potential which makes the grid current increase ϵ times. The authors cited show that $\frac{1}{\beta}$ is approximately $9 \times 10^{-5}\,T$ volts, where T degrees K is the temperature of the filament of whatever material; it is therefore about 0·10 V for oxide-coated, 0·18 V for thoriated, and 0·25 for tungsten, filaments. For grid rectification of weak signals, the modern dull-emitter filaments are superior

4. CLOSER ANALYSIS OF THE GRID RECTIFIER

to bright emitters; but for strong signals, one triode is substantially as good as another*.

The grid characteristic i_g, e_g of Fig. 227 (at $e_a = 150$ V) is replotted with log i_g as ordinate in Fig. 232. It is not a straight line, but over

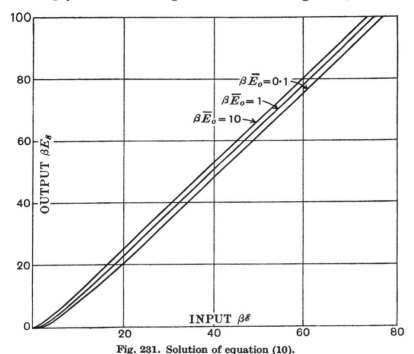

Fig. 231. Solution of equation (10).

the range $0{\cdot}4 - 0{\cdot}8$ V lies fairly close to the dotted straight line whose equation is

$$i_g = \alpha \epsilon^{\beta e_g},$$

where $\quad \alpha = 0{\cdot}000224\ \mu\mathrm{A}$

and $\quad \beta = 11{\cdot}9\ \mathrm{V}^{-1}$.

For the same rectifier (Fig. 226), with these values of α and β, equation (10) in place of the corresponding portions of our old formula (5) yields the calculated response curve shown dotted in Fig. 233. The observed curve is plotted again for comparison. It

* The latter conclusion, here deduced from analysis of the exponential characteristic, was anticipated on general grounds in VI-5 (a).

will be seen that the agreement is fairly close with weak signals; and with stronger signals the divergence is much less violent than in Fig. 228, where the old formula (5) was used.

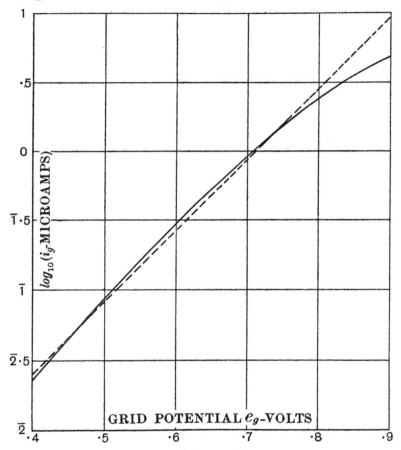

Fig. 232. A grid characteristic plotted semi-logarithmically.

4 (c). *Quartic characteristics.* The failure of equation (8) for numerical computations when the characteristic simulates exponential form, and the sensible departure from exponential form exemplified in Figs. 232 and 233, suggest an attempt to represent the rectifier characteristic by a finite number of terms such as

$$i = f(v) = \alpha + \beta v + \gamma v^2 + \delta v^3 + \epsilon v^{4}*.$$

* See S. E. A. Landale—"An analysis of triode valve rectification," *Proc. Camb. Phil. Soc.* Vol. xxv (1929), p. 355.

4. CLOSER ANALYSIS OF THE GRID RECTIFIER

In equation (6) (p. 333), with any assigned value of $\bar{i}$, E_s is known as $R[\bar{i} - f(v_0)]$, and f, f''', f''''' are known in terms of $\alpha, \beta, \gamma, \delta, \epsilon$, the constants in the equation fitted to the characteristic of the rectifier. Equation (6) is therefore soluble as a quadratic in $\mathscr{E}^2$.

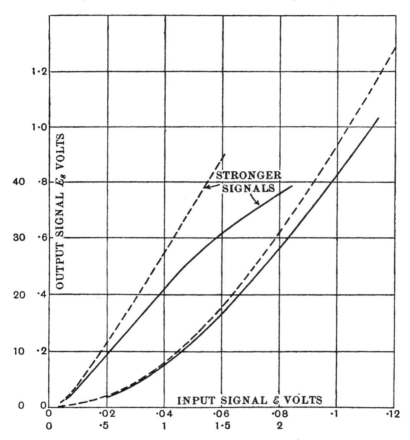

Fig. 233. Exponential treatment of rectifier, Fig. 226.

Landale found that the grid characteristic over a considerable range could be represented very closely by a quartic, and found very good agreement between observed rectified outputs and those calculated from the quartic characteristic. But because the assumed quartic obviously fails at low values of v, for which it would make i negative, it is applicable only when the no-signal value of grid current is impractically large*; or, once more, with very weak signals.

4 (d). *Subsidiary effects*. It has been tacitly assumed in the foregoing examination of the grid rectifier that the grid current is a function of grid potential only. In fact, however, a rise of anode potential without change of grid potential produces a fall of grid current. This is exhibited by the additional i_g characteristics in Fig. 227 for $e_a = 110$ V and $e_a = 190$ V. As

$$i_a = \phi\,(e_a + \mu e_g),$$

where μ is a constant positive number, the amplification factor of the triode, so we may expect to find

$$i_g = f\,(e_g - \nu e_a),$$

where ν is a constant positive number, which may be termed the inverted amplification factor of the triode (i.e. when used with grid and anode interchanged).

In the presence of the condenser C_a in Fig. 224 (p. 327), the rise of anode potential due to the input signal $\mathscr{E}$ is E_s†. The grid current is affected by this change of anode potential (hitherto neglected) as though the mean grid potential had been lowered by νE_s. The effect in the derivation of any rectification formula (e.g. on p. 333) is that $E_s(1 + \nu m)$ takes the place of E_s in the formula. Whether the effect is important depends on whether νm is appreciable in comparison with unity.

The characteristics of Fig. 227 (and others not reproduced) show that in this triode ν is a constant, and is about $\frac{1}{2010}$‡. Since $m = 25$ ($\mu = 38$), νm is only $1{\cdot}2\,\%$. The effect of ν is therefore a slight decrease of the output E_s for any input $\mathscr{E}$§.

---

* E.g. 27μA in Fig. 5 of Landale's paper quoted. $\dfrac{\partial e_g}{\partial i_g}$ had there the impractically low value of 26 kΩ.

† In the absence of C_a it is $(E_s - m\sqrt{2}\,\mathscr{E}\sin pt)$.

‡ I.e. at small grid potentials, as used. Fig. 131 (p. 205) shows that at large grid currents ν is much larger and is not constant.

§ In the absence of C_a the grid current is also affected as though the alternating component of its P.D. had been increased from $\mathscr{E}$ to $(1 + \nu m)\mathscr{E}$. With E_s approximately proportional to $\mathscr{E}^2$, the total effect of removing C_a would then be an increase in E_s for a given $\mathscr{E}$ of $\dfrac{(1+\nu m)^2}{1+\nu m}$ times. In our example this is $1{\cdot}2\,\%$.

5. LOAD ON THE INPUT CIRCUIT

A second manner in which anode conditions may influence the over-all performance of a grid rectifier is by curvature of the anode-current characteristics. If the condenser C_a in Fig. 224 (p. 327) is present, anode current curvature causes a *fall* of anode potential, say E_s', which, according to equation (2) (p. 324), is

$$E_s' = \left(\frac{\mu \mathscr{E}}{2}\right)^2 \cdot \frac{\partial^2 i_a}{\partial e_a^2} \cdot \frac{R_a}{1 + R_a \dfrac{\partial i_a}{\partial e_a}}.$$

The net output signal (for weak signals) is reduced by this amount from the value given by (5), and is accordingly

$$E_s - E_s' = \frac{\mathscr{E}^2}{2} \cdot \frac{\mu R_a}{R_a + \dfrac{\partial e_a}{\partial i_a}} \left[\frac{\partial^2 i_g}{\partial e_g^2} \cdot \frac{R_g}{1 + R_g \dfrac{\partial i_g}{\partial e_g}} - \frac{\partial^2 i_a}{\partial e_a^2} \cdot \mu \frac{\partial e_a}{\partial i_a}\right] \quad (13).$$

An appreciable amount of anode rectification often occurs in grid rectifiers embodying a triode of large amplification factor. In the triode of our example, the anode characteristic ceases to be straight below about 5 mA. Careful examination of the anode characteristic at the no-signal current, 2·97 mA, showed that there $\dfrac{\partial^2 i_a}{\partial e_a^2} = 277 \ \mu\text{A/V}^2$ $\left(\text{and } \dfrac{\partial^3 i_a}{\partial e_a^3} = 80 \ \mu\text{A/V}^3\right)$. The anode rectification occurring was therefore

$$E_s' = 3 \cdot 45 \mathscr{E}^2,$$

which is about 4 % of the grid rectification proper at weak signals. This correction would bring the observed and calculated curves in Fig. 233 closer together.

5. LOAD ON THE INPUT CIRCUIT

5 (a). *Load due to grid current.* Since the grid rectifier is a primary rectifier (according to the classification in Section 1 of this Chapter), it must absorb power from the input circuit. In Fig. 234, whatever the value of R_g, owing to the shunting condenser C_g the alternating P.D. $\mathscr{E}$ to be rectified is applied across grid-filament, and the input circuit LC is shunted at any instant by a leak of the value $\dfrac{\partial e_g}{\partial i_g}$ at that instant.

With weak signals $\dfrac{\partial e_g}{\partial i_g}$ is nearly constant, say r, throughout the cycle. With indefinitely weak signals, if A is the no-signal state-point on the grid characteristic, the leak has the value r, where $\frac{1}{r}$ is the slope of the curve at A. Under the action of the signal, the mean P.D. across grid-filament slides back from OB to OB', and the leakage resistance rises to r' where $\frac{1}{r'}$ is the slope at A'. With strong signals the slide-back BB' (which is the rectified output of the rectifier proper) may be so large that the slope varies widely throughout the cycle, and may indeed be sensibly zero during a

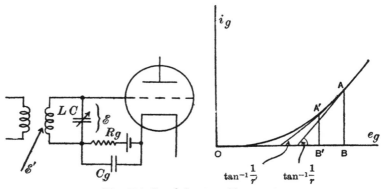

Fig. 234. Load due to grid current.

large portion of the cycle. The condition investigated in VI–5 (a) is then approached. As the applied signal $\mathscr{E}$ is increased in strength, the equivalent damping leakage resistance therefore rises from the value of $\dfrac{\partial e_g}{\partial i_g}$ at A with indefinitely weak signals, towards a limiting value $\dfrac{R_g}{2k}$ with very strong signals, as found in VI–5 (b), where k may in practical cases have a value not far from unity.

The damping of *LC* (Fig. 234) due to the rectifier is thus greatest with weak signals, and may be* very appreciable. As a numerical example, let us suppose that *LC* is a tuned circuit of inherent decrement 0·03, C having a small value such as 100 $\mu\mu$F, for signals of wavelength 1500 m (200 kc/s). With the grid characteristic of Fig. 227 (p. 330), if the grid current were 1 μA, the grid-filament resistance

* Indeed, must be, if the *LC* circuit is proportioned for good efficiency. See VI–3 (b).

5. LOAD ON THE INPUT CIRCUIT

would be about 0·1 MΩ, and so would introduce a decrement of 0·25 into the oscillatory circuit. The P.D. $\mathscr{E}$ produced by a given E.M.F. $\mathscr{E}'$ impressed in L would thus be reduced from $\mathscr{E}' \times \dfrac{\pi}{0·03}$ in the absence of the rectifier to $\mathscr{E}' \times \dfrac{\pi}{0·03 + 0·25}$ in the presence of the rectifier. The circuit would also become very poorly selective*. Moreover, if R_g were (say) 2 MΩ, on passing from weak to strong signals the effective resistance of the rectifier would rise from 0·1 MΩ to something of the order of 1 MΩ; and $\mathscr{E}$ would accordingly rise from $\mathscr{E}' \times \dfrac{\pi}{0·03 + 0·25}$ to $\mathscr{E}' \times \dfrac{\pi}{0·03 + 0·025}$. A form of distortion, sometimes called amplitude distortion, would thus be introduced, strong signals being favoured in comparison with weak.

The choice of a more suitable grid current, such as the 0·16 μA of our numerical example on p. 330, exhibits the grid rectifier in a more favourable light. Here $\dfrac{\partial e_g}{\partial i_g}$ is 0·5 MΩ, and the decrement introduced is now 0·05. If, too, R_g is changed to 1 MΩ, the damping of the circuit will not vary widely however weak or strong the signal, the decrement introduced by the rectifier being about 0·05 with both very weak and very strong signals.

5 (b). *Load due to anode-grid capacitance.* Damping of the input circuit attached to the grid is due not only to grid current—present in the grid rectifier but absent in the anode rectifier—but also to any absorption of power from the grid circuit through the agency of the anode-grid capacitance K (Fig. 235). If the resistance R_a in the anode circuit is shunted by a capacitance C_a large enough to prevent any high-frequency fluctuation of anode potential, the presence of K has no influence on the decrement of the input circuit LC, but merely throws a small capacitance K in parallel with C. In an anode rectifier C_a is always present, but in a grid rectifier it may be present or absent.

When C_a in Fig. 235 is absent, the current i flowing through the source of alternating P.D. E of frequency $\dfrac{p}{2\pi}$ may be evaluated

---

* As may be seen from Fig. 36, p. 72, to halve the P.D. would require distuning by 0·9 % in the absence of the rectifier, and by 8·4 % in the presence of the rectifier.

from first principles; or we may make use of equation (34) on p. 317, writing
$$\alpha = R_a,$$
$$\beta = 0,$$
$$\gamma = \frac{1}{jpK}.$$

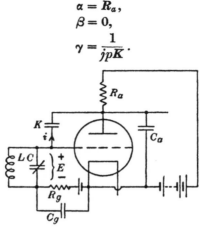

Fig. 235. Load due to anode-grid capacitance.

This gives
$$\frac{E}{i} = \frac{R_a + 0 + 0 + \frac{1}{jpK}(1 + aR_a)}{1 + (g + a) R_a};$$

whence
$$i = E\left(g + a + \frac{1}{R_a}\right) \frac{1 + j\dfrac{a + \dfrac{1}{R_a}}{pK}}{1 + \dfrac{\left(a + \dfrac{1}{R_a}\right)^2}{p^2K^2}}.$$

The power supplied by E is the real part of Ei, viz.
$$E^2 \frac{g + a + \dfrac{1}{R_a}}{1 + \dfrac{\left(a + \dfrac{1}{R_a}\right)^2}{p^2K^2}}.$$

The grid connection therefore absorbs power from the source E as though there were shunted across it a resistance
$$\frac{1 + \dfrac{\left(a + \dfrac{1}{R_a}\right)^2}{p^2K^2}}{g + a + \dfrac{1}{R_a}} = \frac{\rho + \dfrac{\left(1 + \dfrac{\rho}{R_a}\right)^2}{\rho p^2 K^2}}{\mu + 1 + \dfrac{\rho}{R_a}}.$$

5. LOAD ON THE INPUT CIRCUIT

This equivalent shunt resistance may be small enough to constitute a heavy load on the input circuit. Taking the triode conditions as in the example of a complete grid rectifier on p. 330, and with the input circuit of p. 342 tuned to $\lambda = 1500$ m, if K were 5 $\mu\mu$F, the equivalent shunt resistance would be 36·8 kΩ, causing in this circuit a decrement increase of about 0·7.

The price to be paid, in the form of this heavy load thrown upon the input circuit, for any advantage in omitting the high-frequency shunting condenser C_a (Fig. 235), is clearly much too high. The omission of C_a would ordinarily be permissible only if the load we

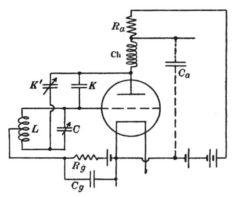

Fig. 236. Removal of load due to rectifier.

have calculated as thereby thrown upon the input circuit were more or less wholly cancelled by some counter-retroaction. This may readily be accomplished as shown in Fig. 236. Ch is a high-frequency choke, which we will assume has a reactance at the frequency in use large enough to suppress all high-frequency current in the resistance R_a. (The presence or absence of the shunt C_a is then, of course, irrelevant as far as high-frequency conditions in the triode are concerned.) This does not remove the load thrown on the input circuit*, but it furnishes a conveniently controlled retroaction through the adjustable condenser K'. It is an application of the neutrodyne principle explained in VIII–7 (c). If the tap on

---

* Indeed, analysis readily shows that if the choke reactance and R_a are both much greater than ρ, and if $\dfrac{1}{pK} \ll \rho$, the load caused by Ch (Fig. 236) is the same as that caused by R_a (Fig. 235), viz. $\dfrac{1}{p^2 K^2 \rho (\mu+1)}$.

L were at the mid point—which is by no means necessary—the load thrown upon LC by the anode-grid capacitance K is just cancelled when K' is made equal to K.

The disposition of Fig. 236 is a very useful one. The heavy damping in LC when K' is below the critical value at which the effect of K is cancelled, and the increase of retroaction up to the self-oscillating point as K' is increased through the critical value, confer a powerful and smooth control of the decrement over a wide range of tuning of the oscillatory circuit LC.

The residual high-frequency P.D. across R_a consequent on the finite value of the reactance of Ch can be rendered negligible by a

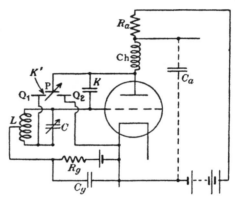

Fig. 237. Modification of Fig. 236.

moderate value of C_a. This facilitates the provision of a small time-constant $C_a R_a$, which is desirable on acoustic grounds when the rectifier is used for telephony*.

As regards the acoustic frequency, R_a is shunted by the capacitance $(C_a + K + K')$. If it is desired to keep this strictly constant despite adjustment of K', the condenser K' may be of the differential type, as in Fig. 237. The sum of the capacitances between the moving vanes P and the two sets of fixed vanes Q_1 and Q_2 is constant. Retroaction is increased on moving P away from Q_2 and towards Q_1.

It is often desirable readily to cause the retroaction to pass through zero and reverse. This is conveniently arranged with the differential condenser in Fig. 237 by connecting Q_2 to the grid instead of the filament.

* See XIII–5 (d).

6. Approach to Linear Performance

The push-pull disposition in XIII–5 (e) is another way of avoiding load on the input circuit due to anode-grid capacitance, but offers no advantage over the use of the shunting condenser C_a in Fig. 235 as regards suppression of high-frequency anode current within the triode.

6. Approach to Linear Performance

If the fluctuations of $\mathscr{E}$ (which constitute the input signal) are to be reproduced to scale by the fluctuations of E_s (which constitute the output signal), a linear relation between $\mathscr{E}$ and E_s over the working range is necessary.

Since no rectifier is known which shows the discontinuity of slope exhibited in the ideal characteristic of Fig. 102, p. 159, the "perfect rectifier" cannot be realised. All actual rectifiers follow the square law $E_s \propto \mathscr{E}^2$ when $\mathscr{E}$ is small enough. Over this range $\dfrac{dE_s}{d\mathscr{E}} \propto \mathscr{E}$. As $\mathscr{E}$ is increased, $\dfrac{dE_s}{d\mathscr{E}}$ ceases to rise as rapidly as $\mathscr{E}$, and there may follow a range of $\mathscr{E}$ over which $\dfrac{dE_s}{d\mathscr{E}}$ is very nearly constant. In examining a rectifier for linear performance, we wish to know the lower limit $\mathscr{E}_1$ and the upper limit $\mathscr{E}_2$ (if any) of the sensibly linear region over which $\dfrac{dE_s}{d\mathscr{E}}$ is constant. For good behaviour in telephony* $\dfrac{\mathscr{E}_2}{\mathscr{E}_1}$ should be upwards of 19 (for 90°/₀ modulation ratio); and to permit of operation with weak signals, $\mathscr{E}_1$ should preferably be small.

Thanks to the slide-back effect in the rectifier, examined in VI–5 (a), any rectifier through which current cannot flow in the reverse direction—as in all high-vacuum thermionic rectifiers—settles down to an approximately linear performance as the signal is strengthened. In an anode rectifier the upper limit is imposed by the greatest anode potential the triode will withstand without damage. Suppose this is 200 V in the "DEH 410" triode used to illustrate anode rectification in Section 2 of this Chapter. Since $\mu = 34$, it can be seen from the characteristic in Fig. 222 (p. 324) that for a no-signal current of 50 μA the lumped potential is

$$(77 - 34 \times 1\cdot 45) = 28 \text{ V}.$$

* See p. 412.

The largest negative grid bias usable is therefore B_g, where

$$200 - 34 B_g = 28,$$

i.e. $B_g = 5\cdot 1$ V.

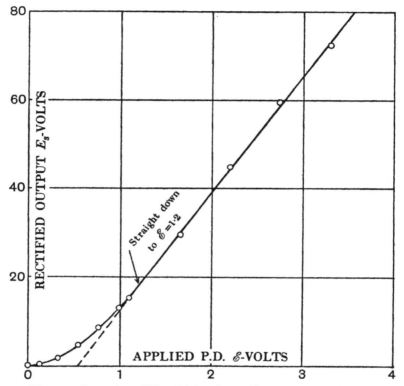

Fig. 238. Extension of Fig. 223 (anode rectifier) to stronger signals.

To avoid grid current $\mathscr{E}$ must therefore not much exceed

$$\frac{5\cdot 1}{\sqrt{2}} = 3\cdot 6 \text{ V},$$

equivalent to $34 \times 3\cdot 6 = 122$ V in the anode circuit.

Fig. 238 is the observed $\mathscr{E} - E_s$ curve for the anode rectifier of Fig. 223 extended to stronger signals[*]. It is seen that the lower

[*] As the grid bias was here only $-1\cdot 45$ V, grid current occurred with the stronger signals. Although this would throw a load on the input circuit, it would not appreciably affect the relation between $\mathscr{E}$ and E_s.

6. APPROACH TO LINEAR PERFORMANCE

limit of rectilinearity $\mathscr{E}_1$ is about 1·2 V, so that to obtain a rectilinear range $\dfrac{\mathscr{E}_2}{\mathscr{E}_1}$ as great as 19, excessively strong signals would be necessary. Measurements of many anode rectifiers have shown that it is seldom possible to have $\mathscr{E}_1$ even as low as in this test.

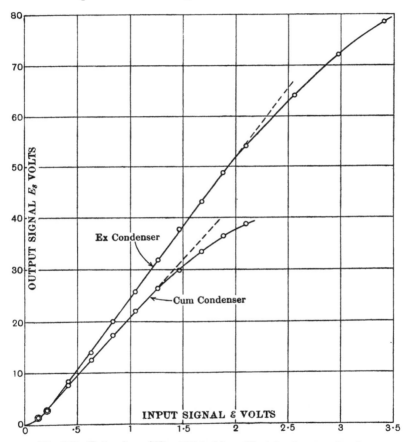

Fig. 239. Extension of Fig. 228 (grid rectifier) to stronger signals.

The need for very strong signals to obtain linearity of the response curve in anode rectifiers is due to the gradual rate of rise of the foot of the anode current characteristic. The grid current characteristic, tending to exponential form, is more propitious. Fig. 239 is the observed $\mathscr{E} - E_s$ curve for the grid rectifier of

Figs. 226 and 228 extended to stronger signals. The lower curve is with the large capacitance C_a (Fig. 226) shunting the anode circuit resistance R_a, and the upper without. Both are straight from about $\mathscr{E} = 0.2$ V; i.e. $\mathscr{E}_1 = 0.2$ V. With the condenser, $\mathscr{E}_2$ is about 1·2 V; and without the condenser, $\mathscr{E}_2$ is about 2·1 V. In this rectifier the upper rectilinear limits are due mainly to the onset of strong anode rectification*. The partial suppression of the alternating component of anode current by omission of the condenser C_a (e.g. in Fig. 224), or by the insertion of a choke (as in Fig. 236), more or less suppresses anode rectification and thereby raises $\mathscr{E}_2$.

Although this grid rectifier is much superior to the anode rectifier both in the smallness of $\mathscr{E}_1$ and in the largeness of $\dfrac{\mathscr{E}_2}{\mathscr{E}_1}$, it still does not meet the desideratum, $\dfrac{\mathscr{E}_2}{\mathscr{E}_1} \not< 19$. To make it do so, it would be necessary to provide a larger no-signal anode current than the 2·97 mA, requiring a larger no-signal anode potential than the 150 V provided.

An arrangement which combines the advantages of relatively sharp rise of characteristic, giving low $\mathscr{E}_1$, possessed by the grid rectifier (B, Fig. 219, p. 321), with the advantage of no upper limit $\mathscr{E}_2$ to the linear performance possessed by the ordinary (secondary) anode rectifier (A, Fig. 219), is that shown at C, Fig. 219. It may be termed a primary anode rectifier. By maintaining the grid of a triode at a suitable fixed small positive potential, and providing a no-signal anode potential of smaller positive value or zero, an $i_a - e_a$ characteristic may be obtained which rises even more sharply than the grid characteristic. Fig. 104 (p. 163), used in Chapter VI to illustrate general rectifier theory, is actually such a characteristic, together with its first and second differential derivatives. It was observed with the same "DEH 410" triode used in the examples of anode and grid rectifiers already given in Sections 2 and 3 of this Chapter, when the grid potential was held fixed at $+3.8$ V. Used as the rectifier shown at C, Fig. 219, the observed response curve was that already given in Fig. 106 (p. 166). In that Figure the ordinate numbers are rectified current in microamperes; they must be multiplied by 0·101 to give the rectified output in volts. It is seen that above $\mathscr{E} = 0.4$ V the response curve is a straight line, ex-

* In the presence of the shunting condenser C_a, when $\mathscr{E}$ was 1·2 V the anode current fell during the cycle as low as about 25 μA.

tending in the observations to $\mathscr{E} = 20$ and, it is to be supposed, indefinitely further*.

The inconveniently low input effective resistance here—about 61 kΩ (see p. 166)—is not a necessary feature of this type of rectifier. The resistance R_a (in C, Fig. 219) might with advantage have been made (say) 1 MΩ instead of 101 kΩ. This primary anode rectifier would then throw no more load on the input circuit than an ordinary grid rectifier (such as at B, Fig. 219).

7. Summary Comparison of Triode Rectifiers

In discussions of the relative merits of crystal, grid and anode rectifiers, with particular reference to their ability to reproduce faithfully the modulated signals of wireless telephony, the grid rectifier has commonly been stigmatised as peculiarly unfitted to follow the rapid rise and fall of amplitude corresponding to the higher acoustic frequencies. The existence of a time-constant $C_g R_g$ (Fig. 224, p. 327) of the grid leak and shunting condenser has been recognised, properly, as introducing a sluggishness of response to change of high-frequency amplitude. But it should be recognised that every rectifier falls under the same criticism. The time-constant CR in Fig. 94 (p. 149) and Fig. 102 (p. 159), for example, has precisely the same significance as $C_g R_g$ in Fig. 224; and the considerations governing the choice of values of C and R are the same whether the asymmetric conductor used as the rectifier is a crystal contact, or the anode-filament path in a diode or triode, or the grid-filament path in a triode. In each rectifier, to keep the sluggishness sufficiently inappreciable, the time-constant CR must be kept sufficiently small in comparison with the shortest acoustic period to be dealt with. The term *cumulative*, sometimes applied to the grid rectifier, is true in that the final state of the rectifier, under the action of (say) a steady input $\mathscr{E}$, is reached—strictly only after the lapse of infinite time—as the result of a charge gradually accumulated in the condenser C_g (Fig. 224). But the same might be said of the high-frequency shunting condenser C in any other rectifier (Fig. 94 or 102). The term cumulative, if

* H. L. Kirke seems to have been the first to publish an account of the use of a triode in this way. See "The diode rectifier," *Wireless World*, Vol. xxiv (1929), p. 32. On trying the powerful triodes (Osram "LS 5" and "LS 5B") with large grid potential there recommended, the present author was unable to obtain such good performance as with the small triodes and lower grid potentials he had previously and independently used. During an extensive search for a thermionic rectifier with a long linear range, for signals of moderate strength, he has encountered none with a better performance than that of Fig. 106.

XI. THE TRIODE AS RECTIFIER

applied to the grid rectifier, should be applied to the others also; and is therefore otiose.

In Fig. 224 (p. 327), a second high-frequency shunting condenser C_a is shown, introducing a second time-constant $C_a R_a$. But this is for the convenience of shunting out high-frequency effects from the subsequent low-frequency stages of the complete receiver. It does not enter into the rectification (except as a second-order effect), and other less crude methods of effecting the separation are feasible and often preferable. The use of a choke as in Fig. 236, or a tuned rejector circuit in place of it, is one way; the push-pull disposition of Fig. 275 (p. 414) is another.

Assuming that the performance desired is that of a telephony receiver, viz. a linear relation between output P.D. and amplitude of high-frequency input to the rectifier, unaffected by the frequency of the acoustic modulation of that amplitude, the three types of triode rectifier may be summarily compared as follows, where:

A = anode rectifier (of ordinary, i.e. secondary type);

B = grid rectifier (complete, i.e. with anode in service);

C = primary anode rectifier.

Output E_s for given input $\mathscr{E}$. About the same in all. (But for weak signals, making $E_s \propto \mathscr{E}^2$, in B much larger than in A, and in A much larger than in C.)

Load on input circuit. In A, nil; in B and C quite appreciable. In B there is an additional heavy load from anode-grid stray capacitance if the high-frequency fluctuation of anode current is prevented.

Lower limit of linear response, $\mathscr{E}_1$. In A, large; in B, less; in C, least.

Upper limit of linear response, $\mathscr{E}_2$. In A and C, none. In B there is a limitation imposed by anode-current curvature, severe if the high-frequency anode current is not suppressed.

Connection to subsequent amplifier (acoustic). In A and C the anode-circuit impedance should be very large, and is therefore preferably a pure resistance. In B the anode conditions are those of an amplifier, and a transformer may be employed.

Convenience for retroaction. In A and C, bad. In B, good; through a condenser if high-frequency anode current is suppressed, and through a mutual inductance if not.

Each form of rectifier has exclusive desirable features, but in the majority of situations the grid rectifier has the most to recommend it.

CHAPTER XII

RETROACTIVE AMPLIFIERS, AND SELF-OSCILLATING RECEIVERS

1. Reduction of damping by retroaction

In examining the conditions for a retroactive circuit to generate oscillation, we saw that in the circuit of Fig. 177 (p. 267) any transient disturbance in the oscillatory circuit LRC is continued as a decreasing or increasing oscillation

$$i = I\epsilon^{bt} \sin pt,$$

according as b is negative or positive respectively; that $\frac{p}{2\pi}$ is a

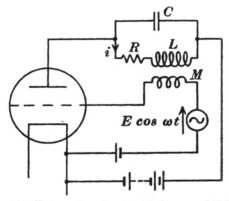

Fig. 240. Retroactive circuit with impressed E.M.F.

frequency nearly equal to the resonance frequency $\frac{1}{2\pi \sqrt{CL}}$; and that

$$b = -\frac{1}{2L}\left(R + \frac{aL}{C} - \frac{gM}{C}\right),$$

where a and g are the conductances of the triode. This shows that the resistance of the oscillatory circuit, owing to the triode connection, is reduced for current of this particular frequency from its inherent value R by an amount $\frac{gM - aL}{C}$, and can be made as nearly zero as desired, for example by merely increasing M.

That the effect of retroaction is a real reduction of resistance, and does not depend on the currents in the circuit possessing the particular frequency at which self-oscillation occurs when the retroaction is pushed far enough, may be seen by extending the differential equation on p. 268 to include an alternating E.M.F. impressed upon the circuit from outside, as in Fig. 240. The equation is now

$$CL \frac{d^2i}{dt^2} + (CR + aL - gM) \frac{di}{dt} + (1 + aR) i = -gE \cos \omega t.$$

This is the equation which would be written down for the simple circuit in Fig. 241, and so proves that the resistance of the *LRC* circuit by association with the triode has been changed from R to $\left(R - \frac{gM - aL}{C}\right)$, whatever the frequency $\frac{\omega}{2\pi}$ of the impressed E.M.F.

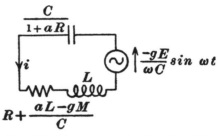

Fig. 241. Circuit equivalent to Fig. 240.

Retroaction in any other circuit capable of generating oscillation can be regarded in the same way as a reducer of resistance when the retroaction is not pushed far enough to generate.

We have, then, in the triode a means of reducing the resistance of any circuit to any extent; and since the current built up in a tuned circuit by a sustained isochronous alternating E.M.F. is inversely proportional to the resistance, it is theoretically possible to produce finite current changes under the action of infinitesimal impressed E.M.Fs. Thus in Fig. 242, *LRC* is an oscillatory circuit in which a small E.M.F.—say 1 mV—is impressed from a sustained C.W. signal of frequency n in the antenna. If the dimensions of the circuit are, for example, $L = 5000$ μH, $R = 100$ Ω, $C = 508$ $\mu\mu$F, and the natural frequency is therefore about 100 kc/s ($\lambda = 3000$ m), as n is changed the steady-state current varies between a small value and a maximum of 0·01 μA, as indicated by the resonance curve marked "$R = 100$ Ω"*. If now the triode circuit within the

* See IV-2 (e).

1. REDUCTION OF DAMPING BY RETROACTION

dotted rectangle is connected to the *LRC* circuit, the resistance of the latter is increased by $\frac{aL}{C}$ owing to the power absorbed by the

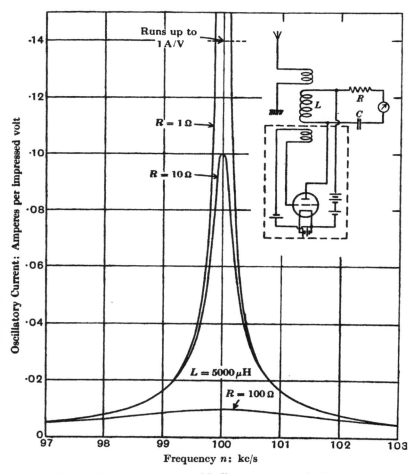

Fig. 242. Resonance curves with divers amounts of retroaction.

anode circuit, and is decreased by $\frac{gM}{C}$ owing to the retroaction. Its net resistance can be reduced to some value much smaller than its inherent 100 ohms, say to 10 ohms or even to 1 ohm. The resulting enhancement of the sharpness of tuning, and the increase in value of the tuned current produced by the same impressed E.M.F., are shown by the resonance curves in Fig. 242 marked "$R = 10\,\Omega$" and "$R = 1\,\Omega$."

Since we are able actually to pass through the zero value of effective resistance of the oscillatory circuit, it might seem at first sight that the engineer has here within experimental grasp the infinities of mathematical conception. But although the device is of extraordinary potence, it is limited in practice for the following reasons:

(i) Slight inherent unsteadiness of triodes, batteries, etc. demands the provision of a margin of stability between the actual effective resistance and the region of self-sustained oscillation lying beyond zero resistance.

(ii) The part of the total resistance introduced by the triode connection is, in the absence of perfect linearity of triode characteristics, itself affected by changes in potentials of grid and anode, and these vary with the current in the oscillatory circuit.

(iii) Extreme constancy of frequency of the incoming signal is necessary if the full benefit from very low resistance is to be obtained.

(iv) Although with theoretically perfect conditions the ratio between steady-state current and impressed E.M.F. becomes indefinitely great, an indefinitely long time is required before the steady state is reached; and in telegraphy the signal is not continued beyond the length of the Morse dot*.

But although infinity cannot be occupied (as a resting place), it may be approached. In the numerical example of Fig. 242, the inherent decrement ($R = 100\,\Omega$) of the *LRC* circuit was 0·1, and this was supposed to be reduced by retroaction to 0·001 ($R = 1\,\Omega$). In a research† requiring the measurement of very low decrements, it proved possible under laboratory conditions to maintain and measure with considerable precision decrements well below 0·001. Observations on the telegraphic signals of commercial stations ranging from 23 km to 4 km in wavelength showed that under slow Morse keying conditions there was no advantage in reducing

---

* At 125 words per min. the Morse dot lasts only $\frac{1}{100}$ second. Much more rapid fluctuations still are, of course, necessary in wireless telephony.

† L. B. Turner and F. P. Best, "The optimum damping in the auditive reception of wireless telegraph signals," *Journ. I.E.E.* Vol. 63 (1925), p. 493.

1. REDUCTION OF DAMPING BY RETROACTION

the decrement below about $\frac{120}{n}$, where n c/s is the frequency. At $\lambda = 3000$ m this gives an optimum decrement of 0·0012. For the short waves now used in long-distance communication, and with the steadiness of frequency now conferred by the fork or quartz-crystal master oscillator, it is probable that much lower decre-

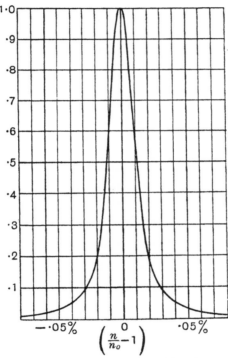

Fig. 243. Chain of three circuits of low decrement.

ments would be advantageous if they could be provided; for the frequency interval between two transmitters, necessary to keep interference at any given receiver to a certain level, decreases without limit as the selectivity of the receiver is increased.

For all signalling processes it is, however, preferable to obtain high selectivity by employing a chain of moderately selective circuits rather than a single circuit of extremely low decrement*. If the successive circuits are associated so as to have negligible

* See XIII-5 (b).

reaction—e.g. if they are coupled together indefinitely loosely; or, more practically, if they are coupled through triodes devoid of all retroaction from anode to grid*—the over-all resonance curve is found by taking for each abscissa the product of the ordinates of the resonance curves of the several circuits taken separately. Three such circuits, each with a decrement of 0·001, would give the over-all resonance curve plotted in Fig. 243. Triode retroaction in each of the circuits would be required to reduce its decrement to the low value of 0·001. Fig. 243 exhibits a degree of selectivity too great to be applied in ordinary signalling processes, but it may be regarded as a not impracticable *tour de force*.

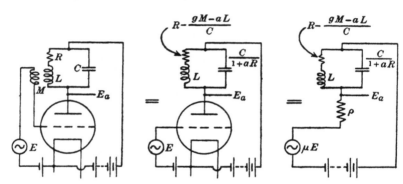

Fig. 244. Amplifier with retroaction.

2. Retroactive amplifiers

Let us regard the disposition in Fig. 240 as an amplifier, in which the input is E and the output is the alternating component E_a of anode potential. The amplification $\frac{E_a}{E}$ can be deduced from Fig. 241 or from equation (22) on p. 315. But as triode amplifiers without retroaction have been treated at length already in Chapter VIII, the clearest picture of the effect of retroaction will be given by embodying the result of the last Section as an elaboration of the circuit equivalence already used in VIII–2 (Fig. 151, p. 231). This is done in Fig. 244.

The effect of the retroaction M on tuned amplification and on selectivity is apparent from the investigation in VIII–6 (b). The

* By a neutrodyne balance, or the use of a screen-grid tetrode; see VIII–7 (c).

low decrement there seen to be necessary for good amplification with short-wave signals (i.e. E of very high frequency) can be secured by retroaction, whether of the mutual-inductance type or any other.

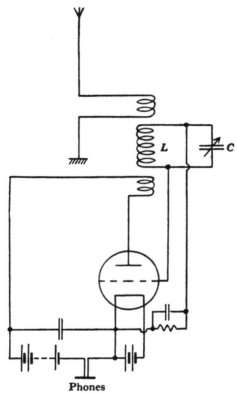

Fig. 245. Retroactive amplifier-rectifier.

3. THE AUTOHETERODYNE

Retroaction from anode to grid can, of course, be provided in a triode rectifier, whether of the grid or anode (ordinary, secondary) type. It is especially easily applicable in the grid rectifier because here the anode conditions are those of an amplifier (with moderate and constant anode resistance ρ). Fig. 245 would be an ordinary grid rectifier if the coil in the anode circuit were absent. The addition of this coil, suitably coupled to L, reduces the decrement of the *LC* circuit in the manner and with the results examined in the previous Sections of this Chapter.

If the retroaction is pushed far enough, the circuit will maintain itself in oscillation irrespective of any incoming signal. By adjustment of C the frequency of the self-oscillation may be made to differ slightly, by n_a c/s, from the frequency of an incoming C.W. signal. The distuning n_a being made an agreeable acoustic frequency, the triode fills the double office of heterodyne oscillator* and rectifier in one. The disposition is called an *autoheterodyne*.

In comparison with a heterodyne receiver embodying an independent oscillator, the autoheterodyne possesses one important practical advantage and suffers from two serious disabilities.

In searching with an independent heterodyne receiver for a signal of unknown wavelength, at least two separate tuning adjustments must be simultaneously correct before the signal is heard: the receiving circuit proper has to be tuned in sufficiently closely to be sensitive to the incoming waves, and the heterodyne oscillator has to be distuned sufficiently slightly to produce a beat of audible frequency. In the autoheterodyne, on the other hand, on merely swinging the condenser C (Fig. 245) across its scale the correct adjustment must be encountered.

If n_a is the beat frequency, the receiving circuit is distuned by n_a from the frequency n of the incoming signal. The signal P.D. applied to the rectifier is therefore less than could have been obtained from the same incoming wave if the receiving circuit had been accurately tuned to it (and an independent heterodyne oscillator had been employed). We may think of n_a as of the order of 1 kc/s. The fractional distuning is therefore of the order of $\frac{1}{n}$, where n is expressed in kc/s. The longer the wavelength the more considerable does this fraction become; with waves exceeding 3000 m, $\frac{1}{n}$ exceeds 1%, and the loss of signal strength is very marked even with circuits of moderately high decrement. What this loss amounts to depends, of course, on the effective decrement of the oscillating receiver, which is considered in the next paragraph; but it is such that for long-wave reception the autoheterodyne is in practice debarred.

In a non-oscillating retroactive amplifier the decrement is re-

* See VI-7.

3. THE AUTOHETERODYNE

ducible at will; and provided the operating range is such that the triode characteristics are sensibly linear, the decrement is constant whatever the strength of the signal. Where curvature is encountered, however, the effective decrement changes* as the amplitude of the oscillation changes. In a steady self-oscillating system the effective decrement for any externally impressed E.M.F. is zero when that E.M.F. is infinitesimal; but any rise of amplitude must increase the effective decrement, since, as we have seen in IX-2 (b), it is this which determines the amplitude at which the generator will maintain itself in steady operation. In heterodyne reception, when the local oscillation of amplitude a and the incoming signal of amplitude b† are syn-phased, the rectifier is subjected to $(a + b)$, and when anti-phased to $(a - b)$. In the autoheterodyne the dependence of effective decrement upon amplitude implies that if a_0 is the value of a in the absence of a signal b, then, on the advent of the signal, a_0 falls to a_1 during syn-phase epochs and *rises* to a_2 during anti-phase epochs. The difference between syn-phase and anti-phase amplitudes of P.D. applied to the rectifier is therefore not $2b$, but $(2b - \overline{a_2 - a_1})$. The great sensitivity and selectivity obtainable by retroaction in a receiver with independent heterodyne cannot be obtained in the autoheterodyne, however small the wavelength.

This disability of the autoheterodyne is minimised—it cannot be made to vanish—by so designing the circuit, *qua* self-oscillator, that the condition of maintenance resembles that of the linear régime (IX-2 (b)) rather than the over-retroacted high-efficiency régime (IX-4 (a)). It should, of course, also be arranged that the curvature which limits growth of amplitude as a self-oscillator is not the curvature on which rectification depends. Thus in the disposition of Fig. 245 used as an autoheterodyne, the dimensions should be such that, while rectification depends on the sharpest obtainable curvature of the grid-current characteristic, amplitude of self-oscillation is limited by the slightest obtainable curvature of anode-current characteristic.

---

* The decrement rises or falls according to the sense of the curvature. *Fall* of effective decrement consequent on a rise of amplitude has been applied in the author's "Oscillatory valve relay: a thermionic trigger device," *Journ. I.E.E.* Supplement to Vol. 57 (1920), p. 50.

† See VI-7.

4. THE SUPERHETERODYNE

In a heterodyne receiver the beat frequency $(n_1 \sim n_2)$ between the local oscillation (n_1) and the E.M.F. impressed by the incoming signal (n_2) is controllable between inaudibly low (hypo-acoustic) and inaudibly high (hyper-acoustic) frequencies*. In VI–7 the beat frequency was an acoustic frequency; but it is obvious that the heterodyne may be used as a frequency reducer without reference to acoustic effects, converting an incoming signal of frequency n_2 and amplitude b into an outgoing signal of frequency $(n_1 \sim n_2)$ and amplitude (assuming linear-response rectification) linearly related to b. In Fig. 109 (p. 169), the output current flowed through a telephone; when $(n_1 \sim n_2)$ is hyper-acoustic, the telephone is replaced by apparatus appropriate to such frequencies as are met in long-wave radiation, as in Fig. 246. Such receivers are called *superheterodyne*.

Fig. 246 includes one intermediate-frequency amplifier stage, the grid and anode circuits $C_1 L_1$, $C_2 L_2$ being tuned to the frequency $(n_1 \sim n_2)$. This stage might be repeated and elaborated with any of the circuit devices already considered in connection with high-frequency amplifiers. Ordinarily the rectifiers are triode rectifiers, and there are low-frequency amplifier stages between the second rectifier and the final indicating instrument (relay or telephone).

If the original incoming signal is of C.W. telegraphy form, there exists across $C_2 L_2$ a steady hyper-acoustic P.D. If this is to produce an audible signal in the telephone, a second heterodyne oscillation (frequency n_3) must be introduced, with a second rectifier. The pitch of the final signal in the telephone is $[(n_1 \sim n_2) \sim n_3]$. If the in-

---

* The difference between n_1 and n_2 cannot be made to vanish altogether, even when (as in heterodyne reception of signals from a remote station) there is no appreciable reaction between the generators of the two oscillations beating together. This is a second-order effect dependent on curvature of triode characteristics, and has been analysed by B. van der Pol ("Erzwungene chwingungen in einem System mit nichtlinearem Widerstand," *Zeitschr. für 'lochfrequenz*, Vol. 27 (1926), p. 124). When the frequency of the external impressed E.M.F. is very close to the frequency n_2 of the heterodyne oscillator, extra damping is introduced causing the condition of maintenance of self-oscillation to be no longer met. The only current then existent is of the impressed frequency n_1. This phenomenon of forced synchronisation is well known to experimentalists.

A practical upper limit to the beat frequency is reached when $(n_1 \sim n_2)$ approaches either of the high frequencies n_1 and n_2, since then it is no longer feasible to segregate the beat-frequency currents in the rectifier and subsequent circuits.

4. THE SUPERHETERODYNE

coming signals are telephonically modulated (see next Chapter), the second heterodyne oscillator n_3 is omitted.

It is convenient to refer to the several frequencies as:

n_h, designating both the high frequencies n_1 and n_2;
n_i, designating the intermediate frequency $(n_1 \sim n_2)$;
n_l, designating the low frequency $[(n_1 \sim n_2) \sim n_3]$.

As an illustration:

$n_h \doteqdot 1000$ kc/s (wavelength 300 m);
$n_i = 30$ kc/s (corresponding with wavelength 10,000 m, given by $n_1 = 1030$ or 970 kc/s);
$n_l = 1$ kc/s (given by $n_3 = 31$ or 29 kc/s).

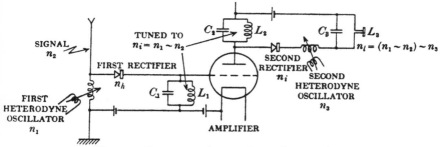

Fig. 246. Frequencies in superheterodyne receiver.

The advantages offered by the superheterodyne method are three:

(i) However high the frequency (n_h) of the incoming signal may be, high-frequency amplification can be carried out at a convenient (lower) frequency (n_i). The difficulties besetting amplification at very high frequencies are in this way entirely avoided.

(ii) A single set of amplifying stages (coils, condensers, etc., for n_i) suffices for dealing with incoming signals of all wavelengths. The n_i circuits may even be tuned up once for all and left without provision for further adjustment; for whatever n_h may be, the correct frequency n_i to suit the amplifying stages is secured by adjustment of the *first* heterodyne oscillator.

(iii) For the same decrement, the selectivity of each tuned circuit is much greater after heterodyne-conversion of frequency than if the original high frequency had been retained. A fractional

XII. RETROACTIVE RECEIVERS

change $\frac{\delta n_h}{n_h}$ of incoming signal frequency becomes $\frac{\delta n_h}{n_i}$ in the intermediate-frequency circuits—an increase of 33 times in our numerical illustration*.

The choice of intermediate frequency is largely conditioned by considerations of the necessary discrimination in various parts of the circuits between currents of one frequency and of another.

Part of circuit (Fig. 246)	Impedance		Reason
	for frequency	should be	
$C_1 L_1$	n_h	small	For good performance of Rectifier No. 1
$C_1 L_1$ and $C_2 L_2$	n_h	small	To avoid introducing strong high-frequency P.D. into intermediate and low-frequency circuits
	n_i	large	For good performance of amplifier
$C_3 L_3$	n_i	small	For good performance of Rectifier No. 2
L_3	n_l	large	The n_l P.D. across L_3 is a measure of the final output
C_3	n_l	much larger than impedance of L_3	To avoid prominent frequency discrimination as n_l changes (in telephony)

These considerations are indicated in the Table. When it is remembered that advantage (iii) above (incoming wavelength selectivity) is the greater the smaller is n_i, that (in telephony) n_i must be too high for audibility, and that n_l wanders (say) between 50 and 5000 c/s, it will be seen that the several requirements may prove

* But if the decrements in the n_i circuit of the superheterodyne receiver and in the n_h circuit of the alternative ordinary receiver are *not* equal, but are in each receiver reduced (e.g. by triode retroaction) as far as the *ringing* of the signals permits, this advantage becomes illusory. Ringing (i.e. long persistence of transients) is determined by the product of decrement and frequency, so that the lowest decrement permissible (with respect to ringing) would be proportionately greater at the n_h frequency than at the n_i frequency. (See L. B. Turner, "The relations between damping and speed in wireless reception," *Journ. I.E.E.* Vol. 62 (1924), p. 192.) In short-wave practice, however, desirable decrements are not determined by the ringing effect, and the gain of selectivity conferred by amplification at frequency n_i instead of n_h is a real one.

to be conflicting*. It may be said, summarily, that for long-wave signals (especially telephony), superheterodyne reception is not advantageous; but that as the wavelength is reduced (i.e. as $\frac{n_h}{n_l}$ gets greater), the difficulties tend to disappear. For short waves, the superheterodyne method is of great utility†.

The invention of the superheterodyne method of reception seems to have been the work of E. H. Armstrong‡, to whom much of the early study of triode retroaction must be credited.

5. SUPER-REGENERATIVE RECEPTION

A triode circuit with retroaction carried so far as to be capable of maintaining a steady state of self-oscillation is not *ipso facto* in a state of self-oscillation. The retroaction merely confers a negative sign on the damping exponent of the circuit, so that any transient disturbance will upset the unstable equilibrium and cause an oscillation to grow until the condition of maintenance is reached, i.e. until any further growth would restore the positive sign to the effective damping.

It was shown on p. 354 that the equation to the flow of electricity in the circuit of Fig. 240 is the same as in that of Fig. 241. When M is sufficiently large, the resistance of the circuit, $R - \frac{gM - aL}{C}$, is negative, say $-W$; and if q is the charge in the condenser we have (as in IV-2(a))

$$L\frac{d^2q}{dt^2} - W\frac{dq}{dt} + \frac{1+aR}{C}q = -\frac{g}{\omega C}E\sin\omega t.$$

If there is resonance between the impressed E.M.F and the circuit, and if the circuit is electrically dead ($q = 0$, $i = 0$) at the instant $t = 0$, the solution, as found already in IV-2(d), is

$$i = \frac{g}{\omega C} \cdot \frac{E}{W}\left(1 - e^{\frac{W}{2L}t}\right)\sin\omega t.$$

The amplitude therefore grows at an ever-increasing rate as time

---
* The situation is eased by the provision of more elaborate filtering circuits for the n_i frequency than are shown in Fig. 246.
† For an example, see XIII-7(b).
‡ See his "The super-heterodyne—its origin, development, and some recent improvements," *Proc. I.R.E.* Vol. 12 (1924), p. 539.

passes, but at any instant it is proportional to the magnitude of the impressed E.M.F*.

Suppose now that after being allowed to grow for several periods the oscillation is made to die away substantially to zero—i.e. is *quenched*—by some extraneous means, e.g. by the temporary introduction of a large resistance into the circuit or of a short-circuit across the condenser; and that the quenching is applied and removed periodically, the impressed E.M.F. being sustained throughout. Assuming that, between quenchings, the amplitude never rises

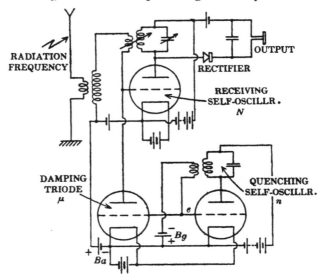

Fig. 247. Super-regenerative receiver.

enough to approach the condition of maintenance—i.e. that the triode performance remains substantially rectilinear—the mean amplitude (i) increases very rapidly† with the duration T of the unquenched epoch, and (ii) is proportional to the magnitude E of the impressed E.M.F. By making T as long as may be (in comparison with the duration of the Morse dot in telegraphy, or in comparison with the shortest significant acoustic period in telephony), owing to (i) the strength of output for a given input E may be made very

* The same result would be found for a transient impressed E.M.F. in the form of a brief shock.

† As $\dfrac{1}{T} \displaystyle\int_0^T \left(1 - \epsilon^{\frac{W}{2L}t}\right) dt$.

5. SUPER-REGENERATIVE RECEPTION

great, and owing to (ii) that output is linearly proportional to the input.

A periodically quenched self-oscillatory receiver operating in this manner has been termed *super-regenerative* by its inventor, E. H. Armstrong*.

The quenching may be effected in many ways. At a slow rate it could be done with a vibrating or rotating switch contact. At the high rates usually desirable (in telephony, hyper-acoustic) a triode circuit oscillating at the quenching frequency can be connected

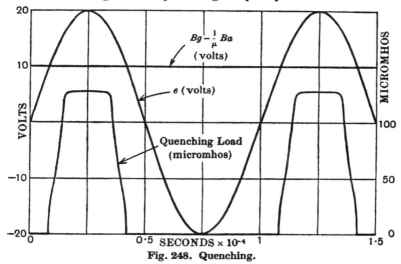

Fig. 248. Quenching.

to the receiver oscillating at the radiation frequency in various ways so as to quench the latter periodically. Fig. 247 illustrates the principle in one simple disposition. The damping triode is without influence on the receiving self-oscillator when the lumped potential of the former, $[B_a + \mu(e - B_g)]$, is negative, i.e. when $e < \left(B_g - \frac{1}{\mu} B_a\right)$; but during epochs when e is (algebraically) more than $\left(B_g - \frac{1}{\mu} B_a\right)$, the anode-filament slope conductance of the damping triode constitutes a load on the grid circuit of the receiving triode, and, when this load is heavy enough, quenches the oscillation in the latter.

* See his "Some recent developments of regenerative circuits," *Proc. I.R.E.* Vol. 10 (1922), p. 244.

As a numerical example, let the quenching oscillator have a frequency 10 kc/s, and its grid potential have an amplitude 20 V, so that (in volts and seconds)

$$e = 20 \sin 2\pi \times 10^4 t.$$

Let the damping triode be an Osram "DEL 410" (Fig. 132, p. 206), $\mu = 13$, $B_a = 130$ V, $B_g = 20$ V. $\left(B_g - \dfrac{1}{\mu} B_a\right)$ is then 10 V, and damping is present during epochs when e exceeds 10 V, i.e. during one-third of the cycle of the quenching oscillator. The quenching conditions are shown in Fig. 248, in which the curve marked "Quenching Load" is the anode-filament slope conductance of the quenching triode as given by the lumped characteristic in Fig. 180, p. 274. The precise duration of the epochs during which this load renders the receiving oscillator incapable of self-oscillation depends upon the dimensions of that circuit, but it might be something only slightly less than $\frac{1}{3}$ of 10^{-4} sec. During $\frac{2}{3}$ of 10^{-4} sec. the receiving oscillator is quite undamped by the quenching connection, and so is free to build up an amplitude under the influence of any incoming signal E according to the factor $E\left(1 - \epsilon^{\frac{W}{2L}t}\right)$ found on p. 365.

When there is no incoming signal, the slight unsteadiness inherent in the receiving triode and its batteries prevents complete quiescence in the circuit; some oscillation, though relatively weak, must grow up during the unquenched epochs even when $E = 0$. This seems to be an unavoidable defect in the super-regenerative method of reception (especially for telephony), since some background of rustling sound is present in the telephone when there ought to be complete silence. The necessity to keep the oscillations which grow up under the action of the signal E much larger than those which grow up under the action of the inherent unsteadiness of the triode imposes a lower limit on the signal strength which can be dealt with. Nevertheless, it proves possible in practice to obtain by this method relatively enormous output signals from input signals too weak to be effective with ordinary heterodyne or with non-self-oscillating reception.

6. EXAMPLE OF TRIODE RECEIVER

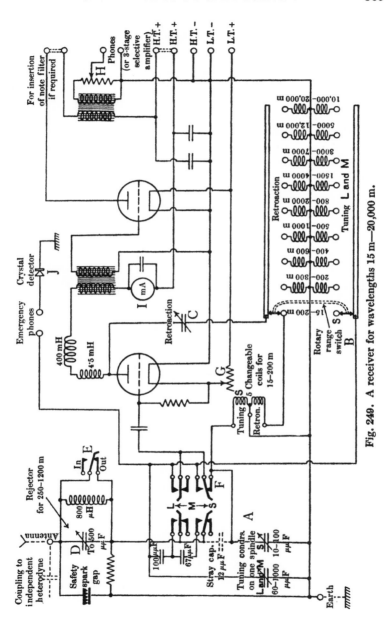

Fig. 240. A receiver for wavelengths 15 m.–20,000 m.

6. Example of triode receiver

As an example of modern construction we take the receiver shown in Plates XXIV and XXV, and in Fig. 249*. This instrument is of simple and robust type, designed for marine service; but it covers the whole range of wavelengths from 15 m to 20,000 m. It comprises only two triodes, one acting as retroactive detector, or actually oscillating as autoheterodyne, and the other as acoustic amplifier. From 20,000 m to 200 m ("L" and "M"), the wavelength is controlled wholly by switch and condenser adjustments; but from 200 m to 15 m ("S"), five changeable plug-in coils are used. These

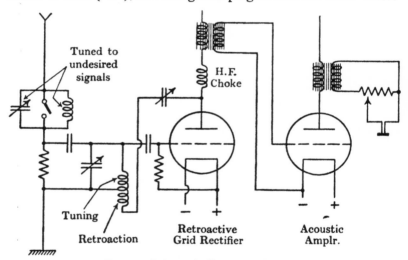

Fig. 250 Schematic diagram of Fig. 249.

are all housed inside the case, and are accessible on raising the flap-door at the top. They are visible in Plate XXV.

With the aid of the schematic diagram in Fig. 250, the complete circuit diagram (with a few approximate electrical dimensions) in Fig. 249 will be easily intelligible. The antenna is not tuned. It is coupled to the tuned grid circuit through a capacitance of $100 \mu\mu F$ (L), $40 \mu\mu F$ (M; $100 \mu\mu F$ and $67 \mu\mu F$ in series), or

* "Receiver Type 352" of the Marconi International Marine Communication Co., to whom I am indebted for the photographs and details. The same letters identifying the several components are used in the Plates and the Figure.

6. EXAMPLE OF TRIODE RECEIVER

$12\,\mu\mu\text{F}$ (S; stray capacitance across switch parts only). The rectifier is of grid type. Retroaction is provided in the manner described with reference to Fig. 236 (p. 345). For Morse signals, at short wavelengths the receiver is used in the self-oscillating condition (autoheterodyne; XII–3); but although it can be brought to self-oscillation throughout, for the longer waves a separate heterodyne oscillator is used, for the reason explained on p. 360. The milliammeter in the anode circuit of the rectifier triode serves to indicate—by decrease of current—the incidence and strength of self-oscillation.

The component parts are mounted on the front panel, which is of aluminium, and on a horizontal wood base fixed thereto. The whole assembly is accessible when drawn out of the cast aluminium containing box.

The wireless cabin of a modern liner, S.S. *Orontes*, is shown in Plate XXVI. A is a transmitter for $\lambda = 600\text{–}800$ m (I.C.W.) and for $\lambda = 200\text{–}2700$ m (C.W.). B is a transmitter working on three wavelengths between 17 m and 36 m, and C is the corresponding short-wave receiver. D is the emergency transmitter (spark) for $\lambda = 600$ m. E is a receiver, in four units placed close together, for $\lambda = 300\text{–}20{,}000$ m. F is two Morse keys. The cupboard G contains the alternator supplying power to the transmitters, and H is its starter.

CHAPTER XIII

TELEPHONY

1. The transmission unit

A method of specifying the relation between the strengths of a signal at one position and at another in a chain of apparatus (or telephone circuit or transmission system of any kind) is that of the *transmission unit**. It originated in line telephony, but has now spread to all branches of electrical signalling and to acoustics. The magnitudes of two powers W_1 and W_2 are said to differ by N transmission units when

$$\frac{W_1}{W_2} = 10^N,$$

i.e. $N = \log_{10} \frac{W_1}{W_2}.$

The two powers to be related need not be rates of passage of energies of the same physical kind. They might, for example, be the electrical power supplied to a receiving antenna and the acoustic power delivered by the associated loud-speaker; the complete receiver would show a rise or gain of N transmission units if $\frac{\text{acoustic power out}}{\text{electrical power in}} = 10^N$. Even when, strictly, the signal is not a flow of energy at all, but is a fluctuation of potential unaccompanied by current (as from grid to grid in a chain of amplifying triodes), the amplification or attenuation is expressed by the number of transmission units equal to the logarithm of the *square* of the ratio of the fluctuations. For when P.D. or current is associated with flow of energy, the power is proportional to the square of P.D. or current.

The basic transmission unit is named the *bel* (after Alexander Graham Bell, the inventor of the telephone sounder); the letter b stands for bel, and db for a tenth of this unit, the *decibel*. Thus the *gain* or *loss* between two points in a transmission system expressed in decibels is 10 times the logarithm to the base 10 of

---

* For a succinct statement, with references to published discussions of the new nomenclature by various writers, see W. H. Martin, "Decibel—The name for the transmission unit," *Bell System Tech. Journ.* Vol. 8 (1929), p. 1.

the ratio between the powers at the two points. The decibel is the customary unit to employ. It is of convenient size in that a change of one decibel in the strength of a sound of medium pitch is about the smallest change which a normal ear can perceive. It corresponds with a power ratio of $10^{0 \cdot 1} = 1 \cdot 26$, and an amplitude ratio of $\sqrt{1 \cdot 26} = 1 \cdot 12$*.

A few illustrations of the use of the decibel are appended. The references prefixed in brackets are to places in this book where an equivalent statement in alternative form may be found.

(P. 6.) The radiation from the London (Oxford Street) broadcasting station is 6·0 db higher at Hitchin than at Huntingdon.

(P. 9.) Over a telephone line with the conventional commercial limit of attenuation there is a loss of 40 db.

(P. 10.) A simple wireless telegraph receiver is estimated to have a total loss of 38 db.

(P. 247, Fig. 161.) The three-stage amplifier showed a gain of 70·0 db at 100 c/s, 70·4 db at 1,000 c/s, and 66·3 db at 10,000 c/s. A change of frequency from 1,000 c/s to 10,000 c/s produced a loss of 4·1 db.

"Good engineering practice specifies that the continuous 'tube noise'"—i.e. noise arising from spontaneous variations within the triodes—"should be more than 40 db below the signal...when using any gain up to the maximum of 84 db." (Bailey, Dean and Wintringham, *loc. cit.* p. 400.)

With reference to the short-wave receiver of Fig. 281 (p. 427): "The intermediate frequency filter has a loss of about 11 db in the pass band and therefore additional gain is supplied in the intermediate frequency amplifier to counteract this, and the overall gain of this amplifier and filter is around 85 db." (A. G. Lee, *loc. cit.* p. 309.)

2. Telephony and Telegraphy Compared

The word *telegraphy*, in its usual restricted sense, signifies the art or process of electrically signalling the letters of the alphabet by the use of a conventional code, wherein the several combinations of signs of only two or a few kinds stand for the several letters to be

---

* An alternative transmission unit, the *neper*, is used in some European countries. In nepers, $N = \frac{1}{2} \log_e \frac{W_1}{W_2}$. The same power ratio is expressed by N nepers as by $8 \cdot 7 N$ decibels.

communicated. In the Morse code the signs consist of *marks* of short and long duration called *dots* and *dashes*, and of *spaces* of short and long duration. The electrical process involved in Morse telegraphy consists, therefore, merely in transmitting and receiving an electrical disturbance during short or long epochs, with short or long intermediate epochs left blank. When the speed of signalling by the Morse code is 125 standard words per minute, the shortest epoch, viz. the duration of the dot and of the short space, is $\frac{1}{100}$ second*.

Telephony, more appropriately named, signifies the reproduction (by electrical means) at one place of sound then† occurring at a far distant place. Since the simplest analysis of sound exhibits it as a mixture of tones of every loudness and pitch (between limits)‡, the electrical processes involved are much more delicate in telephony than in telegraphy. Simultaneously or in quick succession tones covering a very great range of pitch must be transmitted and received (cf. the mere succession of dots, dashes and spaces in telegraphy); and their ever-varying strength relationships must be retained (cf. the mere presence or absence of the disturbance in telegraphy). Telephony does for the ear what the now developing art of television does for the eye.

While, therefore, the broad outlines are the same in both, the smaller details of theory and design take a more prominent place in telephony than in telegraphy. The way of the experimentalist is harder, and his need of guidance from theoretical investigations more acute. Indeed, telephony involves every point of theory and every practical difficulty (except the use of the Morse code) met

---

* The dash, and the space between letters, last three times as long. Signalling speeds in wireless telegraphic practice range about from 10 to 300 words per minute.

† Definition of the precise meaning of "then" might involve us in the theory of relativity. The playing of a gramophone record copied by electrical means from another far distant gramophone record would not be designated telephony. This process bears the same relation to telephony as does electric facsimile reproduction to television.

‡ For speech to be *intelligible* (not to sound *true*), the components having frequencies between about 500 c/s and 2000 c/s must be preserved; 200 c/s–4000 c/s would be highly satisfactory for the telephonic communication of intelligence; and in ordinary conversations the acoustic power changes from moment to moment (disregarding intervals of silence) through a range of about $10^5 : 1$ (50 decibels). For excellent reproduction of music, frequencies covering the much wider range of about 30 c/s–10,000 c/s must be faithfully dealt with; and the range of acoustic power in an orchestral performance may be of the order of $10^{12} : 1$ (120 decibels).

2. TELEPHONY AND TELEGRAPHY COMPARED

with in telegraphy, with many others added. It has been observed that "the difference in degree is not far from that between ruling a dot-and-dash line and making a dry-point etching of an autumn landscape*," and perhaps the analogy is even closer if a photograph is substituted for the etching.

As a subdivision of wireless telegraphy in the larger (and legal) sense of wireless electrical communication, wireless telephony is distinguished by the following features:

(i) The transmitter must be of the continuous wave type; or if a true continuous wave is not produced, the period of fluctuation of amplitude (or the interval between consecutive separate wave trains) must be so small as to correspond with a tone which is preferably of inaudibly high pitch, or which is at least of higher pitch than the highest important constituent frequency of the sound to be reproduced.

(ii) The amplitude or the frequency of the oscillation in the transmitting antenna must be controlled by some form of voice-sensitive microphone, instead of a Morse key.

(iii) The receiver must contain a telephone sounder† as the final indicator of the signal.

(iv) Heterodyne reception, providing a beat tone with C.W. Morse signals, must not be used‡.

The only one of these four features of telephone apparatus which is not familiar to the student of telegraphy as already outlined in this book is (ii). The same C.W. generators—arc, alternator or triode—as are used for telegraphy can be used for telephony. The telegraph receiver practically always includes a rectifier and telephone sounder; and although such sounders are designed rather to be sensitive at about 1000 cycles per second than to give good voice articulation, they are always usable for telephony. Hence any wireless telegraph receiver as used for spark signals—i.e. without a heterodyne—may be used as a wireless telephone receiver,

* A. N. Goldsmith, *Radio Telephony* (1918). This book contains a good descriptive account, very well illustrated, of the early development of wireless telephony.

† The term "telephone sounder" is here used to connote the ordinary Bell's instrument one applies to the ears (or the more powerful substitute called a *loud-speaker*). In Germany and France this is called a telephone, and in England sometimes a telephone receiver; but both these terms are also commonly used in their wider senses, and must be so used in this Chapter.

‡ But the super-heterodyne (XII–4), with its hyper-acoustic beat, may be used. Super-regenerative reception (XII–5) is also possible.

the rectified current in the sounder increasing and decreasing with the microphonic modulation at the transmitter. As regards new processes, therefore, we are concerned in this Chapter mainly with the means by which the transmitting voice is enabled to control or modulate the antenna oscillation.

3. Microphonic modulation at the transmitter

3 (a). *Methods not dependent on the use of triodes.* The ordinary carbon-granule microphone has been used in line telephony very successfully since about 1880 to produce alternating electric currents or P.Ds. corresponding with the alternating air pressures which

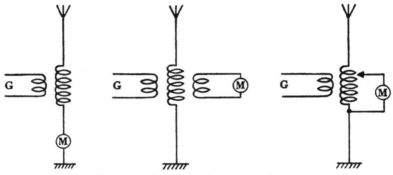

Fig. 251. Microphone in antenna circuit.

constitute sound. The same translating instrument is available for wireless telephonic transmission.

The earliest and simplest method of applying a microphone to control the amplitude of the transmitted waves was to insert it in the antenna itself, or in a circuit coupled to the antenna, as in Fig. 251, where M is the microphone and G stands for the oscillation generator of whatever kind. The resistance of the antenna circuit, and therefore the amplitude of the high-frequency current in it, fluctuates in accord with the voice. The maximum power variation in the antenna cannot then much exceed the mean power consumed as heat within the microphone; but as long as the latter does not exceed a very few watts, excellent results are obtainable in this way with the ordinary microphones of line telephony. A slight modification consists in placing the microphone in an intermediate circuit between the generator and the antenna, as in Fig. 252; but the same limitation of power obtains.

3. MICROPHONIC MODULATION AT THE TRANSMITTER 377

When the microphone is traversed by the high-frequency current, it must, of course, be situated very near to the antenna; and this may be inconvenient or impossible in some circumstances.

To deal with larger powers by this method, unusual microphone arrangements must be employed. The Poulsen arc was for long available as a practicable means of producing relatively large antenna powers, and the only bar to long-distance wireless telephony was the absence of any microphone capable

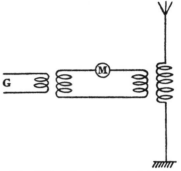

Fig. 252. Microphone in intermediate H.F. circuit.

of controlling large power with good articulation. Great efforts were made by many inventors to produce satisfactory groups of the ordinary small carbon-granule microphones; and various types of microphone were devised not employing carbon contacts at all. But so little success attended these efforts that it may be doubted whether long-distance wireless telephony would ever have become practicable if it had continued to depend on control by means of a high-power microphone.

With the advent of the high-frequency alternator, a form of relay-control became possible, at least in theory. The output from the alternator is entirely dependent on the field magnets, so that a microphonic control of the field would control the amplitude of the antenna oscillation. This method is illustrated diagrammatically in Fig. 253, where F is the field magnet of a high-frequency alternator and M is the microphone. The method is not a very practical one, for in a

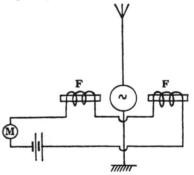

Fig. 253. Microphone in field circuit of alternator.

small alternator the power used to excite the field cannot be made very small compared with the high-frequency output of the machine. In any case, the high-frequency alternator is adapted rather to large powers and very long waves than to small powers and short waves.

Another form of relay-control is that of the magnetic amplifier (Fig. 86, p. 136) adopted by E. F. W. Alexanderson for the Morse-keying of high-power, high-frequency alternators. The principle appears to have originated with L. Kühn and would, it seems likely, have proved a very important advance if the advent of the triode had not transformed the situation*.

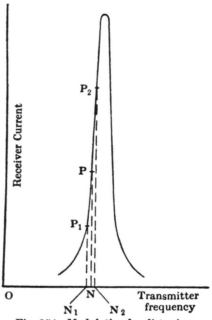

Fig. 254. Modulation by distuning.

When speech control takes the form of variation of wavelength rather than amplitude at the transmitter, the varying wavelength of the transmitter is translated into varying amplitude at the receiver as indicated in Fig. 254. The receiving circuits are adjusted to resonate at a frequency differing slightly from the no-speech trans-

* See L. Kühn, "Über ein neues radiotelephonisches System," *Jahrb. d. dr. Telegraphie*, Vol. 9 (1915), p. 502. As evidence of the revolutionary development since that date, the concluding paragraph of his paper—written, be it noted, by an inventor describing his own very considerable improvement—is worth quoting. "The problem of wireless telephony seems thus to be brought near a practical solution. But it rests with the future to decide whether and when radiotelephony, even if all important technical requirements are met, will take rank amongst other commercial methods of communication."

3. MICROPHONIC MODULATION AT THE TRANSMITTER 379

mitter frequency ON, thus bringing the representative point on the resonance curve to some position P on one of its steep sides. When speech occurs, the transmitter frequency varies between ON_1 and ON_2, and the receiver current amplitude varies between N_1P_1 and N_2P_2.

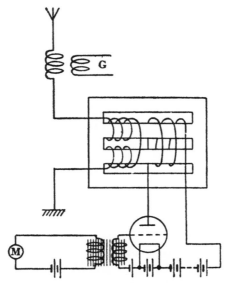

Fig. 255. Magnetic amplifier with triode.

3 (b). *Methods dependent on the use of triodes.* The use of the triode, singly or in groups, extends almost indefinitely the possibilities of controlling large antenna power by means of a feeble microphone. Firstly, its mere amplifying properties could be applied to reinforce the action of the microphone in the arrangements of Fig. 253 and Fig. 86 (p. 136). The microphone could be made to modulate the potential of the grid, where the absorption of power is small or zero, and the anode current controlled thereby be made to flow in the field winding of the alternator (Fig. 253) or the control winding of the magnetic amplifier (Fig. 86). The latter, for instance, then develops, in its simplest form, into the arrangement of Fig. 255*. If the grid is kept at a negative potential, the only current taken from the secondary of the microphone transformer is due to stray

* Modulation is effected in this way in the Lorenz transmitters at Königs Wusterhausen referred to on p. 143.

capacitance; so that the microphone imposes no limit on the power controllable by the anode. A very large triode, or a large group of triodes in parallel, can then be controlled by the ordinary small microphone of line telephony. Moreover, cascade amplification can be resorted to, any type of low-frequency amplifier being interposed between the microphone M and the triode in Fig. 255.

Secondly, the damping effect of a triode with its anode-filament connected across the whole or part of the inductance of an oscillatory circuit may be utilised. In Fig. 256, the LC circuit is damped by the presence of the triode during the whole cycle if the anode-filament slope conductance remains finite throughout the cycle; or, if not, during that part of the cycle when the anode current is neither zero nor saturated. By arranging that this proportion shall vary with the grid potential, a microphone which influences that potential is made to influence the amplitude of the oscillation in the tuned circuit LC produced by a steady high-frequency E.M.F. impressed upon it.

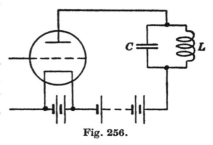

Fig. 256.

One way of accomplishing this is depicted in Fig. 257. Here the anode potential of the damping triode is $(B_a + e_r)$, where e_r is the P.D. of radiation frequency across the coil produced by a C.W. generator G. The grid potential is $(-B_g + e_m)$, where e_m is the P.D. of acoustic frequency, sensibly constant during a few of the radiation-frequency cycles, produced by the microphone M. We will suppose that the lumped battery potential $(B_a - \mu B_g)$ is zero, so that the anode absorbs power only when $(e_r + \mu e_m)$ is positive (and whenever $\overline{e_r + \mu e_m}$ is positive, provided that saturation is never reached). When the microphone is quiescent, $e_m = 0$ (A, Fig. 257), the antenna is damped by the triode during epochs T_A; when e_m has a negative value (B, Fig. 257), the damping epochs are reduced to T_B; and when e_m is positive (C, Fig. 257), they are increased to T_C. This is sufficient to show that rise (and fall) of the acoustic E.M.F. in the transformer secondary under the action of the microphone must reduce (and increase) the amplitude of high-frequency current in the antenna. To calculate the amount it would be necessary to allow for the fact that in B (Fig. 257) the e_r curve should have

3. MICROPHONIC MODULATION AT THE TRANSMITTER

a larger amplitude than in A, and in C a smaller amplitude. We have neglected this above, in order to expound as simply as possible the principle of modulation by triode damping.

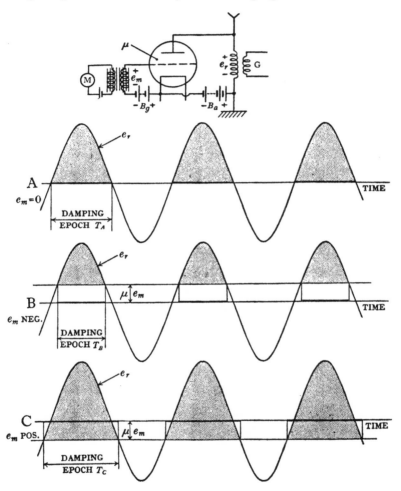

Fig. 257. Modulation by variation of damping.

An example of this type of control is contained in Fig. 258, which is the diagram of connections of the transmitter used by the Marconi Company at Chelmsford during some early telephony experiments with Madrid. There are here two stages of low-frequency amplification interposed between the microphone and the damping group of triodes. The apparatus to the right of the antenna is the un-

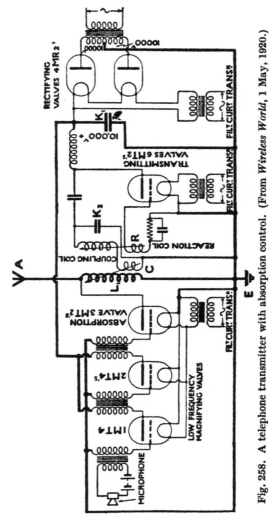

Fig. 258. A telephone transmitter with absorption control. (From *Wireless World*, 1 May, 1920.)

specified oscillation generator G of Fig 257; it is, in this case, an oscillating group of triodes fed through a pair of valves from an A.C. transformer.

In any modulating system where the microphone does not carry

3. MICROPHONIC MODULATION AT THE TRANSMITTER

high-frequency current, the association of a triode amplifier with the microphone not only removes the earlier pressing need for microphones capable of dealing with higher power than are the carbon-granule microphones of line telephony; it goes further, and makes possible the use of other types of microphone capable only of even smaller output, e.g. the electromagnetic microphone and the condenser microphone. A simple example of the former is an ordinary telephone sounder reversed; sound impinging on the diaphragm produces in the magnet coils corresponding E.M.Fs. capable of stimulating the grid of an amplifier. Reversed telephone sounders of the moving coil type* are sometimes used as microphones in broadcasting studios.

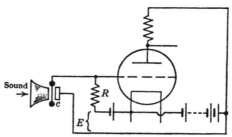

Fig. 259. Condenser microphone.

A condenser microphone consists, ordinarily, of a two-plate condenser with plates very close together, one being in the form of a stretched diaphragm. Impinging sound waves cause the diaphragm to vibrate and the capacitance to vary accordingly. Fig. 259 shows how a condenser microphone may be caused to excite the grid of an amplifying triode. Under the influence of sound of frequency $\frac{\omega}{2\pi}$, let the capacitance of the condenser be

$$c = C(1 + m \sin \omega t),$$

where m is a small number much less than unity, a measure of the sound intensity. If the time constant CR is very large compared with the period $\frac{2\pi}{\omega}$, the charge in the condenser is nearly constant and the P.D. across R is nearly

$$\frac{E}{1 + m \sin \omega t} \doteqdot E(1 - m \sin \omega t)\dagger.$$

* See Subsection 6 (b) of this Chapter.
† Condenser microphones of a form developed by E. C. Wente in the Bell Telephone Laboratories for acoustic measurements develop about 400 μV per

While triodes provide a solution—a very complete one—of the problem of modulating the oscillation produced by any form of C.W. generator, they also constitute the most convenient form of generator; and with a triode oscillator as generator, special methods of modulation are applicable. The amplitude of the oscillation is limited by the representative point on the lumped characteristic running off the steep part as the anode current approaches zero* (see Fig. 180,

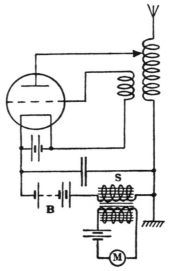

Fig. 260. Microphonic control of mean anode potential.

p. 274). A reduction (or increase) of the static lumped potential ($P_0 = B_a - \mu B_g$, in Figs. 179 and 180) therefore effects a reduction (or increase) in the amplitude of oscillation. The alteration may be made in either B_a or B_g with like effect, provided that the values are not such as to invalidate the lumped characteristic (e.g. by causing the grid potential during the cycle to rise too high).

In Fig. 260 the amplitude of steady oscillation would rise and fall with the P.D. of the battery B; it will rise and fall, at an acoustic rate, under the action of an E.M.F. produced in S by sound

dyne/cm² of pressure on the diaphragm over the whole acoustic range—varying, in a particular microphone, from 275 at very low frequencies to 500 at 8 kc/s. See I. B. Crandall, "Sounds of speech," *Bell System Technical Journal*, Vol. IV (1925), p. 586.

* And/or at the top as it approaches saturation, if there is limited filament emission. In a high-efficiency régime, absorption of power by the grid also may exercise a minor influence in limiting growth.

3. MICROPHONIC MODULATION AT THE TRANSMITTER

falling upon the microphone M. The condenser shunting B and S carries the high-frequency component of anode current, while constituting a negligible shunt for the steady and acoustically varying components.

The arrangement of Fig. 260 is, of course, just as limited with respect to power controllable by a feeble microphone as are those of Fig. 251. But the power controlled by the same feeble microphone can be increased without theoretical limit in either or both of two ways. Firstly, a low-frequency amplifier can be inserted between M and S, precisely as in Figs. 255 and 258. In a simple form, this

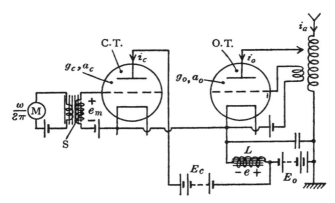

Fig. 261. Anode-choke modulation.

arrangement is shown in Fig. 261. S is now attached to the grid of the control triode C.T., its place in the anode supply circuit of the oscillating triode O.T. being taken by a choke L through which passes the anode current of both triodes. A falling anode current in C.T. under the action of the microphone produces a rise in the P.D. ($E_0 + e$) supplying the oscillator O.T.; and vice versa. This method of modulation often goes by the name of *choke control*. With elaborations it is probably the method most commonly used at the present time; it can be made to give a very nearly linear response for a large depth of modulation* over the whole acoustic range of frequencies. A quantitative analysis of the method is given in the next Subsection.

The second way in which the microphone in Fig. 260, while itself

---

* See Subsection 4(a) of this Chapter.

dealing with very small power, could be made to modulate large power in the antenna is by the insertion of a high-frequency amplifier between the weak speech-controlled oscillator and the antenna, as indicated in Fig. 262. The amplifier may be in several stages and the amplification very large, so that the power directly controlled by the modulating choke may be a very small fraction of the antenna power. In practice the microphone would be relieved of all load by the insertion also of a low-frequency amplifier as in Fig. 261.

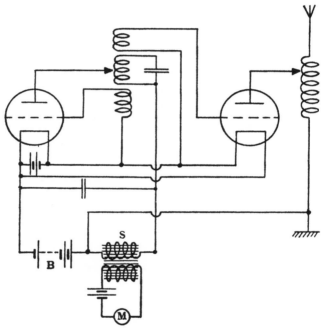

Fig. 262. Amplification after modulation.

Modulation at a low-power level, followed by amplification of the modulated high-frequency oscillation, is a great easement of the difficulties met by the designer of a high-power telephone transmitter in his efforts to obtain at all relevant acoustic frequencies a linear relationship between amplitude of oscillation and microphone E.M.F. In order to modulate faithfully, the choke L with its shunting condenser in Fig. 261 should present an indefinitely large reactance at all acou-

3. MICROPHONIC MODULATION AT THE TRANSMITTER

stic frequencies and amplitudes—a condition which the designer can meet well enough if the choke has a very small volt-ampere capacity, but which becomes progressively harder to meet as the size is increased.

3 (c). *Analysis of anode-choke control.* Let the triode O.T. in Fig. 261 be oscillating in the linear régime studied in IX–2. Let the P.D. across S be the sinoidal alternating quantity e_m of frequency $\frac{\omega}{2\pi}$, viz. the frequency of a tone at the microphone M; and let e be the P.D. (of ω-frequency) which arises across L. Let i_c be the ω-frequency component of anode current in C.T.; and let i_o, i_a be the ω-frequency components of the acoustically fluctuating amplitudes of high-frequency current in O.T. and in the antenna respectively. Let a_c, a_o be the anode-filament conductances of the two triodes.

Then $\frac{\partial i_c}{\partial e} = a_c$. And since there is the linear relation of Fig. 180 (p. 274) between the amplitude of high-frequency oscillation and the high-frequency mean of lumped potential in O.T. (i.e., in Fig. 180, between I_a and P_0),

$$i_o = a_o e \text{ and } i_a = ke,$$

where k is a constant of the oscillating triode circuit.

The quantities e_m, e, i_c, i_o and i_a are alternating quantities of frequency $\frac{\omega}{2\pi}$. They may be treated by the $j \equiv \sqrt{-1}$ notation, where the operator $\frac{d}{dt}$ is placed by $j\omega$.

$$e = -j\omega L (i_c + i_o);$$
$$i_c = a_c e + g_c e_m,$$

where g_c is the mutual conductance of the triode C.T.;

$$i_o = a_o e;$$
$$\therefore \quad e = -j\omega L (a_c e + g_c e_m + a_o e);$$
$$\text{i.e.} \quad e(1 + j\omega L . \overline{a_c + a_o}) = -j\omega L g_c e_m;$$
$$\therefore \quad i_a = ke = k . \frac{-j\omega L g_c}{1 + j\omega L . \overline{a_c + a_o}} e_m.$$

That is, $\dfrac{\text{acoustic amplitude of antenna current}}{\text{amplitude of acoustic P.D. produced by microphone}}$

$$= \frac{k\omega L g_c}{\sqrt{1 + \omega^2 L^2 \cdot \overline{a_c + a_o}^2}} = \frac{k \cdot \dfrac{g_c}{a_c + a_o}}{\sqrt{1 + \dfrac{1}{\omega^2 L^2 \cdot \overline{a_c + a_o}^2}}}.$$

For faithful modulation this ratio must be sensibly independent of ω; that is,

$$2\omega^2 L^2 \cdot \overline{a_c + a_o}^2 \gg 1$$

at the lowest acoustic frequency to be transmitted. This determines the lowest permissible inductance L of the choke*. With this sufficiently large value of L, the ratio becomes sensibly

$$\frac{kg_c}{a_c + a_o} = \frac{k\mu_c}{1 + \dfrac{a_o}{a_c}},$$

where μ_c is the amplification factor of the control triode. With a given oscillating triode (a_o), the most sensitive modulation is obtained by using a control triode (μ_c, a_c) making this ratio as large as possible.

When L is adequate for faithful transmission,

$$e \fallingdotseq -\frac{g_c}{a_c + a_o} e_m;$$

$$i_c = e_m \cdot g_c \left(1 - \frac{a_c}{a_c + a_o}\right) = e_m \cdot \frac{g_c a_o}{a_c + a_o};$$

$$i_o = -e_m \cdot \frac{g_c a_o}{a_c + a_o}.$$

Thus i_c and i_o are equal anti-phased currents, and the total anode current supplied to the two triodes (Fig. 261) is unaffected by the microphone, which merely diverts current from one triode to the other.

Since linear performance is desired in both triodes, it would not be an economical arrangement to use similar triodes for C.T. and O.T., with a common high-tension source. For in O.T. (but not in C.T.) there is a high-frequency oscillation superposed on the acoustic

* The bigger L is the better, provided that its reactance at the highest acoustic frequency to be transmitted remains far less than the reactance of the shunting condenser at that frequency.

fluctuations we have been calculating; the instantaneous value of anode current in O.T. therefore fluctuates through a wider range than does the anode current in C.T. With the aid of the above analysis relating modulator and oscillator, in conjunction with the analysis of the oscillator given in IX–2, for any given oscillator and available high-tension source a suitable control triode C.T. can be chosen, and appropriate values for its grid and anode batteries can be determined.

4. The modulated oscillation

4 (a). *As first produced.* A generator oscillating at the high (radiation) frequency $\frac{p}{2\pi}$, emitting therefore radiation of the form

$$A \sin pt,$$

is a C.W. telegraph transmitter when it is switched on and off under the control of the Morse key. If the amplitude A, instead of being constant (when the key is down), contains a variable portion which copies the changes of air-pressure on a microphone in the presence of sound, the apparatus becomes a telephone transmitter. If the sound is a pure tone of frequency $\frac{\omega}{2\pi}$, the high-frequency currents and the radiation are of the form

$$(A + B \sin \omega t) \sin pt,$$

where B measures the amplitude of the sound. The graph of this quantity as a function of time is shown in Fig. 263. The oscillation is a tone-modulated oscillation, and the ratio $\frac{B}{A}$ is called the *modulation ratio* or *depth of modulation*.

A P.D. of this form applied to a detector giving a rectified output always proportional to the high-frequency amplitude—as detectors subjected to strong signals actually do give*—produces an output of the form of the envelope of the high-frequency oscillation, i.e. the dotted curve in Fig. 263. If $\omega \ll p$ and $B < A$, this is sensibly the curve

$$A + B \sin \omega t.$$

The effect on a telephone sounder is then of the form $B \sin \omega t$, a true copy of the acoustic disturbance at the microphone.

* See VI–5 and 6, and XI–6 and 7.

XIII. TELEPHONY

The function

$$(A + B \sin \omega t) \sin pt,$$

while easy to apprehend pictorially as in Fig. 263, contains a product of two sinoidal oscillations. Its mathematical treatment is facilitated by rewriting it in the alternative form of the sum of a set of sinoidal oscillations. Now

$$(A + B \sin \omega t) \sin pt \equiv A \sin pt + \frac{B}{2} \cos (p - \omega) t - \frac{B}{2} \cos (p + \omega) t.$$

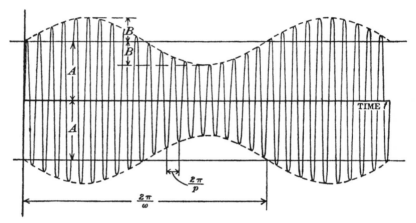

Fig. 263. The modulated carrier.

The tone-modulated wave therefore is[*] the sum of three sinoidal high-frequency oscillations with a special relationship between the three frequencies and a special relationship between the amplitudes of two of these oscillations.

$A \sin pt$ is called the *carrier-wave*;

$\dfrac{B}{2} \cos (p - \omega) t$ is called the *lower side-wave*;

$\dfrac{B}{2} \cos (p + \omega) t$ is called the *upper side-wave*.

[*] There is no room to question this *is*; it is a mathematical identity. A discussion, beginning in *Nature*, 18 Jan. 1930, as to the validity of what is wrongly termed "the wave-band *theory* of wireless transmission," in which even learned persons have taken part, is therefore futile. Those to whom mathematical transformations are not convincing should satisfy themselves, by experiments with filters and a heterodyne oscillator, that in a tone-modulated wave sinoidal oscillations of these three frequencies are indeed present.

4. THE MODULATED OSCILLATION

These are indicated diagrammatically in Fig. 264, which is another way of portraying the same conditions as are shown in Fig. 263.

The tone being transmitted is seen to affect the wave $A \sin pt$ existing during silence by causing the addition to it of the lower and upper side-waves.

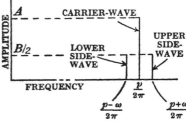

Fig. 264. Carrier-wave and side-waves.

In speech or music, the sound is a complex and changing disturbance containing components of many frequencies. In place of $B \sin \omega t$, we must write

$$B_1 \sin \omega_1 t + B_2 \sin \omega_2 t + B_3 \sin \omega_3 t + \ldots *$$

Fig. 265 portrays, after the fashion of Fig. 264, the conditions if all the components $B \sin \omega t$ have frequencies lying between (for example) 200 c/s and 4000 c/s†, and amplitudes bounded by the curves shown. The areas marked "lower side-band" and "upper side-band" would not be filled at any one instant with vertical lines standing for side-waves; but as the sound changes they may at any moment be occupied by vertical lines in any positions and of any heights within the areas. The tops of the side-band areas are shown

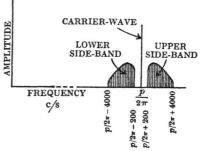

Fig. 265. Side-bands.

---

* The B's and the ω's are treated as though they are constants, or at least retain their values during many of their periods. Actually, of course, the very essence of both speech and music is that they are not continued periodic disturbances. It is true that a vowel sound, or the sound from a pipe or bowed string, owes its character not to its beginning or its end but to what lies between; but a consonantal sound or (less completely) the sound from a percussion instrument, depends for its peculiar psychological effect upon transients, the beginning or end of a quasi-periodic disturbance. Nevertheless it is usual in the theory and practical calculations of telephony—line or wireless—to treat the sound to be transmitted as though it were a mixture of sustained pure tones. It seems to be still open to discussion how far the excellence of a system for transmitting periodic sounds covering a wide range of frequencies is a complete criterion of its excellence for transmitting transients. Owners of good broadcast receivers will reflect on the unrealistic clapping they are accustomed to hear after an excellent orchestral reproduction.

† We will adopt these figures for our numerical illustrations.

sloping downwards as the separation from the carrier-wave increases because in most sounds the amplitudes of the component tones tend to be smaller as the pitch is higher.

For good reproduction the transmitting and receiving antennas, and the other high-frequency circuits carrying the modulated oscillation at the transmitter and at the receiver, must respond substantially uniformly over the range of frequencies $\left(\frac{p}{2\pi} - 4000\right)$ c/s to $\left(\frac{p}{2\pi} + 4000\right)$ c/s. Their selectivity must therefore not be too great; and the longer the wavelength the flatter must be the high-frequency resonance curves if the side-waves corresponding with the

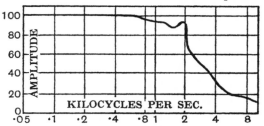

Fig. 266. A case of failure to transmit the higher acoustic frequencies.

higher acoustic frequencies are not to be too much repressed. That the partial cut-off of the higher acoustic frequencies is a real difficulty, even at moderately short wavelengths, is shown by the response curve in Fig. 266*. This refers to the high-frequency transmitting circuits at the broadcasting station "5XX," wavelength about 1600 m. At 4000 c/s the amplitude has fallen to 30 % of its correct value, despite the fact that at this station modulation was performed at high power so that high-frequency amplification of the modulated oscillation was not employed. The effect was due mainly to the selectivity of the rather lowly damped antenna.

It is the practice in those telephone systems where simplicity of apparatus and operating technique is important, and even where it is not if the carrier-wave is not long (e.g. in broadcasting on both accounts), to employ the complete modulated oscillation

$$A \sin pt + \frac{B}{2} \cos (p - \omega) t - \frac{B}{2} \cos (p + \omega) t :$$

* Taken, by permission of the Institution, from Fig. 7 in P. P. Eckersley's "Design and distribution of wireless broadcasting stations for a national service," *Journ. I.E.E.* Vol. 66 (1928), p. 501.

4. THE MODULATED OSCILLATION

the carrier-wave and both side-bands (Fig. 265) are, as far as can be contrived, all transmitted and received. But it is obvious on the one hand that this cannot in practice be fully achieved, since an oscillatory circuit responds differently to E.M.Fs. of different frequencies; and on the other hand that it is not theoretically necessary. Since the carrier-wave bears no imprint from the microphone, it might, as far as reception is concerned, as well be manufactured at the receiver as at the transmitter, with obvious economic advantage*. Deformations, intentional and unintentional, of the modulated oscillation before it is reproduced as a P.D. across the receiving rectifier are the subject of the following Subsections.

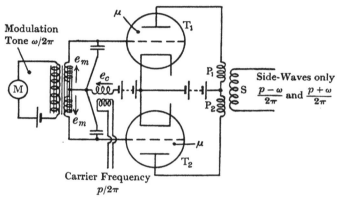

Fig. 267. Suppression of carrier-wave.

4 (b). *Suppression of the carrier-wave.* The suppression of the carrier-wave from the modulated oscillation might ideally be effected by a process of filtering applied between the modulated oscillator and the antenna; but if the frequency-band which may be suppressed is only 400 c/s wide (the band lying between the two side-bands, Fig. 265), this would be impracticable except with waves too long for use in telephony.

The principle of the method† actually used to produce the side-bands without the carrier-wave is explained with reference to Fig. 267. The two triodes T_1 and T_2 are equal, and the circuit is

* Whatever his views on tariffs, no engineer would care to meet a demand for steady alternating current on one side of the Atlantic by generating it on the other.

† The circuit device—not the use of the triodes—is the same as that employed for separating the real-circuit signals from the phantom-circuit signals in line telephony and telegraphy with superimposed circuits.

symmetrical with respect to them. The grid and anode batteries are such that the lumped battery potential is about zero. Since the high-frequency E.M.F. e_c is continuously present, in the absence of a modulation tone, the two triodes are together alternately alive and dead* according as e_c is positive or negative. The E.M.Fs.

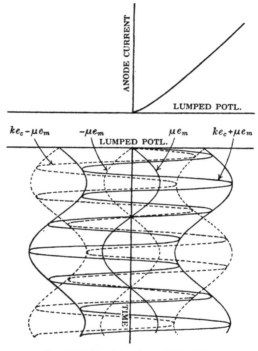

Fig. 268. Explanation of Fig. 267.

induced in the secondary S from the two primaries P_1, P_2 are equal and opposite: the output to the antenna is zero.

The alternating component of lumped potential of each triode caused by e_c is ke_c, where k is some constant depending on the dimensions of the triode and the anode circuit. When the microphone is excited the lumped potentials are

$$ke_c + \mu e_m \text{ in } T_1,$$
and $$ke_c - \mu e_m \text{ in } T_2.$$

These are shown by the full and dotted curves in Fig. 268. While

* For simplicity of explanation we take them as fully alive or fully dead only—i.e. we neglect the curvature of the foot of the triode characteristic.

4. THE MODULATED OSCILLATION

e_m is positive, T_1 is longer alive than dead, and T_2 is longer dead than alive. The action of P_1 on S overcomes the action of P_2 on S and produces a current in the antenna of frequency $\frac{p}{2\pi}$ and amplitude more or less proportional to e_m. The antenna oscillation is therefore of the form

$$B \sin \omega t . \sin pt = \frac{B}{2} \cos (p - \omega) t - \frac{B}{2} \cos (p + \omega) t;$$

that is, the side-waves are both present, but the carrier-wave is not.

The device is a push-pull combination of two rectifiers* excited from the acoustic source, with a central element (the coil marked e_c) added for introducing symmetrically an oscillation of the carrier frequency. It is commonly called a *balanced modulator*.

Re-introduction of the carrier-wave at the receiver consists theoretically in the mere addition of the term $A \sin pt$; or, rather, of a term at the receiver corresponding with the missing term $A \sin pt$ at the transmitter. Before assessing the practical difficulties of this restoration, it is necessary to consider the effects of the phase displacements which occur between the components of the modulated oscillation in the processes undergone between its generation in the transmitter and its reappearance at the rectifier in the receiver†.

4(c). *Propagation between transmitter and receiver.* A transmitted wave $M_1 \sin 2\pi nt$ after propagation across a length l of free space appears as

$$M_2 \sin \left(2\pi nt - \frac{l}{\lambda} . 2\pi\right) = M_2 \sin 2\pi n \left(t - \frac{l}{c}\right),$$

where λ is the wavelength and $c = n\lambda$ is the "velocity of light." Since $\frac{l}{c}$ is independent of n, the changes of phase of the three sinoidal waves constituting the modulated wave are such as would be introduced into the expression

$$A \sin pt + \frac{B}{2} \sin (p - \omega) t - \frac{B}{2} \sin (p + \omega) t$$

---

* Cf. the "push-pull" amplifier disposition, p. 249.

† This recondite subject cannot here be treated with any attempt at completeness, but is sketched in the following Subsections. An excellent and suggestive account of the scope of the full enquiry has been given by R. V. L. Hartley, "Relations of carrier and side-bands in radio transmission," *Proc. I.R.E.* Vol. 11 (1923), p. 34.

by a mere change in the arbitrary instant from which time is measured. They therefore have no physical significance.

The amplitude M_2 is less than M_1; but since the frequencies of the three component waves are nearly equal (assuming $p \gg \omega$), their attenuations are nearly equal and the amplitudes at the receiver are nearly in the same ratio as at the transmitter.

Where transmission is not through free space, but depends on Heaviside-layer effects, serious phase and amplitude distortions may occur. Experience with long-distance, short-wave telephony strongly suggests that such distortions are sometimes very serious.

4 (d). *Change of phase of one side-wave with respect to the other.* This is a change which must occur to some extent in any single oscillatory circuit energised by both side-waves. In practical cases it is sometimes quite large.

It will aid clarity if henceforth we write for the modulated oscillation

$$m = C \sin pt + \frac{S}{2} \cos (p - \omega) t - \frac{rS}{2} \cos (p + \omega) t \ldots(1),$$

where C stands for the amplitude of the carrier-wave (formerly A),

„ S „ „ lower side-wave (formerly B),

„ rS „ „ upper side-wave (formerly B).

m is a high-frequency oscillation whose envelope expresses the acoustic output given by a linear-response detector. When $r = 1$, this expression (1) is that of the undistorted modulated wave

$$m = (C + S \ \omega t) \sin pt.$$

Let us change the arbitrary instant of zero time by τ. Writing $(t + \tau)$ in place of t in (1), we get

$$m = C \sin (pt + p\tau) + \frac{S}{2} \cos (\overline{p - \omega} . t + \overline{p - \omega} . \tau)$$

$$- \frac{rS}{2} \cos (\overline{p + \omega} . t + \overline{p + \omega} . \tau) \ldots(2).$$

Since there is no physically necessary relation between p and ω, τ may be chosen to make $p\tau$ an integral multiple, (say k_1) of 2π, and also $\overline{p - \omega} . \tau$ an integral multiple (say k_2) of 2π, without restricting our freedom to make $\overline{p + \omega} . \tau$ exceed an integral multiple

4. THE MODULATED OSCILLATION

(say k_3) of 2π by any specified angle ϕ. On doing so, (2) becomes

$$m = C \sin(pt + k_1 \cdot 2\pi)$$
$$+ \frac{S}{2} \cos\overline{(p - \omega) \cdot t + k_2 \cdot 2\pi} - \frac{rS}{2} \cos\overline{(p + \omega) \cdot t + k_3 \cdot 2\pi + \phi}$$

$$= C \sin pt + \frac{S}{2} \cos(p - \omega)t - \frac{rS}{2} \cos\overline{(p + \omega) \cdot t + \phi} \quad \ldots\ldots(3).$$

Since ϕ appears in (3) as the result of a mere change of the arbitrary instant from which time is measured, whatever its value it can have no physical significance*. It is to be noticed that this is true whether the amplitudes of the side-waves are equal or not.

4 (e). *Change of phase of carrier-wave with respect to side-waves.* This is a change which may be caused by frequency discrimination in an oscillatory circuit, and (practically speaking) must occur when the carrier-wave is suppressed at the transmitter and re-introduced at the receiver from a local oscillator. Although approximate synchronism is easily achieved, it is not possible in practice to hold the phase of the re-introduced carrier consistently coincident with the phase of the carrier which would have arrived from the transmitter; indeed, slight departure from synchronism is equivalent to a continued wandering between syn-phase and anti-phase†.

Let the phase displacement of the carrier be θ. Equation (1) for the modulated oscillation then becomes

$$m = C \sin(pt + \theta) + \frac{S}{2} \cos(p - \omega)t - \frac{rS}{2} \cos(p + \omega)t \quad \ldots\ldots(4),$$

$$= \left[C \cos\theta + \frac{S}{2}(1 + r) \sin\omega t \right] \sin pt$$
$$+ \left[C \sin\theta + \frac{S}{2}(1 - r) \cos\omega t \right] \cos pt,$$

$$= \left\{ C^2 + \frac{S^2}{4}(1 + r^2) - \frac{S^2}{2} r \cos 2\omega t + \right.$$
$$\left. CS \sqrt{(1 + r^2) + 2r \cos 2\theta} \cdot \sin(\omega t + \beta) \right\}^{\frac{1}{2}} \sin(pt + \alpha) \quad \ldots(5),$$

where
$$\tan \alpha \equiv \frac{C \sin\theta + \dfrac{S}{2}(1 - r) \cos\omega t}{C \cos\theta + \dfrac{S}{2}(1 + r) \sin\omega t},$$

---

* I am indebted to Mr R. R. M. Mallock for suggesting this argument.
† At $\lambda = 300$ m, asynchronism of 1 in 10^6 causes the phase angle to change through $180°$ twice a second.

and
$$\tan\beta \equiv \frac{1-r}{1+r}\tan\theta.$$

If the side-waves are both present and equal, i.e. if $r=1$, (5) becomes

$$m = \sqrt{C^2 + \frac{S^2}{2} - \frac{S^2}{2}\cos 2\omega t + 2CS\cos\theta\sin\omega t} \cdot \sin(pt+\alpha) \quad \ldots(6),$$

where
$$\tan\alpha \equiv \frac{C\sin\theta}{C\cos\theta + S\sin\omega t}.$$

We may distinguish five particular cases, (i)–(v), of equation (6).

(i) If $\theta \neq 0$, (6) shows that an infinite series of harmonics of $\frac{\omega}{2\pi}$ is produced.

(ii) If $\theta = 0$, (6) reduces to the original modulated oscillation
$$m = (C + S\sin\omega t)\sin pt,$$
giving the correct acoustic signal $S\sin\omega t$.

(iii) If $C^2 \gg S^2$ and $\theta \neq 0$—a condition realised without difficulty when the carrier is re-introduced by an oscillator at the receiver—(6) reduces to
$$m \doteqdot (C + S\cos\theta\sin\omega t)\sin(pt+\alpha) \ldots\ldots\ldots\ldots(7),$$
giving the acoustic signal
$$S\cos\theta\sin\omega t \quad\ldots\ldots\ldots\ldots\ldots\ldots\ldots(8).$$

(iv) If $C^2 \gg S^2$ and θ is constant, (8) is a signal of the correct constant frequency $\frac{\omega}{2\pi}$, but of amplitude reduced from S to $S\cos\theta$.

(v) If $C^2 \gg S^2$ and θ is not constant, but is changing slowly—e.g. owing to the inexact synchronism of the re-introduced carrier—(8) can be regarded as a single note of the correct frequency but of amplitude slowly and continuously varying between S and zero, or as the sum of two beating tones of a single constant amplitude and of constant frequencies displaced equally above and below the correct frequency $\frac{\omega}{2\pi}$.

The practical precept conveyed by this examination is that if both side-bands are transmitted, the carrier must not be suppressed at the transmitter and re-introduced at the receiver, unless a continuous watch can be kept on the performance of the receiver and a phase correction of the re-introduced carrier can be effected empirically

4. THE MODULATED OSCILLATION

from moment to moment. This might be feasible with long waves; but, as we shall see in the next Subsection, there is a better way of avoiding the ill-effect of changing carrier phase.

4(f). *Suppression of one side-wave.* The effect of any reduction of the amplitude of one side-wave—say the upper—can be calculated from (5), by assigning to r the appropriate value between 1 and 0. The formula is, however, too complicated for any general conclusions to be drawn.

The case of the complete suppression of one of the side-bands is given by putting $r = 0$ in (5), which then reduces to

$$m = \sqrt{C^2 + \frac{S^2}{4} + CS \sin(\omega t + \theta)} \cdot \sin(pt + \alpha) \quad \ldots\ldots(9),$$

where
$$\tan \alpha \equiv \frac{C \sin \theta + \frac{S}{2} \cos \omega t}{C \cos \theta + \frac{S}{2} \sin \omega t}.$$

We may distinguish the same five particular cases (i)–(v) of equation (9) as we did in the last Subsection with reference to equation (6). As before let θ be a phase displacement of the carrier.

(i) If $\theta \neq 0$, (9) shows that an infinite series of harmonics of $\frac{\omega}{2\pi}$ is produced.

(ii) If $\theta = 0$, an infinite series of harmonics of $\frac{\omega}{2\pi}$ is still produced. (Contrast the undistorted performance when both side-waves were present.) But if $C^2 \gg S^2$, (9) reduces to

$$m \doteq \left(C + \frac{S}{2} \sin \omega t \right) \sin(pt + \alpha) \quad \ldots\ldots\ldots(10),$$

giving the correct acoustic signal (of half the amplitude existing when both side-waves were present).

(iii) If $C^2 \gg S^2$ and $\theta \neq 0$, (9) reduces to

$$m \doteq \left[C + \frac{S}{2} \sin(\omega t + \theta) \right] \sin(pt + \alpha) \quad \ldots\ldots(11),$$

giving the acoustic signal

$$\frac{S}{2} \sin(\omega t + \theta) \quad \ldots\ldots\ldots\ldots\ldots(12).$$

(iv) If $C^2 \gg S^2$ and θ is constant, (12) is a signal of the correct frequency and of constant amplitude $\dfrac{S}{2}$. The carrier phase displacement θ has now produced no weakening of the acoustic signal.

(v) If $C^2 \gg S^2$ and θ is not constant, but is changing at the constant rate ω', (12) still has constant amplitude and frequency, but the frequency is $\dfrac{\omega + \omega'}{2\pi}$ instead of the correct frequency $\dfrac{\omega}{2\pi}$. (Contrast the periodic passage of the acoustic amplitude through zero when both side-waves are present.) Since the change of frequency $\dfrac{\omega'}{2\pi}$ is the same for all components $\dfrac{\omega_1}{2\pi}, \dfrac{\omega_2}{2\pi}, \ldots$ of (say) a musical sound, the occurrence of ω' (i.e. of a changing θ) produces discord, and does not merely change the key as when a gramophone record is played too fast*.

The practical precept now apparent is that when one of the side-waves is suppressed, the carrier-wave may be suppressed at the transmitter and re-introduced at the receiver without prejudicing good acoustic reproduction, provided that the re-introduced carrier is in adequate strength $(C^2 \gg S^2)$ and that its frequency, while not necessarily exactly equal to that of the suppressed carrier, differs only slightly therefrom $(\omega' \ll \omega)$.

4 (g). *Single-side-band transmitter.* It has been shown in the preceding Subsections that the original sound actuating the microphone at the transmitter can be reproduced at the receiver although both carrier-wave and one of the side-bands are not propagated between the stations; and some of the conditions for faithful reproduction have been examined. The chief advantages of single-side-band transmission over the simpler process in which the complete modulated wave is radiated (e.g. in broadcasting) are (i) and (ii) as follows. For the sake of concrete illustration we will suppose that the carrier-frequency is 60 kc/s ($\lambda = 5000$ m) and that the acoustic frequencies to be reproduced range from 200 c/s to 4000 c/s†.

---

* Experience has shown that $\omega' = 20$ c/s is large enough to cause appreciable decrease of speech intelligibility, and that the effect on speech or music of as little as 2 c/s can be detected. See Bailey, Dean and Wintringham, "Transatlantic radio telephony," *Proc. I.R.E.* Vol. 16 (1928), p. 1682.

† The long-wave transatlantic telephone service between Rugby and New York is conducted with a carrier-wave of about 60 kc/s, and is designed to deal faithfully with an acoustic range of about 400 c/s to 2600 c/s. The power supplied to the antenna reaches some 150 kW.

4. THE MODULATED OSCILLATION

(i) In the complete modulated oscillation most of the power pertains to the carrier-wave. Its suppression at the transmitter implies economy in coal and size of apparatus. The power in the transmitter is proportional to the sum of the squares of the amplitudes of the component sinoidal oscillations. Hence for the complete modulated oscillation

power in carrier-wave $= C^2$ (see (1), p. 396);

power in two side-waves $= 2 \times \left(\dfrac{S}{2}\right)^2 = \dfrac{S^2}{2}$;

total power $= C^2 + \dfrac{S^2}{2}$.

And with single-side-band transmission,

total power $= \left(\dfrac{S'}{2}\right)^2$,

where $\dfrac{S'}{2}$ is the amplitude of the single side-wave giving reception equal to that got with carrier A and both side-waves $\dfrac{S}{2}$; i.e. $S' = 2S$ (see (ii) on p. 399). Hence the ratio of transmitted powers is

$$\frac{S^2}{C^2 + \dfrac{S^2}{2}} = \frac{2\left(\dfrac{S}{C}\right)^2}{2 + \left(\dfrac{S}{C}\right)^2}.$$

Even with the largest possible depth of modulation, $\dfrac{S}{C} = 1$, this ratio is $\tfrac{2}{3}$; and since, during the course of speech or music, the mean square of the depth of modulation is very much less than the maximum*, the saving in power is much more than is represented by the ratio $\tfrac{2}{3}$. Thus when the sound being transmitted is such that $\dfrac{S}{C} = 10\%$, the ratio of powers is only about 1%.

(ii) The band of frequencies occupied by the single-side-band transmitter is narrower than the complete modulated oscillation. Thus (see Fig. 265, p. 391) the occupied range in the latter is

56·0–59·8, and 60·0, and 60·2–64·0 kc/s,

i.e. practically the whole of 56–64 kc/s. With single-side-band transmission this is reduced to (say) 56–59·8 kc/s, i.e. practically halved.

* See footnote on p. 374.

The advantages of the restriction of the frequency band are two. First, there is room for more channels of communication in one geographical area without mutual interference—a point of the greatest practical importance, so severe is the congestion of wireless telegraph and telephone stations. Secondly, more selective high-frequency circuits are usable—are, indeed, necessary to obtain the first advantage. Now, except with very short waves, the radiation resistances of antennas of practicable height are low, and it is therefore not feasible to provide an efficient antenna of high damping. Hence it is an important advantage to raise the permissible selectivity

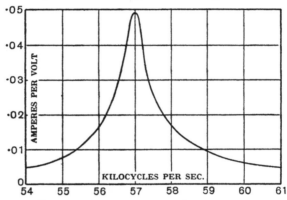

Fig. 269. Resonance curve of an antenna.

of the antenna. Let us suppose, for example, that the acoustic requirements permit of a frequency discrimination by the antenna not more than (say) 2 : 1 in amplitude. If the antenna is required to pass the complete modulated wave it would be tuned to resonate to the carrier frequency, 60 kc/s, and must give a response not less than half the resonant response at frequencies of 56 and 64 kc/s, i.e. when distuned by 7 %. Its decrement must therefore be not less than 0·24 (see Fig. 36, p. 72). But to deal with the lower side-band only, the antenna might be tuned to about 58 kc/s, and need possess only half the former decrement, viz. 0·12. The difficulty of providing economically a system tuning sufficiently flatly is in any case a severe one; it is reduced by the suppression of one of the side bands*.

* The difficulty, and the attempts to meet it, at the New York transmitter of the long-wave transatlantic telephone are described by A. A. Oswald and J. C. Schelleng, "Power amplifiers in transatlantic radio telephony," *Proc. I.R.E.* Vol. 13 (1925), p. 313. Fig. 269 (copied, by permission, from Fig. 20 of that paper) shows the resonance curve of the antenna alone.

4. THE MODULATED OSCILLATION

To effect the suppression of one of the side bands by a direct filtering process, the filter* would be required to discriminate very sharply between frequencies close together. It would be required to admit (sufficiently uniformly) all frequencies within the band 56·0–59·8 kc/s, and to reject (sufficiently completely) all frequencies outside this band, in particular those lying within the other side-band, viz. 60·2–64 kc/s. Because $\frac{59·8}{60·2}$ is so near to unity, such a filter would be very costly to construct. The difficulty lies in the largeness of the carrier frequency. It is eased by a process of double (i.e. repeated) modulation. In the first modulator the acoustic signals modulate a carrier of lower than the radiation frequency; and in the second modulator the output from the first modulator modulates the final carrier (which is suppressed).

The process is explained by reference to Fig. 270. The top part is a frequency diagram constructed after the fashion of Fig. 265, and the bottom part is a schematic diagram showing the relations of the pieces of apparatus corresponding with the several stages of the complete process from the microphone to the antenna. This figure shows how an acoustic band A occupying the range 500–3200 c/s is transmitted from New York as a single-side-band S_2†. Full-line areas indicate wave bands actually produced in the process, shaded if they are wanted, left unshaded if they are filtered away as unwanted. The dotted lines C_1, C_2 indicate suppressed carriers.

The first modulator is fed with the acoustic input A and a steady oscillation of 33·7 kc/s. The carrier C_1 is balanced out in the modulator, and the side-band S_1' is filtered out, leaving only S_1. A second balanced modulator is fed with the side-band S_1 and a steady oscillation of 89·2 kc/s. The carrier C_2 is balanced out in the modulator, and the side-band S_2' is filtered out, leaving the side-band S_2 to excite the antenna through a series of amplifying stages.

The first filter has to accept 33·2 kc/s and to reject 34·2 kc/s; it therefore does not require so sharp a cut-off as if there were a single modulation, in which case the filter would have to accept 55·5 − 0·5 = 55·0 kc/s and to reject 55·5 + 0·5 = 56·0 kc/s.

A further important benefit is conferred by the double modulation.

---

* See XVI.

† Fig. 270 is prepared from information in R. A. Heising, "Production of single sideband for transatlantic radio telephony," *Proc. I.R.E.* Vol. 13 (1925), p. 291. A somewhat similar diagram is given there.

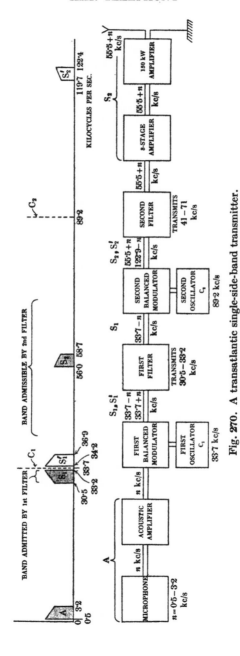

Fig. 270. A transatlantic single-side-band transmitter.

The second filter may have a wide band of admissible frequencies, since S_1' and C_2 are far removed in frequency from S_2. Without any change of filters, therefore, the position of the radiated band S_2 may be shifted by merely changing the frequency of the second oscillator.

Successful application* of single-side-band transmission in the wireless field has been facilitated by its earlier use in the field of carrier-wave multiplex line telegraphy and telephony ("wired wireless"). It is a development of prime importance for long-wave wireless telephony. As yet it has not been used in short-wave working, where the advantages it offers are less pronounced and the difficulties in the way of its application are greater.

5. Distortions in the Receiver

5 (a). *General.* Although the receiving dispositions of wireless telephony do not exhibit the counterpart of the specialised features which distinguish the telephone from the telegraph transmitter, certain aspects of telephonic reception require special quantitative consideration. In telegraphy the receiving circuits can, as a first approximation, be treated as though the signal is either present in the steady state or wholly absent†. In telephony no such treatment is possible.

At its simplest, during the transmission of a steady single tone, the high-frequency oscillation is continuously changing in amplitude. When the modulation frequency is very low in comparison with the frequency of the oscillation modulated—i.e. $\omega \ll p$, in the last Section—the amplitude remains substantially constant during many of the high-frequency cycles. As regards high-frequency phenomena, the oscillation may then be treated as a sine function of time. Expressed in the terms of the last Section, the components of the complete modulated oscillation may be regarded, over a long epoch, as sensibly equal in frequency. Thus with short waves—but not, as we saw in the last Section, with long—the discrimination

* By the American Telephone and Telegraph Co., and the Western Electric Co.
† In analysing the possibilities of high-speed Morse signalling, and the disturbance caused by it, this treatment is, of course, inadequate. See, e.g., L. B. Turner, "The relations between damping and speed in wireless reception," *Journ. I.E.E.* Vol. 62 (1924), p. 192. A more thorough procedure is doubtless to treat telegraphic conditions as a particular form of the modulation studied in telephony. See, e.g., R. V. L. Hartley, *lot. cit.* p. 395.

exercised by an oscillatory circuit between the carrier and side-waves, because of their unequal frequencies, may be negligible*. Then, likewise, in the detector, where high-frequency and low-frequency meet, it is possible to employ high-frequency bye-pass condensers (such as C_g, C_a in Fig. 224, p. 327) possessing negligible reactance at the high-frequency and negligible admittance at the low. With long waves this is not possible†, and further examination is necessary.

Apart from the effects peculiar to single-side-band transmission examined in the last Section, the more fundamental origins of acoustic distortion in the receiver may be classified as follows:

(i) In the high-frequency circuits, due to frequency discrimination. (See Subsection 5 (b).)

(ii) In the detector, due to non-linear performance. (See Subsection 5 (c).)

(iii) In the detector, effects of finite modulation frequency. (See Subsections 5 (d) and (e).)

(iv) In the acoustic amplifier.

(v) In the telephone sounder (loud speaker) and triode feeding it. (See Section 6.)

As regards item (iv), the implications of Chapter VIII are sufficiently obvious without further remark. The other items are discussed in the remainder of this Section, and in the next.

5 (b). *In the high-frequency circuits.* Effects of unequal changes of phase and amplitude amongst the components of the modulated oscillation at the transmitter have been examined in the last Section; and, of course, similar discriminations in the receiving circuits produce similar results.

In Fig. 167 (p. 254) frequency-response curves were given for alternative designs of an amplifier stage tuned to signals of 50 kc/s. If these were telephonic signals, according as the more selective or the less selective design were adopted, side-waves corresponding with a tone of 5 kc/s would be reduced in amplitude to about 16 °/₀ and 99 °/₀ of their correct values in relation to the carrier-wave, and to low-pitched side-waves. This is with a single stage. If there were three such stages—it might be, for example, one at the trans-

* At $\lambda = 30$ m, $\dfrac{p}{2\pi} = 10^4$ kc/s; and even for a side-wave of $\dfrac{\omega}{2\pi} = 5$ kc/s, $p - \omega$ and $(p + \omega)$ differ from p by only 0·05 °/₀.

† As adumbrated in VI–5 (c).

5. DISTORTIONS IN THE RECEIVER

mitter and two at the receiver—in the more selective design, tones of the high acoustic frequencies would be almost entirely cut off, those of 5 kc/s being only $0{\cdot}16^3 = 0{\cdot}4\,\%_0$ of the correct amplitude*. In the less selective design, the high tones would be reproduced at almost their full strength; but such an arrangement would be quite impractically at the mercy of interfering stations.

A telephone receiver would give at once the most faithful response to the desired signals, and possess the greatest immunity from other signals, if its circuits could be so designed that the overall frequency-response curve were a rectangle, as in Fig. 271. Everything within the band of desired frequencies would be uniformly admitted, while everything outside this band would be wholly

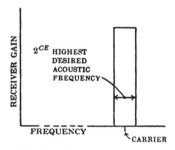

Fig. 271. Ideal frequency characteristic.

rejected. For high musical quality the accepted band might be 10 kc/s–20 kc/s wide, and for good speech intelligibility it might be 6 kc/s wide. This performance the designer tries to approach as nearly as engineering exigencies will permit.

In obtaining any required degree of selectivity, the superiority of a chain of moderately damped circuits over a single circuit of very low damping (referred to on p. 357) is illustrated by Fig. 272. The full-line curve shows the response of a chain of three non-reacting circuits, each of decrement 0·031 and each tuned to a carrier-wave of 1000 kc/s ($\lambda = 300$ m). Side-waves corresponding with a tone of 5 kc/s, i.e. with frequencies of 995 kc/s and 1005 kc/s, are depressed 8·4 db below the carrier-wave. The distuning has divided their amplitude by 2·65. The whole acoustic band would be received with very satisfactory uniformity. Signals distuned by more than (say) $2\,\%_0$ from the carrier (i.e. signals of wavelengths

---

* They would be depressed by 47·8 db.

less than 294 m or more than 306 m) would be depressed by more than 37 db (i.e. divided in amplitude by more than 69). The dotted curve in the same figure shows the response of a single circuit of decrement 0·0123. This decrement has been chosen because it gives the same depression of the 5 kc/s tone as was given by the chain of three more highly damped circuits. The inferiority of the single circuit in cutting out unwanted signals in clear. 2 °/₀ distuning, for instance, now produces a loss of 20 db instead of the previous 37 db.

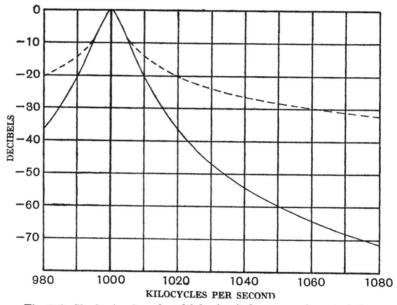

Fig. 272. Single-circuit and multiple-circuit frequency characteristics.

Fig. 273 gives the response curves of the same circuits as does Fig. 272, plotted with a scale of relative amplitudes instead of decibels.

The full-line curves of Figs. 272 and 273, although much nearer to the ideal rectangle of Fig. 271, are still far from it. Much better approximations to the rectangle are obtained by the use of chains of reacting circuits known as *filters**. Fig. 337 (p. 505) is the response curve of the filter shown in Fig. 338 (p. 506), and should be compared in shape with the curves of Fig. 272.

* See XVI.

5. DISTORTIONS IN THE RECEIVER

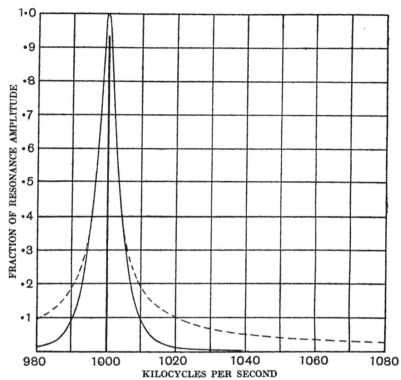

Fig. 273. Another expression of Fig. 272.

5 (c). *Non-linear rectification.* A detector giving an output P.D. proportional to the square of the input P.D.—as with weak signals all rectifiers do (see VI-2)—must obviously produce false amplitude relations. But when stimulated by an acoustically modulated high-frequency P.D., such a detector is the cause of another form of distortion too. A modulated high-frequency input of the form

$$(C + S \sin \omega t) \sin pt,$$

i.e. corresponding with a pure tone of frequency $\dfrac{\omega}{2\pi}$, appears after rectification as an acoustic signal containing an intruder of the harmonic frequency $2\dfrac{\omega}{2\pi}$.

This is seen from the rectifier analysis in VI-2 on writing

$$e = C \sin pt + \frac{S}{2} \cos \overline{p - \omega} t - \frac{S}{2} \cos \overline{p + \omega} t.$$

Taking for the rectifier characteristic the parabola
$$i = \alpha + \beta v + \gamma v^2,$$
we have
$$i = \alpha$$
$$+ \beta \left[v_0 + C \sin pt + \frac{S}{2} \cos \overline{p - \omega} t - \frac{S}{2} \cos \overline{p + \omega} t \right]$$
$$+ \gamma \left[v_0^2 + \frac{C^2}{2} (1 - \cos 2pt) + \frac{S^2}{8} (1 + \cos 2 . \overline{p - \omega} t) \right.$$
$$+ \frac{S^2}{8} (1 + \cos 2 . \overline{p + \omega} t) + 2v_0 C \sin pt + v_0 S \cos \overline{p - \omega} t$$
$$- v_0 S \cos \overline{p + \omega} t + \frac{CS}{2} (\sin \overline{2p - \omega} t + \sin \omega t)$$
$$\left. - \frac{CS}{2} (\sin \overline{2p + \omega} t - \sin \omega t) - \frac{S^2}{4} (\cos 2pt + \cos 2\omega t) \right].$$

The component currents are tabulated below in order of frequency.

Frequency $\times 2\pi$	Amplitude	Remarks
0	$\alpha + \beta v_0 + \gamma v_0^2 + C^2 . \tfrac{1}{2}\gamma + S^2 . \tfrac{1}{4}\gamma$	Steady current. Inaudible. The last term is that used in Morse telegraphy.
ω	$CS . \gamma$	The desired current. $\propto CS$.
2ω	$S^2 . \tfrac{1}{4}\gamma$	Intruding audible harmonic. $\propto S^2$.
$p - \omega$	$S(\tfrac{1}{2}\beta + \gamma v_0)$	
p	$C(\beta + 2\gamma v_0)$	
$p + \omega$	$S(\tfrac{1}{2}\beta + \gamma v_0)$	
$2(p - \omega)$	$S^2 . \tfrac{1}{8}\gamma$	Inaudible
$2p - \omega$	$CS . \tfrac{1}{2}\gamma$	
$2p$	$C^2 . \tfrac{1}{2}\gamma + S^2 . \tfrac{1}{4}\gamma$	
$2p + \omega$	$CS . \tfrac{1}{2}\gamma$	
$2(p + \omega)$	$S^2 . \tfrac{1}{8}\gamma$	

The ratio between the amplitudes of the intruding audible harmonic and the desired current is $\tfrac{1}{4}\frac{S}{C}$. This distortion introduced by the rectifier tends to vanish as the modulation ratio $\frac{S}{C}$ is lowered*.

* Although weak modulation always favours acoustic faithfulness in both transmitter and receiver, broadcasting authorities are urged, by considerations of economy and of interference with neighbouring stations, to modulate so deeply that during loud passages $\frac{S}{C}$ approaches unity.

5. DISTORTIONS IN THE RECEIVER

The distortion due to intruding audible frequencies is much exaggerated by the coexistence of several modulation frequencies, $\frac{\omega_1}{2\pi}, \frac{\omega_2}{2\pi}, \frac{\omega_3}{2\pi}\ldots$. In addition to the constituents of rectifier current found above with a single ω, other combination frequencies now appear. With two simultaneous modulations, making

$$e = (C + S_1 \sin \omega_1 t + S_2 \sin \omega_2 t) \sin pt,$$

similar analysis shows that there are now 18 high-frequency terms (e.g. $2p - \omega_1 + \omega_2$), and 6 acoustic-frequency terms. The last are

Frequency $\times 2\pi$	Amplitude	Remarks
ω_1	$CS_1 \cdot \gamma$	True
ω_2	$CS_2 \cdot \gamma$	
$2\omega_1$	$S_1^2 \cdot \frac{1}{4}\gamma$	Intruder*
$2\omega_2$	$S_2^2 \cdot \frac{1}{4}\gamma$	
$\omega_1 + \omega_2$	$S_1 S_2 \cdot \frac{1}{2}\gamma$	
$\omega_1 - \omega_2$	$S_1 S_2 \cdot \frac{1}{2}\gamma$	

When the rectifier is in series with a large output impedance, so that its mean P.D. slides back under the action of a signal, the foregoing results are, of course, modified. The analysis of the behaviour of the detector with a modulated input signal is very troublesome. It has been carried a long way by F. M. Colebrook† for an exponential rectifier characteristic; and S. E. A. Landale‡ has shown that with a parabolic characteristic an effect of the impedance is to substitute an infinite series of harmonics of ω for the single intruding acoustic term (2ω) we have found on p. 410.

It is probable that distortion of this type—the intrusion of harmonics of the modulating frequencies, and of sum and difference frequencies—is often severe in actual practice; but its importance is the less in that, by the choice of proper detector dimensions and the use of adequately strong signals, a substantially linear detector performance may, and always should, be obtained. (See VI-5, and XI-6 and 7.) If the response curve of the detector (i.e. high-frequency input P.D. plotted against acoustic output P.D.) is

---

* Suppression of one of the side-bands at the transmitter eliminates three of the four intruder terms, and halves the amplitudes of the two desired terms and of the remaining intruder term.

† *Loc. cit.* p. 384.

‡ *Loc. cit.* p. 338. In Landale's treatment the signal must be so weak that the rectifier current never falls to zero.

substantially straight between the extreme input amplitudes*, the intruder frequencies do not arise. It was for this reason that $\frac{\mathscr{E}_2}{\mathscr{E}_1} \not< 19$ was taken in XI–6 as a criterion of satisfactory telephonic detector performance; for with the high modulation ratio $\frac{S}{C} = 90°/_\circ$, $(C + S) = 19(C - S)$.

5 (d). *Rectifier effects of finite modulation frequency.* Even if the detector gives an output P.D. linearly proportional to the high-frequency input P.D. when that input is maintained of constant amplitude (or is modulated indefinitely slowly), it does not follow

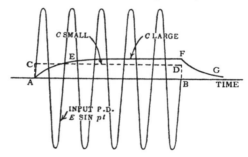

Fig. 274. Acoustic sluggishness in the detector.

that the proportionality will remain unimpaired when the amplitude is modulated at an acoustic frequency. The analysis of VI–5 (a) (p. 158) must be reconsidered to find how far it is vitiated on changing the constant E into the acoustically varying $E \sin \omega t$.

If C is large enough (Fig. 102, p. 159), as we assumed before, to render negligible the fluctuation of its P.D. during a high-frequency cycle, the former analysis does give the steady rectified output E_s reached on continued application of the steady input $E \sin pt$. This value E_s of the condenser P.D. would, however, be reached gradually on the application of the input $E \sin pt$, and would die away gradually on its removal. In Fig. 274, the rectified output is CD with a small value of C, and some such curve as AEFG with a large value of C. The rate of building up (AE) depends mainly on the product of Cr (assuming R is much greater than r), and the rate of dying away (FG) depends on the product CR. When the input is the modulated P.D. $E \sin \omega t . \sin pt$, the effect of an indefinitely large C would

* I.e. $(C - S)$ and $(C + S)$ when the modulation is a single tone.

5. DISTORTIONS IN THE RECEIVER

clearly be to smooth away from the output P.D. all the acoustic fluctuation; and with finite values of C, the degree of smoothing out effected must increase with the modulation frequency $\frac{\omega}{2\pi}$.

This form of distortion is the more completely avoided the smaller is the time constant CR in comparison with the smallest acoustic period to be dealt with. If, for example, R were 1 MΩ or less and C were 100 $\mu\mu$F, CR would be $\frac{1}{10{,}000}$ second or less; and probably the distortion would be inappreciable. With carrier frequencies of 1000 kc/s and upwards, such a value of CR is at once small compared with modulation periods (so keeping the acoustic distortion small), and large compared with the carrier period (so giving substantially the full output from the detector, as calculated in VI-5 (a)). With much lower carrier frequencies it is necessary either to abandon one of the two desiderata, or to compromise between them. The correct course would seem to be to keep CR small (say 10^{-4} sec.), and to put up with reduction of output resulting from CR not being very great in comparison with $\frac{k}{(1-k)n}$ (see p. 162). It is to be noted that when C is reduced to zero, the output of the detector with the unmodulated signal E falls from kE (where k is less than unity, e.g. 0·9 when $R = 100r$) only to $\frac{1}{\pi}\frac{R}{R+r}E$ (e.g. 0·3E when $R = 100r$). How the output depends upon E with intermediate values of C does not seem to have been investigated*.

5 (e). *Push-pull grid rectifier.* When the rectifier employed is of the ordinary grid type, and special attention is being paid to faithfulness of reproduction over a wide range of modulation frequencies, the ill-effect of the finite time constant in the anode circuit ($C_a R_a$ in Figs. 235–237, p. 344) must not be disregarded. The further distortion introduced by the high-frequency shunting condenser C_a can be removed only by the omission of the condenser, with the consequent inconveniences of (i) a heavy load thrown on the input circuit (see XI-5 (b)), and (ii) penetration of high-frequency oscillations into the post-detector, acoustic-frequency circuits†.

---

* Scanty experiments by the writer suggest that the relationship, linear with large C and with zero C, is also (as would be expected) linear with intermediate values.

† (i) is worse the shorter the wavelength, because anode-grid capacitance

XIII. TELEPHONY

By the use of a push-pull disposition of two equal grid rectifiers, however, the condenser may be wholly omitted without causing either of these inconveniences. In Fig. 275, if the two triodes are equal and the anode currents are in the rectilinear region of the characteristics, the high-frequency components of anode current are at every instant equal and opposite. But rectification at the grids makes their mean potentials slide back together and equally. Consequently R_a is traversed by the high-frequency component of anode current in neither triode, but is traversed by the rectified or signal component of anode current in both triodes. Since no high-frequency current passes between the bunched anodes and the filaments, no

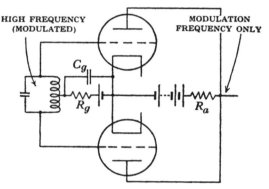

Fig. 275. Push-pull detector.

condenser need shunt R_a (however low the carrier frequency); and since the anode potentials have no fluctuation of carrier frequency, anode-grid stray capacitance throws no load on the input circuit (however high the carrier frequency).

The transformer disposition in Fig. 276 theoretically replaces that of Fig. 275 without prejudicing these advantages of the push-pull device, provided that the halves of the primary winding are oppositely wound and without magnetic leakage.

The push-pull rectifier should be compared with the push-pull amplifier (p. 249), and with the carrier-suppressing modulator (Fig. 267, p. 393). In the language of carrier-wave telegraphy and telephony, a device like that of Fig. 267 is often called a *balanced modulator*, and of Figs. 275 and 276 a *balanced demodulator*.

is then more effective; and (ii) is worse the longer the wavelength, because the carrier frequency is then closer to the high acoustic frequencies.

6. ELECTROMECHANICAL RELATIONS

6(a). *General.* In the preceding pages we have inspected, and in some instances fairly fully analysed, the long chain of purely electrical processes which occur between the microphone and the sounder in a complete wireless telephone installation. At every stage distortions of the acoustic signal may be introduced, and in some instances cannot theoretically be wholly avoided. These electrical processes are susceptible to fairly complete mathematical analysis, and the distortions prove to be reducible to an extent governed mainly by economic rather than technical considerations.

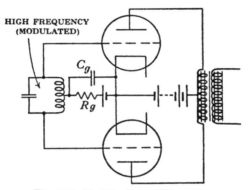

Fig. 276. Modification of Fig. 275.

There remains the electromechanical transformations, *from* sound in air at the microphone and *to* sound in air at the speaker. In any review of telephony in which acoustic faithfulness is important, these processes stand out from the purely electrical processes as being less well understood, less easily measured, and leading to problems less amenable to satisfactory engineering solutions. It would seem that engineers are better equipped, both theoretically and empirically, to deal with electrical vibrations of (say) 50–5000 kc/s than with mechanical vibrations of say 50–5000 c/s*. The weakest points in the working solutions which have so far been achieved undoubtedly lie more in the mechanical than in the electrical stages of the composite process known as telephony.

The practical difficulties reside, not in producing mechanical

---

* Is this because in experimental mechanics mass is less separable from stiffness than is inductance from capacitance in real electric systems?

systems which will deal satisfactorily with vibrations of even high acoustic frequencies such as 5 or 10 kc/s, but in producing a single mechanical system which will deal satisfactorily with both high and low acoustic frequencies. For example, the moving system of a moving-iron loud-speaker (or ordinary telephone sounder of the Bell type), if made small enough to deal effectively with tones of high pitch will fail to respond properly to tones of low pitch. Conversely, in an instrument made on a large scale to suit sounds of low pitch, parts which behave as substantially rigid bodies at low frequencies may become wave-transmission systems at high frequencies; and

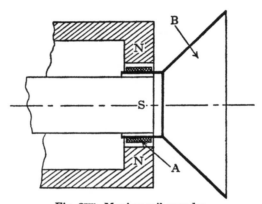

Fig. 277. Moving-coil sounder.

acute resonance phenomena may supervene where impartiality towards frequency is desired. To the difficulties thus met in the vibrating solid components of the mechanical system are added those associated with the reactions between a vibrating fluid (sound-carrying air) and the solid objects immersed therein.

The theory of acoustic vibrations is an abstruse subject standing by itself. It must enter into any complete study of telephony, but is beyond the scope of this book*. We will, however, sketch the theory of one electromechanical system, the moving-coil loud-speaker, making simplifying assumptions which remove the difficulties besetting a fuller analysis.

* The student is recommended to consult I. B. Crandall, "Theory of vibrating systems and sound" (Van Nostrand Co., 1926); and Harvey Fletcher, "Speech and hearing" (Macmillan and Co., 1929). Both these books derive from that far-sighted and thorough research organisation, Bell Telephone Laboratories, New York.

6. ELECTROMECHANICAL RELATIONS

6 (b). *Moving-coil loud-speaker.* The moving-coil type of sounder is particularly interesting in that it is more calculable in its action, and is capable of giving better performance, than the moving iron type; and it can be used, reversed, as the microphone (with the simplification then that the coil carries no current).

The moving system consists of the cylindrical coil A (Fig. 277), attached to a stiff disc or cone* B, suspended in a co-axial cylindrical air-gap between the poles N, S of a powerful magnet. The acoustic-frequency current is made to pass through the coil; it reacts with the steady radial magnetic flux in the air-gap, and so drives the cone back and forth along the axis. The moving system is easily

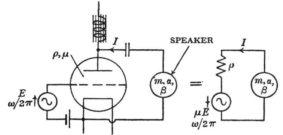

Fig. 278. Moving-coil sounder and output triode.

suspended with very little axial restraint, so that substantially the only axial force at any instant is proportional to the current in the coil at that instant†. The moving disc or cone impels the air in its neighbourhood‡; it therefore acts as a source of sound of frequency equal to the frequency of the current, and of strength dependent on the amplitude of the current.

The speaker is connected, by the use of a transformer or of a large condenser and anode-feed choke (Fig. 278), to the output triode of the acoustic-frequency amplifier, so that the alternating component I of the anode current flows through the coil. We have to find the motion of the coil and cone when a steady acoustic E.M.F. E of frequency $\frac{\omega}{2\pi}$ is applied to the grid of the triode. The equivalent single circuit is shown in the Figure.

---
* The familiar cone form is commonly adopted on account of its remarkable stiffness (as regards relative motion between its particles in the axial direction) for any given mass and diameter.

† Unlike the moving-iron type of instrument, it has no (or negligibly small) natural frequency of oscillation.

‡ The use of a large disc, in place of the small diaphragm of the Bell sounder, makes the use of a trumpet unnecessary.

We shall at first make the mechanical approximations that the cone is a rigid body, and that the resistance to its motion is always a constant times its velocity. Let the dimensions be:

m = mass of moving system;

s = displacement of moving system;

$\alpha \dfrac{ds}{dt}$ = resisting force of air;

β = flux-turns cut by coil per unit displacement,
  = $2\pi a TB$, where a = radius of coil,
  T = no. of turns,
  B = flux-density in gap where coil is located;

l = inherent (static) inductance of coil*;

ρ, μ = anode resistance, and amplification factor, of triode feeding the coil.

The deflecting force due to the current of instantaneous value i is βi. Hence the equation of motion of the coil is

$$\beta i - \alpha \frac{ds}{dt} = m \frac{d^2s}{dt^2} \quad \ldots\ldots\ldots\ldots(13).$$

The E.M.F. generated in the coil by its motion is $\beta \dfrac{ds}{dt}$. Hence if e is the instantaneous value of the grid potential, the equation of electric flow is

$$\mu e - \rho i - l \frac{di}{dt} - \beta \frac{ds}{dt} = 0 \ldots\ldots\ldots\ldots(14).$$

Restricting the analysis to the steady state, when E, S, I are vector quantities all of frequency $\dfrac{\omega}{2\pi}$, and replacing the operator $\dfrac{d}{dt}$ by $j\omega$, where $j \equiv \sqrt{-1}$, we can write (13) and (14) in the form

$$I = \frac{S}{\beta}(-\omega^2 m + j\omega\alpha) \quad \ldots\ldots\ldots\ldots(15);$$

$$\mu E = I(\rho + j\omega l) + j\omega\beta S \quad \ldots\ldots\ldots\ldots(16).$$

Eliminating I from (15) and (16),

$$\frac{\mu\beta E}{S} = -\omega^2(\rho m + \alpha l) + j\omega(\rho\alpha + \beta^2 - \omega^2 ml)\ldots\ldots(17).$$

* The inherent resistance r of the coil is omitted for simplicity. Its effect would be wholly accounted for by writing $(\rho + r)$ in place of ρ.

6. ELECTROMECHANICAL RELATIONS

Hence if E_m, S_m are the amplitudes of E and S,

$$\mu\beta \frac{E_m}{S_m} = \omega \sqrt{(\rho\alpha + \beta^2 - \omega^2 ml)^2 + \omega^2(\rho m + \alpha l)^2} \quad \ldots\ldots(18).$$

Let us suppose that the amplitude E_m impressed on the grid is proportional to the product of the amplitude and the frequency of the sound impinging on the transmitting microphone*; i.e. that

$$E_m = k\omega A_m,$$

where A_m is the sound amplitude at the microphone and k is a constant of the whole installation. Then

$$\frac{kA_m}{S_m} = \frac{1}{\mu\beta} \sqrt{(\rho\alpha + \beta^2 - \omega^2 ml)^2 + \omega^2(\rho m + \alpha l)^2} \quad \ldots(19).$$

For faithful reproduction, $\frac{A_m}{S_m}$ must be independent of both A_m and ω. Equation (19) shows that $\frac{A_m}{S_m}$ is independent of A_m†; but it is not independent of ω unless l and m, the electrical and mechanical inertias, are both negligible.

Theoretically l can be made vanishingly small (and in practice can be much reduced) by providing a short-circuited secondary coil in the form of copper rings let into the pole faces‡. To see more easily the significance of the other terms we will now omit the l terms: whereupon equation (19) becomes

$$\frac{kA_m}{S_m} = \frac{1}{\mu\beta} \sqrt{(\rho\alpha + \beta^2)^2 + \omega^2 \rho^2 m^2} \quad \ldots\ldots\ldots\ldots(20).$$

Despite all ingenuity, it seems to be impracticable to construct a moving system (with a tolerably stiff cone) in which the mechanical inertia is small enough to be unimportant at the higher frequencies; but equation (20) shows how its influence is to be kept as small as is feasible. In striving for faithful reproduction, the aims of the designer must be:

(i) Small mass m.

---

* As it would be if the microphone were a reversed moving-coil speaker without inertia ($m=0$), and if the intermediate transmitting and receiving apparatus gave the linear performance treated as the desideratum in the foregoing Sections.

† A relation which does not obtain in moving-iron speakers.

‡ Even if copper rings are not fitted, eddy currents in the solid poles go a long way towards removing the inductance of the coil.

(ii) Large β: i.e. for a given length of wire in the coil, large flux-density B.

(iii) Large α: i.e. large (and stiff) cone.

(iv) Small resistance ρ.

These aims are, of course, to a large extent mutually destructive. Moreover the magnitude of the output from the speaker, and not its good quality only, is an important practical consideration. At frequencies low enough to make the mass term negligible, a given signal at the grid of the output triode produces a motion whose amplitude S_m is proportional to $\dfrac{\mu\beta}{\rho\alpha + \beta^2}$. This indicates that, although any increase of flux-density in the gap ($\propto \beta$) is always profitable in reducing distortion due to mechanical inertia, it could theoretically be pushed so far as to reduce output. In practice, the highest obtainable flux-density is profitable*.

The foregoing theory is inaccurate chiefly in that, if the effect of the reaction between cone and air is written as the force $\alpha \dfrac{ds}{dt}$, α is actually far from constant. With rising frequency, α settles down to an approximately constant value at frequencies exceeding a border frequency which is the smaller the larger is the disc (radius a); but with discs of usual size (e.g. $a = 10$ cm.), this border frequency appears to be about 1000 c/s†. Furthermore, the mass m is not a constant, for it includes a term representing the "adherent air." With rising frequency, this term also settles down to an approximately constant value at frequencies exceeding a border frequency which is the smaller the larger is the disc. With discs of usual size the variable term in the mass is to a large extent swamped by the inherent or static mass, especially above about 1000 c/s.

With a sufficiently large disc, therefore, it would seem that the above simple analysis (i.e. taking α and m as independent of frequency) would be approximately true, provided that the disc were sensibly rigid, i.e. if all its particles moved in phase. Since, however, the wavelength of sound vibration of 5 kc/s is only

* Saturation in the steel, much intensified by leakage flux outside the air-gap occupied by the coil, restricts the working flux-density to some 8000–10,000 C.G.S. units.

† See C. R. G. Cosens, "Moving Coil Loud Speakers," *Experimental Wireless*, Vol. 6 (1929), p. 353. Cosens deduces his values of α and m from the theory of the motion of a rigid disc given by Lord Rayleigh in his classical "Theory of Sound."

6. ELECTROMECHANICAL RELATIONS

about 6 cm. for propagation in the atmosphere, and is probably much less for propagation radially across the solid disc from the point of application of the driving force, it cannot be supposed that the disc does move as though it were sensibly rigid. Indeed Cosens* calculates for actual loud-speakers that if the cone did vibrate as a rigid body at the high acoustic frequencies, the moving mass would be too great to allow the speaker to be nearly as effective as it is observed to be†.

6 (c). *Impedance of coil movable with elastic restraint in magnetic field.* The effect of its motion on the impedance of the moving coil is of interest‡. We will make the investigation more general by including an elastic restoring force γs, giving the mechanical system *per se* a resonance frequency $\frac{\omega_0}{2\pi}$, where $\omega_0^2 = \frac{\gamma}{m}$; and as there is no longer the triode series resistance ρ in our circuit, we will not now omit the inherent resistance r of the coil. The other symbols have the same meaning as in the last Subsection.

When a current i is flowing in the coil,

$$\beta i - \alpha \frac{ds}{dt} - \gamma s = m \frac{d^2s}{dt^2} \quad \ldots\ldots\ldots\ldots\ldots(21);$$

and if the P.D. across the coil is e,

$$e - ri - l\frac{di}{dt} - \beta \frac{ds}{dt} = 0 \quad \ldots\ldots\ldots\ldots\ldots(22).$$

Restricting the investigation to the steady state as in the last Subsection, and eliminating S between equations (21) and (22), we find

$$\frac{E}{I} = r + j\omega l + \frac{j\omega \beta^2}{m(\omega_0^2 - \omega^2) + j\omega\alpha} \quad \ldots\ldots\ldots(23).$$

Now the impedance of the compound circuit in Fig. 279 is

$$r + j\omega l + \frac{j\omega RL}{R(1 - \omega^2 CL) + j\omega L} \quad \ldots\ldots\ldots(24).$$

---

* *Loc. cit.* p. 420.

† Lack of rigidity of the cone is treated quantitatively in an important research published after this Chapter was in type: P. K. Turner, "Some measurements of a loud-speaker *in vacuo*," read before Inst. E.E. 4 Feb. 1931.

‡ This analysis need not be confined to moving-coil loud-speakers. It may be applied to systems generally in which a current in a wire produces a motion of any part of the system whereby an E.M.F. is generated in the wire: e.g. to an oscillograph, an Einthoven galvanometer, or a moving-iron speaker.

Equations (23) and (24) are identical if

$$R = \frac{\beta^2}{\alpha}, \quad C = \frac{m}{\beta^2}, \quad L = \frac{\beta^2}{\gamma}*.$$

Hence an impedance equal to that of the three elements R, C, L connected in parallel† is added to the impedance $(r + j\omega l)$ which the coil would show if it were held fixed, or if there were zero flux-density in the gap. It is called the *motional impedance* of the coil. It is capacitative or inductive according as the impressed electrical frequency $\frac{\omega}{2\pi}$ is less or greater than the mechanical resonance frequency $\frac{\omega_0}{2\pi}$‡.

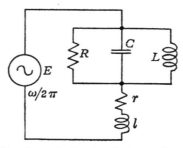

Fig. 279. Circuit equivalent to electro-mechanical system.

6 (d). *Moving-iron speakers.* In an ordinary Bell telephone sounder, or other moving-iron sounder with an elastic diaphragm, there are several well-marked mechanical resonance frequencies. Consequently, as the frequency of the terminal P.D. is raised, the reactance component of the motional impedance passes through a succession of maxima and minima, with change of sign at each resonance; and the resistance component passes through maxima and minima (without change of sign).

The performance of moving-iron, as compared with moving-coil,

---

* Notice that R, C, L are functions of the dimensions of the instrument only. If m, α, β, γ are constants, R, C, L do not vary with the frequency of the signal.

† viz. $\dfrac{\omega^2 L^2 R}{R^2 (1 - \omega^2 CL)^2 + \omega^2 L^2} \left[1 + j(1 - \omega^2 CL) \dfrac{R}{\omega L} \right].$

‡ In the moving-coil loud-speaker without appreciable elastic restraint (last Subsection), $\gamma = 0$ and $L = \infty$. The motional impedance is therefore that of a capacitance $\dfrac{m}{\beta^2}$ shunted by a resistance $\dfrac{\beta^2}{\alpha}$.

speakers is complicated not only by the existence of one or more mechanical resonances, but also by a non-linear relation between current in the coils and the deflecting force. On both theoretical and empirical grounds, it seems that any moving-iron speaker must cause severe acoustic distortion in a variety of ways. The severity of the distortion grows with the strength of the signal dealt with, but it cannot vanish however weak the signals*.

7. Examples of telephone installations

7 (a). *A broadcasting transmitter.* The high-power dual transmitter put into service in 1929 at Brookman's Park is an example of the highest development in telephone transmitters where perfection of acoustic quality is a prime consideration. Descriptions of this station have been widely published†. A review of its essential parts and their relationship will exemplify the engineering application of principles discussed in this book.

Fig. 280 shows a complete transmitter divided into parts according to function. It is a skeleton schematic diagram only. It includes no duplication of parts; none of the many switches and meters of various kinds; none of the auxiliary connections designed to prevent the generation of parasitic oscillations of very high frequency; and none of the neutrodyne connections at each triode overcoming undesired retroaction. The dimensions given are rough only.

For convenience of reference, the triodes shown in the diagram are numbered T_1, T_2, ... Coils acting as chokes, i.e. serving to pass a direct current but to reject alternating current, are marked CH.; and grid biasing connections are marked G.B.

T_1 is the master oscillator, generating a steady oscillation of the carrier frequency. Great care is taken to ensure a high degree of constancy of frequency in this generator.

T_2 is a quasi high-frequency amplifier. Its function is, however, primarily not to amplify, but to separate the master oscillator from the modulated portion of the chain, in order to ensure that the carrier frequency shall be quite unaffected by the modulation. T_2 passes on the unmodulated carrier from T_1 to T_3.

---

* Some analysis of the action of moving-iron speakers will be found in the chapter entitled "Speech apparatus" in E. Mallett, *Telegraphy and Telephony* (1929). The carbon-granule microphone is also briefly discussed.

† See especially P. P. Eckersley and N. Ashbridge, "A wireless broadcasting transmitting station for dual programme service," *Journ. I.E.E.* Vol. 68 (1930), p. 1149. Also *The B.B.C. Yearbook 1931*, p. 269.

T_3 is an acoustic-frequency amplifier. It is fed by cable from the studio, through an amplifier with variable gain located in a control

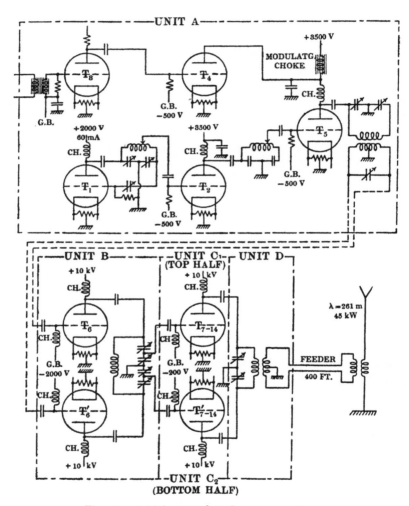

Fig. 280. A high-power broadcast transmitter.

room. There the various signals arriving from the studio, before being passed to T_3, are adjusted to a strength giving the desired depth of modulation.

7. EXAMPLES OF TELEPHONE INSTALLATIONS

T_4 is the control triode, operating on the anode-choke method. Its anode, and the anode of the controlled high-frequency triode T_5, are fed through the same iron-cored choke.

T_5 is the modulator. Its grid input from T_2 is the carrier oscillation of constant amplitude; and its anode output has an amplitude linearly proportional to the mean anode potential—i.e. the potential of the lower end of the modulating choke. Except that T_5 is an externally excited oscillator instead of a self-oscillator, T_4 and T_5 behave precisely as do C.T. and O.T. respectively in Fig. 261 (p. 385).

T_1–T_5 are all air-cooled glass triodes. The output from T_5 is a modulated weak oscillation. There remains but to amplify it and impress it on the antenna.

T_6, T_6' is a push-pull amplifier of intermediate power, comprising two water-cooled metal-glass triodes.

T_7–T_{14}, T_7'–T_{14}' is the final amplifier stage, also arranged in the push-pull disposition, comprising two groups of eight water-cooled triodes. All earlier stages are designed to operate nearly linearly without regard to efficiency; but in this stage, where the power consumed is large, some departure from rectilinear performance is tolerated for the sake of higher efficiency. The push-pull disposition mitigates the distortion introduced.

The antenna is stretched between two towers 200 feet high and 600 feet apart. It is situated as far as possible from the building and from the second antenna, and is fed from the final amplifier over a transmission line.

The anode power for the eighteen water-cooled triodes, some 160 kW at 10 kV, is supplied by two D.C. dynamos connected in series. These are driven by a 220 V D.C. motor taking power from Diesel-driven dynamos and/or a 2000-ampere-hour battery.

Plate XXVII shows the actual layout of all the apparatus included in Fig. 280. It is housed in five well-spaced metal cabinets, whereby interaction between the units is avoided. Reading from left to right, the cabinets contain the groups of apparatus marked units A, B, C_1, D, C_2 in Fig. 280.

7 (b). *An inter-continental short-wave receiver.* As an example of another branch of wireless telephony practice, of even more rapid and recent growth than broadcasting, we take one of the group of short-wave transoceanic beam receivers at Baldock, constructed and

operated by the British Post Office. These receivers, apart from the use of highly directive antenna arrays* which do not here concern us, are especially characterised by the shortness of wavelength, viz. 15–50 m. As in broadcasting, though for a different reason, the complete modulated wave is used. But in this service the emphasis is not on faithful musical reproduction within a few scores of miles, but on good intelligibility of speech for communication over as many thousands of miles†. The chief difficulties to be surmounted are the great and often rapid fading of signal strength from time to time (always exhibited in transmissions relying upon rays reflected from the Heaviside layer), and the interference from other stations.

Owing to the shortness of wave, resort is had to superheterodyne reception. Owing to the fading, very large and widely variable amplification must be available for weak epochs, and the gain should preferably be (and here is) automatically controlled. To avoid interference, close filtering is resorted to; i.e. the overall gain of the receiver is made to fall away very rapidly on each side of the narrow wanted band of frequencies.

A schematic diagram of a complete receiver is given in Fig. 281‡. It will be seen that there are two stages (A) of amplification at the radiation frequency (either 20,000, 15,000 or 10,000 kc/s§), each comprising a push-pull disposition of screen-grid tetrodes. A superheterodyne oscillation (B), with a screen-grid separator stage (C) for constancy of frequency, is then introduced; and the combination is applied to a balanced rectifier (D). The carrier is thus converted from (say) 10,000 kc/s to 300 kc/s.

Since the intermediate frequency of 300 kc/s would be given by incoming waves of two frequencies, say 10,000 and 10,600 kc/s if the superheterodyne oscillator is of 10,300 kc/s, it is a function of

* See XIV–8.

† Transatlantic telephony is effected with received field strengths of the order of 1 μV/m. For good broadcast reception, such as is to be had about 100 miles from Brookman's Park, the B.B.C. consider that the field strength should be 5 mV/m.

‡ See A. G. Lee, *loc. cit.* p. 309. The present author is indebted to the Engineer-in-Chief of the Post Office for additional information, and for the photograph, Plate XXVIII.

§ It is usual to make provision for using at least two alternative wavelengths on any short-wave channel, and to change from one to another as found profitable. One will often fade away completely by day, and the other by night.

7. EXAMPLES OF TELEPHONE INSTALLATIONS

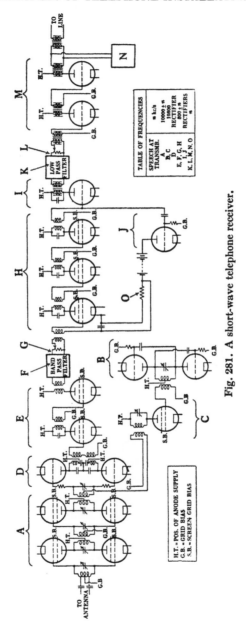

Fig. 281. A short-wave telephone receiver.

the first two stages (A) to discriminate strongly between 10,000 and 10,600 kc/s. Actually the selectivity of these two amplifying stages is such that while the gain for the desired signal is some 25 db, an undesired signal 600 kc/s away suffers a loss of 35 db—a discrimination of 60 db.

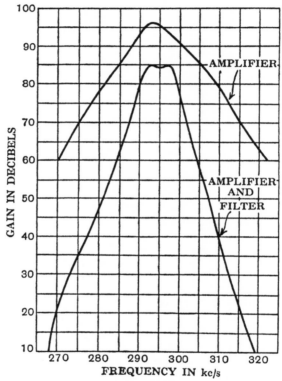

Fig. 282. Frequency characteristics of part of Fig. 281.

It is the intermediate-frequency stages of the superheterodyne receiver which give scope for great amplification and selectivity, especially when, as here, screen-grid tetrodes are used. In the present receiver, after the first detector (D) come two stages of intermediate-frequency amplification (E), then a band-pass filter (F), a potential divider (G) for controlling the power level*, and three more stages (H) of intermediate-frequency amplification. The performance of the

* The potential divider (actually two in cascade) introduces a loss up to 72 db, adjustable in steps of 2 db.

7. EXAMPLES OF TELEPHONE INSTALLATIONS 429

intermediate-frequency amplifier alone (E and H), and of the amplifier and filter together (E and H and F) are given in Fig. 282*.

The wanted part of the signal now consists of an acoustically modulated carrier of 300 kc/s. It is led to the second detector (I), whereafter it is of the acoustic frequency n kc/s. Retroaction from I to H in the automatic gain controller (J) (see below) maintains the signal at appropriate strength. Next comes a low-pass filter (K) designed to cut out frequencies above 3 kc/s, since nothing higher is needed for intelligibility. Again a potential divider (L) to control the power level; then two stages (M) of acoustic amplifier; and finally the signal passes out, to be introduced into the general telephone system of Great Britain. A volume indicator (N) guides the operator controlling the wireless receiver in adjusting the potential divider (G); signals of appropriate strength are passed to the land telephone system whatever (within wide limits) occurs in the transmitter on another continent or—and more especially—in the upper atmosphere which intervenes.

The automatic gain controller (J) is a very simple and effective device. It is a grid detector fed with the same input as the second detector (I). Its rectified anode current flows through a large resistance (O), and the P.D. across this resistance provides the grid bias for the first tube of the intermediate-frequency amplifier (H). A rise of signal strength, caused anywhere prior to the second detector, reduces the anode current of the automatic gain controller and so raises the mean grid potential of the amplifier tube. The operating conditions of this tube are thereby modified in the direction of reduced amplification. By this device, a fluctuation of 50 db in the signal as it leaves the intermediate-frequency filter (F) produces a fluctuation of only about 5 db in the acoustic-frequency signal as it enters the low-frequency filter (K)†. The receiver, being set to work with badly faded incoming radiation, is protected from serious overloading in the acoustic-frequency stages even though the propagation conditions should change suddenly, with an increase in the received field strength of as much as 300 times.

The whole receiver occupies but little space. The assembly is

* From Fig. 22 of the paper by A. G. Lee, *loc. cit.* p. 309, by permission of the Institute of Radio Engineers.

† I.e. carrier power fluctuation of 10^5 : 1 is automatically reduced to about 3 : 1.

shown in Plate XXVIII. The parts are mounted in copper boxes on vertical panels, and only control knobs, meters, etc. can be seen from the front. Careful screening is, of course, necessary, on account of the extremely high frequency in the early stages, and the very large gains introduced in the intermediate-frequency stage. Nevertheless, with the exception of the separate anode battery of the automatic gain controller, all anode and filament supplies (both of this and other like receivers) are taken from common batteries.

This elaborate piece of apparatus receives signals (perhaps from the antipodes) in the form of oscillations at many millions of cycles per second and of enormously varying strength; after treatment by no less than twenty variously functioning thermionic tubes, it passes them out converted and tamed to the likeness of the telephonic signals of the ordinary land-line practice.

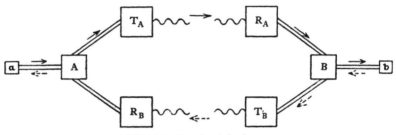

Fig. 283. Duplex telephony.

Since in any practical telephone service it is necessary for either subscriber to become the speaker at any instant, duplex terminal circuits are required. Four-wire working* is adopted, and the complete installation for communication between two sets of land-line terminal apparatus takes the form shown in Fig. 283.

a and **b** are subscribers' instruments in (say) Wolverhampton and Washington. **a** is joined by the ordinary 2-wire circuits to apparatus A located in London. When **a** is speaking, the signals from A pass along a 2-wire line to the wireless transmitter T_A located at Rugby. When **b** is speaking, the radiated signals are received in the wireless receiver R_B (at Baldock if short-wave, at Cupar if long-wave), and pass along a 2-wire line to A and thence to **a**. To prevent interaction between the go and return circuits where they meet at A and B, and by direct radiation from T_A to R_B and from T_B to R_A, voice-operated

* Separate go and return circuits. See text books on line telephony, e.g. E. Mallett, *loc. cit.* p. 423, p. 236.

7. EXAMPLES OF TELEPHONE INSTALLATIONS 431

thermionic devices* are provided which prevent the go and the return circuits from being simultaneously operative. The occurrence of voice currents from a is caused to remove a large loss (about 100 db) normally existent in the go circuit, and to insert simultaneously a like loss in the return circuit. Cessation of speech at a restores the former condition.

* In England, akin to the automatic gain controller of Fig. 281.

CHAPTER XIV

ANTENNAS* AND ANTENNA COMBINATIONS

1. TRANSMITTING ANTENNAS

1 (a). *Long waves.* The antenna in its earliest form, the Hertz doublet, consisted of a straight wire, with or without plates or balls on the ends to localise the capacitance there, with a spark-gap at the middle. The standing waves on this wire, when set into oscillation by the breakdown of the spark-gap, have an antinode of current at the middle and nodes at the ends; one end reaches its maximum positive potential at the instant the other end reaches its maximum negative potential, and at the same instant the current is zero. This is indicated at A, Fig. 40 (p. 77). In the case of the plain wire, the wavelength of the radiation is about twice the length l of the doublet.

G. Marconi's vital contribution from an engineering standpoint to the early scientific work of H. R. Hertz and O. J. Lodge was the enlargement of the short Hertz oscillator into an extensive structure supported high above the ground on masts. He also dispensed with the lower half of the oscillator by joining the potential node to earth, thus arriving at the simple vertical antenna (B, Fig. 40) with a wavelength (unloaded) of about four † times the height h, and a distribution of current falling roughly sinoidally along the wire from a maximum at the bottom to zero at the top. Loading the antenna by an inductance inserted at the bottom lengthens the wave, and modifies the distribution of current and potential as at C, Fig. 40.

* There has been some tendency to use the word *aerial* for an elevated open radiator, i.e. a wire terminating high in the air; and (in this country) the word *antenna* for a long low wire, or a closed wire loop which is usually not spread so obviously wide and high in the air. But any form of antenna depends for its effect on its widespread distribution away from conducting ground; and it therefore seems desirable to use a single word for all forms of the element uniting space radiation with metallic conduction. *Antenna* is preferred to *aerial* because it is easier to pronounce and is common to many languages.

† The exact figure has been the subject of much mathematical discussion. It seems to be about 4·2. It would be 4 if the inductance and capacitance per unit length were the same all along the wire.

1. TRANSMITTING ANTENNAS

It was shown in II-4(c) that the strength of the radiated field is proportional to the product of the magnitude and the vertical length of the current. If, therefore, the whole current at the base of the antenna could be made to flow all the way up to the top, for the same base current the power radiated would be increased; the antenna would have a larger radiation resistance, and the higher the antenna the larger this radiation resistance would be. This is the primary object in attaching to the top of the vertical wire a more or less horizontal system of large capacitance to earth. The distributions of current and potential in the vertical wire then become those of D, Fig. 40. When the capacitance of the top has been made so large compared with the effective capacitance of the uplead that the current in the latter is sensibly uniform, the length of that current is sensibly the full height h of the uplead; and this is the greatest available height with the given masts. For low-power transmitters there is then no advantage in further increasing the capacitance.

With unchanged antenna and wavelength, an increase of power radiated implies larger current in the uplead and a proportionally larger charge CE, where C is the capacitance of the flat top and E its peak potential. When such increase brings E to so high a value (e.g. in the neighbourhood of 200 kV) that corona loss sets in, it is necessary to increase C further. With given height of antenna, large power requires, therefore, an extensive, more or less horizontal, top, carried on a number of masts or towers.

High supporting masts or towers are very costly structures, and the cost rises so rapidly with height that about 250 m has been fairly generally adopted in the designs of recent high-power stations. A wire, or uniform "sausage" of wires, of length h suspended from the top of a mast (say) 150 m high would have a natural wavelength about $4h = 1000$ m. If the antenna were operated at any such short wavelength, for a given current $\mathscr{I}$ amperes the power radiated $\left(\text{viz. } 1580 \dfrac{h^2}{\lambda^2} \mathscr{I}^2 \text{ watts—see II-4 (c)}\right)$ would be conveniently large; moreover, for its capacitance C farads, the antenna potential $\left(\text{viz. } \dfrac{\mathscr{I}}{2\pi n C} \text{ volts}\right)$ would be conveniently small.

However, it is not the power radiated which measures the effectiveness of the transmitter, but the field strength developed by it at the remote receiving station; and in determining the best wavelength to adopt with a given height of transmitting antenna, a dominant consideration is the relation between wavelength and the attenuation between the transmitting and receiving stations. For wavelengths above (say) 1000 m, this consideration calls for greater wavelength the greater the distance of the receiving station. The flat top is added to the antenna upload—at great expense for the supporting masts—with the object first of increasing the effective height of the antenna, and then of reducing the voltage reached. The addition of the flat top itself increases the wavelength*. But attenuation considerations usually call for still longer waves; in which event the antenna is loaded by inductance inserted at the bottom.

As a numerical example, we may consider the high-power Rugby transmitter with the whole antenna (already referred to on pp. 33 and 82); $h = 0.175$ km, $C = 0.045$ μF, $\mathscr{I} = 700$ A. Using the Watson-Eckersley formula for received field strength (p. 42), and taking $a = 0.0013$ and $H = 40$ km, the received field strengths at 100 km, 3600 km and 6400 km are plotted in Fig. 284 for values of wavelength between 4 km and 36 km. A curve is added showing the R.M.S. potential at the top of the antenna. This Figure illustrates the conflict between increase of wavelength to give strong field at great distances, and decrease of wavelength (or reduction of current) to keep the antenna potential below the values causing corona discharge†.

The choice of wavelength, even with any given antenna, is of course influenced by other factors too. For high-speed telegraphic signalling very long waves are unpropitious, and for telephony are ruled out; the strength of atmospheric disturbance depends upon wavelength (see XVII–2); and wavelengths used by other powerful transmitters within range obviously must be avoided.

1 (b). *Short waves.* In recent years communication over very great distances has been found possible, though with interruptions due to fading, by the use of short waves—below 50 m‡. Such short wavelengths make it practicable to use antennas high enough

---

* At Rugby, the natural wavelength of the whole antenna, unloaded, is about 7500 m. The antenna of Plate I, which is carried by masts 150 m high, has a natural wavelength of 925 m.

† This transmitter was actually used at about 16 kc/s (i.e. $\sqrt{\lambda \text{ km}} = 4.34$), with antenna currents up to about 740 A. Brush discharge prevented further increase of the current. The field strength was measured at Houlton, Maine, some 5000 km away, as 580 μV/m.

‡ See III–3, and footnote on p. 426.

1. TRANSMITTING ANTENNAS

to have radiation resistances approaching, and even exceeding, the 40 Ω or so of a vertical grounded wire oscillating at its natural wavelength. The need for a flat top is then no longer felt.

Let us consider a transmitter of frequency n, wavelength λ, working with a vertical antenna about a quarter of a wavelength

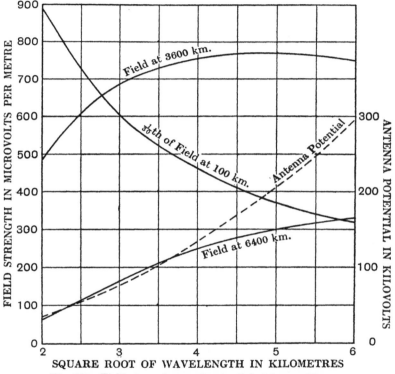

Fig. 284. Field at receiver and potential at transmitter as function of wavelength.

high. The radiation resistance is about 40 Ω. A current $\mathcal{I}$ amperes at the bottom then radiates $40\mathcal{I}^2$ watts; 27 A (R.M.S.) radiates 29 kW*.

If h is the height of the wire, and a its radius, the peak potential E of the antenna is found approximately as follows. Assuming a sine distribution of potential and uniform capacitance per unit

---

* Cf. the long-wave transmitter of II–4(d), where the same power, 29 kW, was radiated by 725 A.

length, the maximum charge in the antenna is $E \times \frac{2}{\pi} C_0$, where C_0 is its static capacitance.

$$\therefore E = \frac{\pi}{2C_0} \int_0^{\frac{1}{2n}} \sqrt{2}\, \mathscr{I} \sin 2\pi nt\, dt = \frac{\mathscr{I}}{\sqrt{2}C_0 n}.$$

Now if h, a, λ are in cm,

$$C_0 \doteqdot \frac{1 \cdot 11 h}{\log_e \frac{\pi h}{4a}} \mu\mu\text{F}^*,$$

and

$$n = \frac{3 \times 10^{10}}{\lambda};$$

whence, in volts and amperes,

$$E = \mathscr{I} \times 85 \log_e \frac{\lambda}{5 \cdot 1 a},$$

$$\doteqdot \mathscr{I} \times 200 \log_{10} \frac{\lambda}{5a}.$$

If $\lambda = 300$ cm, and $a = 0 \cdot 36$ cm (e.g. 7/S.W.G. 18 stranded wire),

$$\log_{10} \frac{\lambda}{5a} = 3 \cdot 2,$$

and $\qquad E = 640 \mathscr{I}.$

When E reaches the breakdown value, e.g. 200 kV,

$$\mathscr{I} = \frac{200{,}000}{640} = 313 \text{ A},$$

and power radiated $= 3900$ kW.

It is seen from this calculation that in short-wave transmitters, even with a single vertical antenna of thin wire, no practical restriction on power is imposed by the potentials reached in the antenna.

1 (c). *Harmonic excitation.* With short waves it is feasible to provide a vertical wire even several quarter-wavelengths in height. Increasing the length from $\frac{\lambda}{4}$ to $2 \cdot \frac{\lambda}{4}$ changes the distribution from that of B to that of A, Fig. 40 (p. 77)†. With the $2 \cdot \frac{\lambda}{4}$-antenna, both the quarter-wavelengths conspire in producing horizontal

---

* See W. H. Eccles, *Wireless telegraph and telephony* (1918), p. 120.

† By appropriate manner of coupling the generator—which is always a triode oscillator, or some form of master oscillator followed by triode amplifiers—the generator can still be situated at the ground even with a distribution approximating to that of A, Fig. 40.

1. TRANSMITTING ANTENNAS

radiation, since the current is in the same direction throughout the wire at every instant. But on further lengthening the wire there are more than one node of current. Then the downward current in one part of the wire produces at any distant point in the horizontal plane a field opposing that produced by the upward current at that instant in another part of the wire. Hence, with a given strength of current antinode in the wire, on lengthening the wire beyond $2 \cdot \frac{\lambda}{4}$, a decrease of horizontal radiation ensues*.

In II-4 we calculated the radiation from a vertical wire carrying a current uniform along the wire; the field strength was found to vary as the cosine of the angle of elevation of the ray, being a maximum for the horizontal and zero for the vertical direction. In the short-wave antennas now under discussion, the current varies along the wire. Examination shows that the interference effects between the components of radiation from the several parts of the wire may cause maximal and vanishing radiation along elevations other than the horizontal and the vertical.

Although the mathematical computation is troublesome, it is easy to understand how this directed distribution of radiation in a vertical plane comes about. Fig. 285 shows the current distribution in a vertical antenna of height $3 \cdot \frac{\lambda}{4}$. Let the shaded area $I_1 \cdot \delta h_1$ (positive) equal the shaded area $I_2 \cdot \delta h_2$ (negative). Then at any remote spot away along a horizontal direction, the field strengths due to the two elements δh_1, δh_2 of the wire are equal in amplitude but of opposite senses, and their resultant field is zero. Along another direction AC, however, the phases of the two components of field differ, not by π, but by

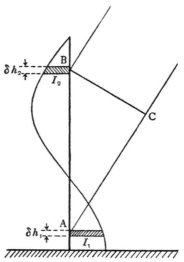

Fig. 285. Interference effects in vertical plane.

---

* It does not follow, as will be seen below, that *for a given total radiated power*, there can be no increase in horizontal radiation.

$\left(\pi - \dfrac{AC}{\lambda}\right)$; and in a direction making $AC = \dfrac{\lambda}{2}$ (as drawn in Fig. 285), the resultant field from the two elements is twice that from either element alone.

Fig. 286* shows the distribution of power (square of field strength)

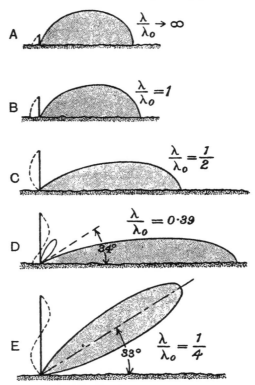

Fig. 286. Vertical-plane directivity curves.

over all elevations around a vertical wire, for the ratios $\dfrac{\lambda}{\lambda_0} \gg 1$, and $\dfrac{\lambda}{\lambda_0} = 1, \tfrac{1}{2}, 0.39$, and $\tfrac{1}{4}$, where λ is the wavelength used and λ_0 is the natural fundamental wavelength of the wire (about $4h$). The power density along any direction is the length of the intercept within the shaded area of the line drawn in that direction from the

* From Figs. 2 and 3 of S. Ballantine, "On the optimum transmitting wavelength for a vertical antenna over perfect earth," *Proc. I.R.E.*, Vol. 12 (1924), p. 833; by permission of the Institute.

foot of the antenna. The diagrams are drawn for the same total power radiated in each case. It is seen that the existence of nodes along the wire may cause concentration of radiation in the horizontal direction; or at an angle of elevation, with even complete cut-off along the horizontal.

It has been held by some experimenters that for obtaining long ranges with short waves the Heaviside layer is used to best advantage when the radiation is directed at some considerable angle of elevation above the horizontal. A distribution such as that in E, Fig. 286, might then be desirable. More recent investigations seem to indicate that concentration in or near the horizontal direction is best. This would favour D, Fig. 286.

A slightly modified method of obtaining vertical-plane horizontal concentration is now commonly used in antenna arrays. (See Subsection 8 (e) of this Chapter.)

With the return to fashion of very short waves—they were used by Hertz in 1888—has come a return to the use of non-earthed antennas too; though now without spark-gap, but excited with triode generators. It has been shown recently that the proximity of the ground may have a large influence on the vertical-plane directivity of a vertical doublet (such as A, Fig. 40 (p. 77)); and curves in variety similar to that of Fig. 286 are given by the Hertz doublet according to the height at which it is suspended above the ground. The effects are complicated by the fact that with very short waves the ground acts by no means as though it were a perfect conductor *.

The vertical-plane directional effects we have been discussing with explicit reference to transmitting antennas can, of course, be utilised in receiving antennas too.

2. Receiving antennas.

The efficiency of a transmitting antenna is the ratio between the power radiated and the power delivered to it from the generator; it is $\frac{r_r}{r_t}$, where r_r is the radiation resistance and r_t the total resistance of the antenna. The transmitting antenna is made high in the endeavour to make r_r large. The efficiency of a receiving antenna, similarly defined, is the ratio between the power it passes to the amplifying and detecting circuits (which we will call the detector power) and the power it absorbs from the radiated field in the

* See the important paper, T. L. Eckersley, "Short-wave wireless telegraphy," *Journ. I.E.E.* Vol. 65 1927\ p 600.

surrounding space. But this ratio is of no practical interest. We desire merely to make the detector power as large as possible with any given strength of received field. If, for example, by readjusting the detector coupling we increase the power delivered to the detector, it is irrelevant whether or not we also produce a change in the power absorbed by the antenna from space.

Let the components of the total antenna resistance r_t be (as in VI-3(b)):

r_r, due to radiation, viz. $1580 \frac{h^2}{\lambda^2}$, where h is the equivalent flat-top height;

r_d, due to the detector, variable at will;

r_h, due to losses (heat).

The E.M.F. produced in the receiving antenna by a given oncoming field is proportional to h, and therefore to $\sqrt{r_r}$. Hence the antenna current is

$$\mathscr{I} = k \frac{\sqrt{r_r}}{r_t},$$

where k is a constant while the transmitting conditions remain unchanged. The useful power, viz. the rate of working on the detector circuits, is therefore

$$\mathscr{I}^2 r_d = \frac{k^2 r_r r_d}{r_t^2}.$$

This is a maximum for choice of r_d (which is entirely under control) when $r_d = \tfrac{1}{2} r_t$, and is then

$$\frac{k^2 r_r}{2 r_t}.$$

On increasing the height of an antenna we effect a proportionate increase in r_r much larger than the proportionate increase in $r_t = r_r + r_d + r_h$*. Hence the higher the receiving antenna, the stronger the signals received.

There are, however, two distinct reasons why such an increase is often of no practical advantage. Firstly, reception often fails not directly from weakness of the incoming signal but from the presence of relatively strong interference from atmospherics or other stations. Nothing is then gained by increasing together the signals and the disturbances. The almost unlimited sensitiveness of modern triode am-

* A rise in height may actually cause a decrease in r_h, by reducing the density of the currents in the earth just below the antenna. The function of the buried earth wires and plates is to reduce this loss.

plifiers strongly reinforces this consideration, and it may be taken that reception nowadays normally fails by interferences of one sort or another rather than by intrinsic weakness of the incoming signal. The need, therefore, is for further discrimination rather than for the increase of receiver sensitivity which would be given by elevating the antenna.

Secondly, the use of retroactive triode circuits makes it possible to reduce the resistance of any circuit to any extent (see XII–1); and this profoundly modifies the conditions governing the design of the receiving antenna. The useful power, with optimum adjustments, was found above as $\frac{k^2 r_r}{2 r_t}$. Let r_t be reduced by triode retroaction to a fraction $\frac{1}{m}$ of the radiation resistance r_r. The detector power is then $m\frac{k^2}{2}$, and is independent of the height of the antenna. It can be stated generally that any two antennas of any sorts or sizes for which the ratios $\frac{r_r}{r_t}$ are equal are equally effective in delivering power to the detector.

This fact renders it unnecessary to employ high receiving antennas, with their costly masts, and leads even to the use of relatively very small closed loops of extremely low radiation resistance, on account of incidental advantages conferred by their directive qualities* and their portability. It can be shown that the radiation resistance of a plane loop of any shape consisting of T close turns enclosing an area A is $31,000 \frac{T^2 A^2}{\lambda^4}$ ohms†. Thus, for example, a loop antenna in a portable receiver for a wavelength of 300 m (1000 kc/s) might consist of 10 turns of wire ½ m square. Its radiation resistance, calculated from the above formula, would be $2 \cdot 4 \times 10^{-5}\,\Omega$. An alternative open antenna consisting of a vertical wire 10 m high would have a radiation resistance of $0 \cdot 7\,\Omega$. Although the heat loss component r_h of the total resistance would in practice be much smaller in the loop antenna circuit than in the open antenna circuit (perhaps $3\,\Omega$ and $30\,\Omega$ respectively), without resort to triode retroaction the signals obtainable with the loop would be vastly weaker than with the open antenna. Nevertheless theoretically, if by the

* See Section 4 of this Chapter.
† See E. B. Moullin, *loc. cit.* p. 11. This assumes that stray capacitance is small enough for the current throughout the wire to be sensibly uniform.

use of retroactive triode connections the total resistance of these two antenna circuits were reduced respectively to say R and $R \times \dfrac{0\cdot7}{2\cdot4 \times 10^{-5}}$, the same power would be delivered to the detector by both antennas.

There are, however, two practical limitations preventing us from pushing this principle indefinitely far. One is the difficulty of maintaining the extremely steady retroactive adjustments necessary for keeping the total antenna resistance almost zero but not negative. The other is that the time required to reach any specified fraction of the final steady amplitude of oscillation (which is approached asymptotically) increases as the resistance is decreased, and may become so great as to restrict the speed of signalling in telegraphy, or to introduce serious acoustic distortion in telegraphy. These limitations have been referred to in XII-1 and in XIII-5 (b).

3. Directive antennas

Any antenna symmetrical about a vertical axis, such as a single vertical wire or an umbrella antenna, obviously radiates (and absorbs) equally in all horizontal directions. Even asymmetric open antennas such as an inclined single wire, an inverted L* or a T antenna show very slight directive effects. The non-directive property of such antennas is a desirable feature in some applications of wireless, e.g. in transmitters used for broadcasting, or in receivers on ships desirous of receiving signals from shore stations or other ships scattered over a wide area.

For many applications, however, directivity would be advantageous. The directional effect may have one or both of two primary features: (i) a concentration of radiated or absorbed power along a particular direction; (ii) an absence or marked diminution of radiation or absorption along a particular direction. In communicating between two fixed points, feature (i) is clearly beneficial, both

* See Fig. 42 (p. 79). An extreme case of the inverted L, however, in which the horizontal top is very much longer than the vertical upload, shows well-marked directive properties, radiating most strongly in the direction tip-to-upload, and absorbing most strongly in the direction upload-to-tip. This type has been much used by the Marconi Co. as an aid in obtaining duplex working with separated transmitting and receiving stations. An example is the transmitting antenna at Caernarvon, where the length of the horizontal top is about nine times the height. There 10 masts 400 feet high support an antenna $\tfrac{2}{3}$ mile long and 500 feet wide. The capacitance is about 0·04 microfarad.

3. DIRECTIVE ANTENNAS

at the transmitter and at the receiver. With a given total power radiated, the desired signals are stronger, the undesired signals and atmospherics arriving at the receiver from other directions are weaker, and neighbouring receivers with which communication is not desired are less subjected to interference from the transmitter, than they would be if non-directive antennas were used.

Advantage of feature (ii) may be taken, both at the transmitters and at the receivers, to assist duplex working with four suitably placed stations; e.g. as in Fig. 283 (p. 430), where a transmitter T_A and receiver R_B near (say) London communicate with a receiver R_A and a transmitter T_B near (say) New York. Feature (ii) at a receiver is beneficial also for eliminating or reducing interference from some particular transmitter or source of atmospherics situated off the line or lines of desired communication.

Finally, if it is possible to swing round at will the direction in which radiation or absorption is concentrated or cut out—as the beam from a lighthouse, or the telescope of a theodolite, is swung round—systems of wireless direction finding can obviously be arranged.

With the small antennas required for very short waves, it is feasible to provide the equivalent of a mirror in optics, in order to direct a proportion of the radiation into a more or less well-defined beam. Fig. 287 shows diagrammatically the geometrical arrangement of an actual transmitter of this sort*. A is the plan view of a vertical wire about half a wavelength long, serving as the antenna of a ½ kW spark transmitter with a wavelength of about 6 m. The dotted line is the trace of a suitable reflector, which might theoretically be a sheet of copper forming part of a parabolic cylinder with focal axis at A. For

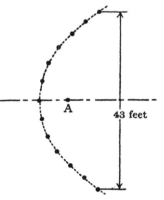

Fig. 287. Antenna with parabolic reflector.

obvious mechanical reasons the sheet is actually replaced by a number of equal vertical wires, indicated in the Figure by dots.

* The rotating beacon on Inchkeith, referred to in Section 6 of this Chapter. See N. Wells, "The Marconi wireless beam reflector on Inchkeith," *Engineering*, Vol. CXIX (1925), p. 309.

444   XIV. ANTENNAS AND ANTENNA COMBINATIONS

These, being of length appropriate to the wavelength used, reflect the radiation incident upon them in much the same way as would the copper sheet. The intensity of signals from this transmitter, measured in a receiver two miles away, gave the polar directivity curve in Fig 288*. In the absence of the reflector, the curve would have been a circle. The concentration of radiation along the axis of parabola is seen to be very marked indeed.

Except with short waves, for transmitting it is seldom economically feasible to use very directive antennas, since here high efficiency $\dfrac{r_r}{r_t}$ is required (see p. 439). Long-wave and medium-wave transmitters

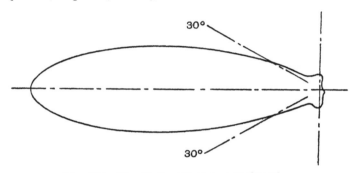

Fig. 288. Directivity of beam at Inchkeith.

ordinarily radiate substantially equally in all directions, even when it is desired to signal only in one direction. For reception, however, various more or less highly directive forms of antenna, and especially of combinations of antennas, can be contrived. The several types are briefly discussed in subsequent Sections of this Chapter.

4. THE BEVERAGE ANTENNA.

A form of antenna which shows high directivity and is suitable for reception is the so-called *wave antenna* of H. H. Beverage, much used since 1923 by the Radio Corporation of America. In its simplest form this consists of a very long wire carried on posts—like a telephone wire on the road side—extending from the receiver in the direction towards the transmitting station. The absorbing property of this peculiar antenna is not due to the height of its

* Copied, by permission, from Fig. 6 of the description in *Engineering, loc. cit.* p. 443. The polar radii are presumably proportional to *square* of field strength.

4. THE BEVERAGE ANTENNA

uplead: indeed, the long wire must not be raised too high above the ground; some 20 ft.–30 ft. is usual. The E.M.F. is introduced into the antenna circuit along the horizontal top, owing to the tilt of the wave-front near the ground (by a degree or so ahead of the vertical) which necessarily accompanies dissipation of power in the ground. With a copper earth the Beverage antenna would not work; it depends for its effectiveness on what, in the functioning of other receiving antennas, is only a defect, although in practice a trivial defect.

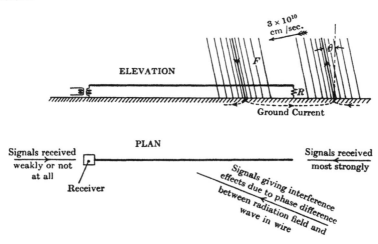

Fig. 289. Beverage antenna.

Fig. 14 (p. 30) shows the plane-polarised wave with vertical wave-front which would travel over the surface of perfectly conducting plane ground. In the upper part of Fig. 289 the electric field is similarly indicated, but the wave-front, incident in the direction of the wire and travelling from right to left, is shown tilted forward through an angle θ owing to the dissipation of power in the imperfectly conducting real ground. Owing to the tilt, where the field strength is F, there is impressed in the horizontal wire an E.M.F. $F \sin \theta$ per unit length. A travelling wave of electric flow is set up in the wire; and if the velocity of propagation in the wire were exactly the same as in free space, as the wave travels along the wire from right to left there would be a building up of amplitude in it which would continue indefinitely with increase of length of the wire. A wave incident at an angle α with the wire affects it as though there were a misfit-ratio $\cos \alpha$ between the wavelength of the incident field and the length of wave set up in the wire; the larger is α, the shorter is the length of wire over which the building up can occur. Actually there is an appreciable difference in the velocities*, so that even when the direction of incidence is along the wire ($\alpha = 0$), on increasing the length of wire beyond a definite point the amplitude at the receiver end would get smaller again. In practice a wire about a wavelength long is adopted, but no exact relationship between the length of wire and the wavelength of the signals is required.

In any transmission line with an open end, a wave travelling in the wire towards that end is there completely reflected; but if the line is terminated on an impedance equal to the *characteristic impedance*† of the line, no reflection occurs. If the antenna were insulated at the right end, signals arriving from the left (e.g. from potentially interfering transmitters or sources of atmospherics situated on the line of the antenna but to the left of the receiver) would build up along the wire from left to right, and would be reflected back to the receiver. To avoid this, the resistance R is connected between the right end and earth, and R is made approximately equal to the characteristic impedance of the line.

Fig. 290‡ is the directivity curve of an ideal simple Beverage antenna one wavelength long. The intercept OP measures the amplitude of the signal produced at the receiver end of the antenna by any given strength of field incident in the direction PO.

The Beverage antennas actually put into service comprise various

---

* The phase velocity (see III–5(a)) at the frequency $\dfrac{p}{2\pi}$ along a line with uniformly distributed inductance L, capacitance C, resistance R and leakance G is $\dfrac{p}{\beta}$; where, when G is negligible, $2\beta^2 = pC(pL + \sqrt{p^2 L^2 + R^2})$. The effect of the resistance of the wire is to increase β above the value $\dfrac{p}{c}$, where c is the velocity of light. In actual Beverage antennas the phase velocity is around 80 % of the velocity of light.

† The characteristic impedance is $\sqrt{\dfrac{R+jpL}{G+jpC}}$. This is approximately $\sqrt{\dfrac{L}{C}}$, and is ordinarily about 600 Ω.

‡ Copied, by permission of the Institution, from Fig. 36 of Beverage, Rice and Kellog, "The wave antenna, a new type of highly directive antenna," *Trans. A.I.E.E.* Vol. XLII (1923), pp. 258, 372 and 510. See also Bailey, Dean and Wintringham, "The receiving system for long-wave transatlantic radio telephony," *Proc. I.R.E.* Vol. 16 (1928), p. 1645.

4. THE BEVERAGE ANTENNA

complications on what has been described above, giving improved directivity and convenience in use; and they are used in groups of several spaced antennas. They appear to be very successful in

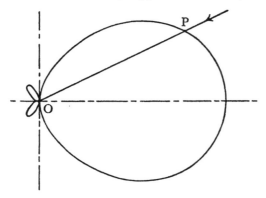

Fig. 290. Directivity of ideal Beverage antenna one wavelength long.

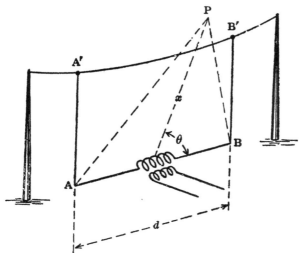

Fig. 291. Two spaced anti-phased antennas.

America; but in England, perhaps owing to differences in the soil, better results have been obtained with combinations of loop antennas[*].

[*] See A. G. Lee, *loc. cit.* p. 309.

448   XIV. ANTENNAS AND ANTENNA COMBINATIONS

5. Loop antennas

Very marked directive effects are obtainable by associating two or more vertical antennas spaced apart but arranged to oscillate isochronously with definite phase relation. A particularly simple and interesting configuration is that shown in Fig. 291. AA' and BB' are two vertical wires hung from insulators at A' and B', distant d apart, containing oscillations of the same amplitude and with

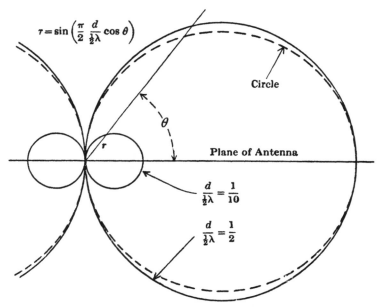

Fig. 292. Directivity curve of Fig. 291.

phase difference 180°. P is a remote point distant x in a horizontal direction inclined at θ to the plane of the antennas. At any instant the electric or magnetic field at P is the sum of the fields due to the two radiators at A and B. Since the radiators are anti-phased, the components at P differ in phase by

$$\pi - \frac{\text{PA} - \text{PB}}{\lambda} \cdot 2\pi,$$

$$= \pi \left(1 - \frac{d}{\frac{1}{2}\lambda} \cos \theta \right) \text{ since } x \gg d.$$

5. LOOP ANTENNAS

If a is the amplitude of each component, the resultant amplitude at P is

$$2a \cos \frac{\pi}{2}\left(1 - \frac{d}{\frac{1}{2}\lambda}\cos\theta\right),$$

$$= 2a \sin\left(\frac{\pi}{2} \cdot \frac{d}{\frac{1}{2}\lambda} \cdot \cos\theta\right).$$

Assuming that $d \not> \frac{1}{2}\lambda$, this varies from zero when $\theta = 90°$ to a maximum $2a \sin\left(\frac{\pi}{2} \cdot \frac{d}{\frac{1}{2}\lambda}\right)$ when $\theta = 0°$; it is then $2a$ if $d = \frac{1}{2}\lambda$, i.e. if the antennas are half a wavelength apart. The shape of the polar curve showing amplitude as a function of θ depends on the ratio $\frac{d}{\frac{1}{2}\lambda}$; when this is small it approximates to a pair of equal tangent circles of diameter $2a \cdot \frac{\pi d}{\lambda}$. Fig. 292 shows the curve for the cases $\frac{d}{\frac{1}{2}\lambda} = \frac{1}{10}$ and $\frac{1}{2}$. The former is indistinguishable from, and the latter is very near to, a pair of circles of diameter $2a \cdot \frac{\pi d}{\lambda}$.

In the converse case of the antenna receiving instead of transmitting, the same function of θ gives the amplitude of the E.M.F. produced in the antenna circuit by horizontally incident radiation which makes an angle θ with the plane of the antenna.

A vertical loop antenna possesses the same directive property, the vertical portions of the loop—or, if the loop is not rectangular, the vertical components of the two halves of the loop—behaving as the pair of vertical antennas just examined.

Directivity curves like those of Figs. 290 and 292 have blunt maxima $\left(\frac{dr}{d\theta} \text{ small near } \theta = 0\right)$ and sharp minima $\left(\frac{dr}{d\theta} \text{ large near } \theta = 90°\right)$. They exhibit feature (ii) of p. 442 in more marked degree than feature (i), and therefore lend themselves specially to the exclusion of unwanted signals in a particular direction. Sharper maxima are obtainable by the use of a parabolic reflector behind a single antenna (Figs. 287 and 288), and as the combination effect of a number of spaced antennas (Section 8 of this Chapter).

The curve for the Beverage antenna in Fig. 290 exhibits sense discrimination along the direction of the wire: signals arriving from

XIV. ANTENNAS AND ANTENNA COMBINATIONS

the right are a maximum, while those from the left are zero. The figure-of-eight directivity curve is not uni-directional in this sense.

In Fig 293 a receiver at R receiving signals from a transmitter at T is protected against interference from a transmitter at I by the use of a suitably orientated loop as the receiving antenna*; but it cannot be protected against interference from I'. But a Beverage antenna with the directivity curve of Fig 290 would cut out I' while receiving T at maximum strength. It is therefore said to be uni-directional.

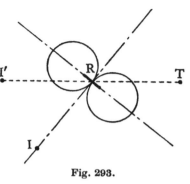

Fig. 293.

By combining the E.M.Fs. received in a loop antenna (figure-of-eight directivity) with the E.M.F., suitably adjusted in phase and

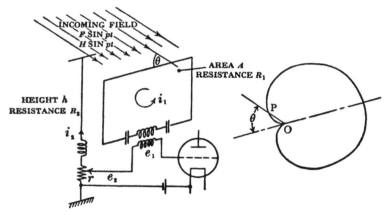

Fig. 294. Cardioid combination.

magnitude, received in a non-directive antenna (circle directivity), a cardioid directivity curve can be obtained. The principle will be readily understood with reference to Fig. 294, which shows what is, perhaps, the simplest out of many possible circuit dispositions. The loop and the open antenna are located close together (in com-

* Although if the angle TRI is (say) less than 45° or more than 135° there is serious reduction of strength of the desired signal also.

5. LOOP ANTENNAS

parison with the wavelength), and are both tuned to the incoming signal. The current in the loop is

$$i_1 = \frac{1}{R_1} \cdot \frac{d}{dt}(AH \cos \theta \cdot \sin pt),$$

where A is the area of the loop (supposed of width small compared with the wavelength);

$$\therefore e_1 = -\frac{p^2 MAH}{R_1} \cos \theta \cdot \sin pt.$$

The current in the open antenna is

$$i_2 = \frac{1}{R_2} \cdot hF \times 3 \times 10^{10} \sin pt^*,$$

where h is the height of the antenna;

$$\therefore e_2 = r \cdot \frac{hF \times 3 \times 10^{10}}{R_2} \sin pt.$$

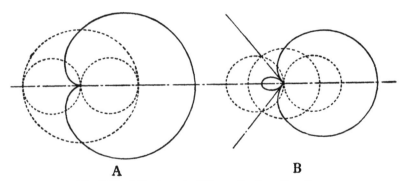

Fig. 295. Adjustment of the extinction directions.

Let the potential divider be adjusted so that

$$r = \frac{p^2 MAH}{R_1} \cdot \frac{R_2}{hF \times 3 \times 10^{10}} = \frac{p^2 MAR_2}{hR_1 \times 3 \times 10^{10}};$$

then
$$e_1 + e_2 = \frac{p^2 MAH}{R_1}(1 - \cos \theta) \sin pt.$$

The plot of the amplitude of $(e_1 + e_2)$ against θ as a polar curve is the cardioid directivity curve in Fig. 294, in which OP measures the strength of signal received along the direction and sense PO.

It will be obvious that this cardioid can be constructed by adding the polar radii of the two separate directivity curves, as at

* Using E.M. units throughout for every quantity except the electric field F. Following II-3, $F = H$.

A, Fig. 295. By merely reducing the value of r on the potential divider, the curves are changed into those of B, Fig. 295, showing that pairs of extinction directions may thus be obtained and adjusted in position.

Fig. 296 shows a modification of the circuit of Fig. 294 which eliminates the separate non-directive antenna. The two halves of the loop serve in parallel as the open antenna.

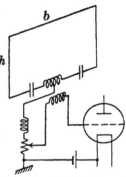

Fig. 296. Modification of Fig. 294.

In all applications of loop antennas involving good extinction effects or measurements which depend on the shape of the directivity curve, it is an urgent, and often difficult, task to avoid the superposition of open-antenna effects. If the two sides of the loop (and any apparatus connected therewith) are not in symmetrical electric relation to earth—notably in their unavoidable stray capacitances—

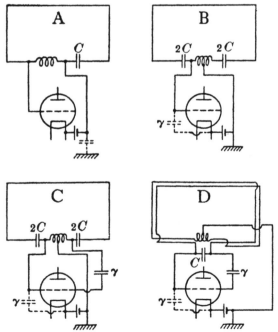

Fig. 297. Stray capacitance with loop antennas.

the equal E.M.Fs. set up in the two vertical sides by an incident ray perpendicular to the loop will cause unequal stray currents to flow, and a resultant circulatory effect in the loop must ensue. The loop will be excited by the radiation falling upon it even when the direction of the radiation is perpendicular to the loop. It is easy to show that in Fig. 296, if the breadth b of the loop is much smaller than λ, the open-antenna E.M.F. impressed by the wave on each vertical side of the loop is $\frac{\lambda}{b}$ times the maximum circulatory E.M.F. round the loop. This is usually a large number; in practical instances it may range (say) from 10 to 1000. If the receiver is to register the loop E.M.F. only, it must be protected against any out-of-balance effects from the much larger open-antenna E.M.Fs. in the sides. The illustrations in Fig. 297 will make this clear. In the highly unsymmetrical disposition A, capacitance between filament and earth would vitiate the balance; even in B open-antenna effect would be present if there were appreciable stray capacitance γ between grid and earth. Symmetry must be maintained by some such device as that in C, Fig. 297; or if the loop has an even number of turns, in D.

6. Direction Finding

6 (a). *Rotating beacons.* Wireless direction finders fall into two primary classes according to whether, in determining the compass bearing of the line joining a transmitting and a receiving station, the directional effect is produced at the transmitter or at the receiver. Installations of the latter class are much more numerous than the former. We will take the former, the D.F.* transmitter, first.

A transmitter which emits radiation distributed in the horizontal plane according to any such curves as those in Figs. 288, 290, 292, and 293, and which is made to rotate about a vertical axis, produces in any receiver signals of strength passing through a succession of maxima and minima determined by the shape of the directivity curve of the transmitter. Thus, if this were the figure-of-eight curve of Fig. 292, the strength of the received signals would change from maximum to minimum (zero) and back to maximum twice per revolution of the transmitter. Since the minimum is well defined

---

* D.F. is the customary abbreviation for direction finder, or direction finding.

454 XIV. ANTENNAS AND ANTENNA COMBINATIONS

and the maximum is ill defined, the position of minimum strength is the more sharply identifiable at the receiver—and is in practice preferred. If the receiving operator is informed of the orientation of the transmitter at the instant when the observed signal strength is a minimum, he can draw on a map a line through the known position of the transmitter, somewhere on which line his ship must be lying. A second like observation taken on signals from another known rotating transmitter provides an intersection fixing the position of the ship on the map.

Ships and aeroplanes do in fact find their bearings by observing the signals from rotating directive transmitters—wireless lighthouses whose beams, immune from fog, are caused to sweep round the compass at a uniform rate. The requisite knowledge of the orientation of the transmitter at any instant is conveyed to the receiving operator by the *quality* (Morse code signs) of the signals whose *quantity* (strength) he is observing. The transmitter emits specified Morse signs at specified positions during the rotation.

A rotating beacon of this kind, erected by the Air Ministry at Farnborough*, transmits on a wavelength of 707 m from a 6-turns loop 5 feet square† rotating at 6° per second (1 revolution per minute). Transmission is maintained throughout the rotation, the signal being in the form of interrupted C.W. (i.e. giving an acoustic note without heterodyning), keyed into specified Morse signs twice per revolution at positions 90° apart when

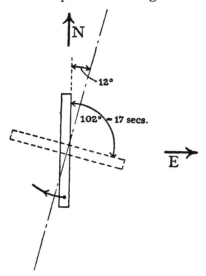

Fig. 298. Reception of beacon signals.

* See T. H. Gill and N. F. S. Hecht, "Rotating-loop transmitters, and their application to direction finding and navigation," *Journ. I.E.E.* Vol. 66 (1928), p. 241.

† A loop of this size at this wavelength has, of course, a very small radiation resistance (see p. 441). But the whole resistance of the loop circuit is kept very low—the cable has 1458 separately insulated strands of S.W.G. 40 wire (see XV-4)—and carries a current of 72 A from a triode oscillator consuming 2·5 kW. Even so, the power radiated must be only about 0·12 W.

6. DIRECTION FINDING

the loop lies exactly N-S and E-W. If the receiving operator starts a stop-watch on hearing (say) the north signal, and observes that the middle of the period during which signals are too weak to be heard occurs (say) 17 seconds later, he knows that his bearing is 12° east of north (or west of south) from the transmitter (Fig. 298).

An example of a different type of transmitting direction finder, much more closely resembling the lighthouse with its sharply concentrated optical beam, is the very-short-wave rotating beacon on Inchkeith in the Firth of Forth, already referred to on p. 443. The whole transmitter with its reflector—in duplicate, for there are two facing in opposite directions—is mounted on a platform which carries a tower 32 ft. high and rotates at $\frac{1}{2}$ revolution per minute. It emits automatically a specified sequence of Morse signals, each of which corresponds with the orientation of the reflector at the instant. As the beam has so sharply defined a maximum (see Fig. 288 (p. 444)), receiving operator on a ship ascertains his bearing by observing which signal is the loudest.

6 (b). *D.F. receivers.* Most direction finding at present practised falls into the second class, i.e. depends on the directivity curve of a receiving antenna. A ship or an aeroplane may be equipped with directional receiving apparatus, and may use it to determine the bearing of the line joining the ship with some known shore transmitter[*]; or a ship not so fitted may request a shore station provided with a D.F. receiver[†] to make the measurement and to communicate the result. Although the elimination of instrumental errors in D.F. reception presents more difficulty on board ship than on land, the technical processes involved are alike.

A receiving antenna in the form of a loop rotatable about a vertical axis can clearly be used to find the orientation of the incident radiation, the signals being strongest when the plane of the loop contains the rays and vanishing when the plane is perpendicular to the rays, according to the figure-of-eight directivity curve. By observing two positions of approximately equal intensity, one on each side of the position of zero or minimum intensity, the operator

---

[*] There are 20 automatic wireless beacons in use or under construction around the coasts of Great Britain and Ireland entirely devoted to this service. Eighteen nations have signed an agreement that from 1 July 1931 all passenger vessels of 5000 tons or over must be fitted with D.F. apparatus.

[†] The British Post Office operates 7 such stations.

is able to determine the orientation (but not the sense) of the incoming rays with remarkable accuracy.

For mechanical reasons it is not practicable to use large rotatable loops, particularly when determinations have to be made rapidly. A development in principle of the rotatable loop method, though historically antecedent, is due to E. Bellini and A. Tosi, with improvements by the Marconi Co. In Fig. 299, A_1, A_2 are two equal loop antennas fixed at right angles; being fixed, they may be large single-turn loops carried on masts instead of small multi-turn loops on movable frames. Each is connected to one or two equal primary windings P_1, P_2 of a *goniometer*. The windings have axes which intersect at right angles, and within them a central secondary coil S is rotatable. The two circuits $A_1P_1C_1$, $A_2P_2C_2$ are made precisely similar, and are set precisely at right angles so that they have no mutual inductance. If each is exactly tuned to the incoming signal (or equally distuned) by means of the condensers C_1, C_2, the oscillations produced will be in phase; and if the width of the loops is small compared with λ, the amplitudes will be proportional to $\cos \theta$ and $\cos \left(\frac{\pi}{2} - \theta\right)$, where θ is the inclination of the incident rays with the plane of one of the antennas (say A_1). The two E.M.Fs. induced in the secondary S by reason of its mutual inductances with P_1 and P_2 will balance for the two positions of S given by

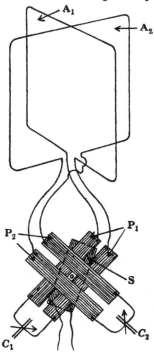

$$\cos \theta \cos \phi + \cos \left(\frac{\pi}{2} - \theta\right) \cos \left(\frac{\pi}{2} - \phi\right) = 0,$$

where ϕ is the angle between S and P_1*;

i.e. $\tan \theta = -\cot \phi,$

$$\theta = \phi \pm \frac{\pi}{2}.$$

Fig. 299. Bellini-Tosi system.

---

* This implies that the mutual inductance between primary and secondary in the goniometer is proportional to the cosine of the angle between them, as it would be if the coil S were very small. But empirical calibration of the goniometer scale makes unimportant whether the cosine relation obtains or not.

6. DIRECTION FINDING

Thus by reading on a scale the position ϕ for zero signals, the orientation (but again not the sense) of the incident radiation is found.

By either of these methods—the rotating loop antenna, or the two fixed loops with a rotating goniometer coil—it is practicable to determine true bearings to within 1°*; but only if the apparatus has been very carefully set up, and various precautions have been taken. These precautions are mainly connected with the avoidance of effects due to the loop antennas acting partly as open antennas as already described with reference to Fig. 297. In the Bellini-Tosi system there is the added complication that the two antennas must be tuned so closely alike that a field incident (say) at 45° to both will produce currents sensibly equal, not only in amplitude but also in phase. Inequality of amplitude would give balance at a wrong angle on the goniometer; inequality of phase would prevent balance at any angle. When the receiver has to be used with signals of various wavelengths, since it is impossible quickly to adjust the two antenna circuits to exact equality, they are sometimes tuned accurately together at a wavelength near that of the signals to be received, and so left. Changes of incoming wavelength then change the strength of the signal but do not destroy the minimum. More often, however, the antennas are used with the tuning condensers C_1, C_2 short-circuited. They are then much less sensitive, but the equalising adjustments are avoided.

With the increase of high-frequency amplification rendered feasible by the advances in triode construction, and especially by use of the screen-grid tetrode, the loop antennas of Bellini-Tosi receivers have been reduced in size. This is particularly advantageous for installations on board ship, where the difficulty of rigging well-balanced large loops was acute. Plate XXIX shows a pair of metal-shrouded Bellini-Tosi loops of modern pattern for ship installations. To aid in combatting the influence of the metal of the ship, the two loops are of unequal sizes; the smaller is set up parallel to the keel, and the larger perpendicular to the keel. Equalisation is effected in the circuits (non-tuned) with which the loops are used. Elimination of the effects of the open-antenna E.M.F. is much facilitated by the very thorough screening of the antennas themselves. The copper tubes seen in the Plate are continuous except for short insulated transverse gaps, without which the screening tubes would act as

---

* Apart from night effect; see the next Subsection.

short-circuited secondary windings. These small rigid screened loop antennas are a striking contrast with the large stretched-wire loops which were formerly used.

The simple rotating loop D.F., and its electrical equivalent the simple Bellini-Tosi D.F., are both subject to the ambiguity of 180° in their indications. This can be removed by the introduction of a non-directive antenna to give the cardioid directivity curve, as described in the last Section. In some installations the direction is first found with precision by a figure-of-eight observation, and the sense in that direction is then ascertained by a rough cardioid observation obtainable when the circuit is modified by the movement of a switch. But the instrumental ambiguity of 180° is often of no moment, for the bearing may be already known to within less than 90°. Moreover, although an aeroplane (for example), desirous of guidance towards an aerodrome where the transmitter is located, may be satisfied by knowledge of its *bearing*, it is more often a fix of *position* which is required. Most directional determinations are therefore made in pairs on two known stations; and the ambiguity of 180° in each direction determination is irrelevant when the position is being fixed as the intersection of two directions.

6 (c). *Night effect*. Direction determinations are subject to many possible errors. By careful design, the purely instrumental errors are reduced under good conditions to less than 1°. But there are other less tractable aberrations depending upon vagaries in the propagation between the two stations. It has been the experience of every operator using a rotating-loop or Bellini-Tosi D.F. that sometimes he gets very well defined bearings (i.e. with well defined minima of signal strength as he swings his coil), and sometimes very ill defined; often so ill defined as to suggest that the signal is arriving about equally in all directions. Even when a sharp bearing is found it may be spasmodically erroneous by very large angles. These aberrations are much more frequent and severe by night than by day, and in general they are much worse with short waves than with long. They are commonly referred to as *night effect* or *night errors*. They have received a great deal of study on account both of the utilitarian importance of combatting them and of the light they throw on the mysteries of the upper atmosphere.

In our examination of the directive quality of the vertical loop antenna, we assumed that the received radiation was incident horizontally, and was plane-polarised with the electric field entirely

6. DIRECTION FINDING

vertical. But where the upper atmosphere is closely involved in the propagation, this ideal condition may be far from the reality*. In what may be called the normal long-wave propagation (as investigated in III–4), except for slight deviations at mountain ranges, rivers, sea-coasts and so on, the course of the ray joining the transmitter to the receiver is the great circle of Earth between the two places. In making a direction determination, we desire to obtain zero E.M.F. in the receiving loop (or its equivalent, the search coil of the goniometer) when its plane is perpendicular to this great circle. The condition for zero E.M.F. is that the line integral of the electric

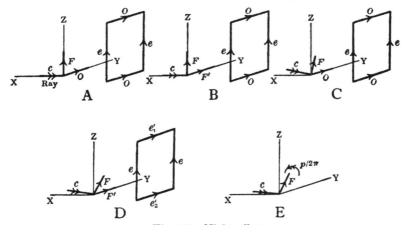

Fig. 300. Night effect.

field round the loop is zero at every instant. If the incident ray is horizontal, we are successful† (A, Fig. 300). If the ray is deflected down from the Heaviside layer, it still lies in the same great-circle plane‡, but the ray at incidence is inclined to the horizontal. If there had been mere bending, with no rotation of the plane of polarisation around the ray, there would still be no horizontal component of electric field in the plane of the loop. Balance would be given with the loop in the same position as before (C, Fig. 300).

Rays which penetrate the upper atmosphere are, however, liable to be changed in polarisation. (See III–4 (d).) If the deflected ray

* See III–3 and 4.

† It would be the same whatever were the polarisation; e.g. B, Fig. 300. Actually, near the surface of the ground, waves travelling horizontally must be almost completely vertically polarised.

‡ Lateral deviations due to what we may think of as Heaviside mountains, etc. seem to be slight.

460　XIV. ANTENNAS AND ANTENNA COMBINATIONS

at incidence possesses a horizontal component F' of electric field (D, Fig. 300), the E.M.Fs. e in the vertical sides still balance, but there are now E.M.Fs. e_1', e_2' in the horizontal sides which do not. In order to obtain a balance, the loop must be rotated round its vertical axis until the E.M.Fs. in the vertical sides are sufficiently unequal to balance the inequality in the horizontal sides. An erroneous bearing is given, the error increasing with the inclination of the ray and with the ratio $\dfrac{F'}{F}$ between the components of the electric field*. Finally, if the deflected ray is circularly polarised (E, Fig. 300), no rotation of the loop about its vertical axis can produce balance.

These aberrations in D.F. reception have in the past more or less completely destroyed its utility at night; but in recent years methods of countering them have been successfully developed. In the above described cases of failure to obtain a balance at the correct position, or any balance at all, the trouble would be removed if the E.M.F. in the horizontal sides of the loop could be eliminated. Theoretically this could be done by tilting the loop about a horizontal axis, so as to bring it perpendicular to the ray; or by using a suitably associated fixed horizontal loop. But since a fresh adjustment would be needed on every change of inclination or of polarisation of the incident ray, it would be no practical solution.

It should be noticed that the screening of the loops as in Plate XXIX does not confer on them the required immunity.

The solution is reached on abandoning the loop in favour of a pair of spaced vertical antennas. Either the horizontal connection between each vertical antenna and the goniometer at the centre is screened from the influence of the waves; or the antennas are so interconnected as to balance out the E.M.Fs. generated in the horizontal parts†. Fig. 301‡ shows an arrangement of screening which has given very good results with short waves; and Fig. 302§

---

* A similar effect is given by the downward ray from an aeroplane transmitter flying within a few miles of the D.F. receiver. The polarisation is not easily controlled, and changes with the movements of the aeroplane.

† These methods were proposed by F. Adcock in 1916, but have been successfully developed only recently.

‡ From Fig. 2 of T. L. Eckersley, "An investigation of short waves," *Journ. I.E.E.* Vol. 67 (1929), p. 992; by permission of the Institution.

§ From Figs. 4 and 5 of R. L. Smith-Rose and R. H. Barfield, "The cause and elimination of night errors in radio direction-finding," *Journ. I.E.E.* Vol. 64 (1926), p. 831. Fig. 303 is from Fig. 6 of the same paper. By permission of the Institution.

6. DIRECTION FINDING

shows an Adcock system for medium wavelengths in which the balancing of the horizontal E.M.Fs. is meticulously effected. In each of the circuit diagrams, only one half of the crossed-antenna system with one field-coil of the goniometer is shown. The other

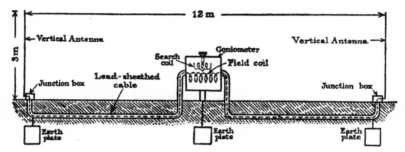

Fig. 301. A short-wave D.F. receiver.

half is exactly similar, and disposed at right angles. The principle will be obvious in each arrangement.

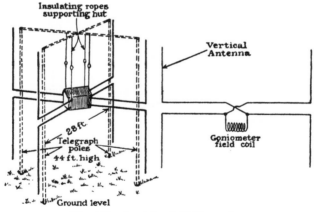

Fig. 302. An Adcock D.F. receiver.

An example of how violent night effect sometimes is, and how much it can be reduced by an appropriate antenna system, is given in Fig. 303. This shows two simultaneous series of determinations at Slough, soon after sunset, of the apparent direction of signals from the broadcasting station at Bournemouth ($\lambda = 386$ m). One series was taken on a rotating loop D.F. receiver, and the other on the Adcock D.F. receiver of Fig. 302.

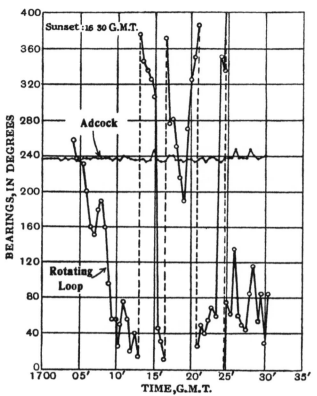

Fig. 303. D.F. aberrations at night.

7. Two spaced antennas

The directive antenna of Fig. 291 (p. 447) is a single oscillator with its ends turned upwards. As a transmitter it is equivalent to the disposition of Fig. 304, if there the generators exciting the two separate antennas produce in them currents of equal frequency and amplitude but displaced 180° in phase. In Fig. 304, however, this particular relation between amplitudes and phases* can be modified

---

* We do not here consider inequality of frequencies in the two antennas; although it might perhaps be of interest as a possible means of obtaining a rapidly rotating or oscillating beam with stationary antennas. In the applications contemplated here, the antennas would in practice be excited from a single source—a local generator if for transmission, the remote transmitter if for reception.

7. TWO SPACED ANTENNAS

at will, with a variety of interesting effects on the directivity curve of the combination.

As the combination of two spaced antennas seems to be more useful in reception than transmission, we will examine it from that aspect. We will confine the investigation to the case of equal antennas, equally and oppositely distuned from the frequency of the incoming radiation, and equally and oppositely coupled* to a combining circuit in which no current flows.

Let the two antennas A, B (Fig. 305) be excited by incoming horizontal radiation of wavelength λ incident at an angle θ with the

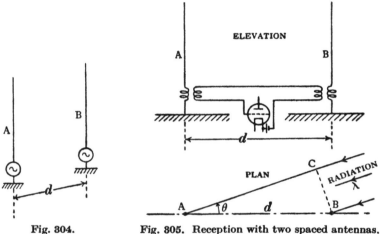

Fig. 304.   Fig. 305. Reception with two spaced antennas.

base line AB. E.M.Fs. $E \sin pt$ and $E \sin (pt + \alpha)$ are impressed in A and B respectively, where $\alpha = 2\pi \dfrac{d \cos \theta}{\lambda}$. Let the antennas be distuned equally in opposite directions so that the currents are equal but lag on the respective E.M.Fs. by $\pm \beta$. Then the currents are

$$I_1 \cos \beta . \sin (pt - \beta) \quad \text{and} \quad I_1 \cos \beta . \sin (pt + \alpha + \beta),$$

where I_1 is the current which would be obtained in each antenna if tuned†. The received signal (i.e. the fluctuation of grid potential in the circuit shown in Fig. 305) is

$$S = S_1 \cos \beta [\cos (pt - \beta) - \cos (pt + \alpha + \beta)],$$

---

* It can be shown that only by this kind of symmetry can an extinction direction be obtained.

† See p. 71.

where S_1 is the signal which would be obtained from one antenna if tuned. That is

$$S = 2S_1 \cos \beta \cdot \sin \frac{\alpha + 2\beta}{2} \cdot \sin \left(pt + \frac{\alpha}{2}\right).$$

We are concerned only with the amplitude, and may drop the factor $\sin \left(pt + \frac{\alpha}{2}\right)$. Accordingly the signal is

$$S = 2S_1 \cos \beta \cdot \sin \frac{\alpha + 2\beta}{2}.$$

The signal vanishes for radiation incident at the angle $\theta = \theta_0$, i.e. for $\alpha = \alpha_0 = \frac{2\pi d}{\lambda} \cos \theta_0$, when

$$\frac{\alpha_0 + 2\beta}{2} = 0 \text{ or } \pi.$$

Then $$\cos \beta = \pm \cos \frac{\alpha_0}{2} = \pm \cos \left(\frac{\pi d}{\lambda} \cos \theta_0\right).$$

Hence for any direction of incidence θ,

$$S = \pm 2S_1 \cos \left(\frac{\pi d}{\lambda} \cos \theta_0\right) \cdot \sin \frac{\pi d}{\lambda} (\cos \theta - \cos \theta_0).$$

With the antennas fixed at any two positions A, B, by distuning them to appropriate values of β between 0 and 45° (thereby reducing the currents to values between I_1 and $\frac{1}{\sqrt{2}} I_1$), the combination can be made completely insensitive to radiation incident at any angles $\pm \theta_0$ whatever. Fig. 306 gives the polar directivity curves for the two separations of the antennas, $d = \frac{1}{10} \lambda$ and $d = \frac{1}{5} \lambda$, when the extinction directions are made $\theta_0 = \pm 90°, \pm 60°, \pm 30°$ and 0. In each diagram the dotted circle is the directivity curve for one antenna alone when tuned to the signal.

It is, of course, possible to combine in similar manner the E.M.Fs. received by two loop antennas spaced apart. The effects obtainable are much more various, since each antenna is by itself directive, and the loops may be set up at various inclinations to the line joining them. Extinction is now caused in two separate ways, viz.: (i) because neither of the loops (being parallel) receives any E.M.F.; and (ii) because the E.M.Fs. in the two loops are balanced in virtue of the special relation between their separation, the angle of in-

7. TWO SPACED ANTENNAS 465

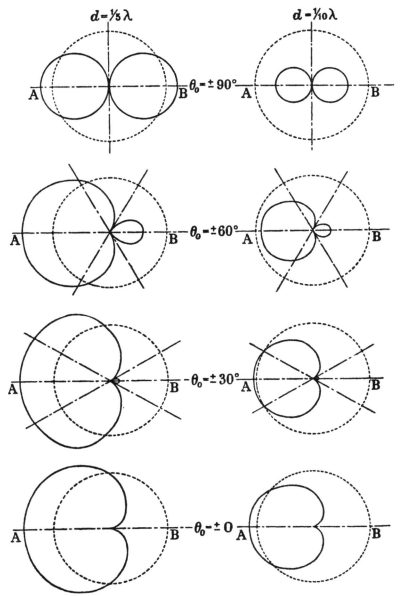

Fig. 306. Directivity curves of Fig. 305.

466 XIV. ANTENNAS AND ANTENNA COMBINATIONS

cidence of the radiation, and the tuning (or other phase-changing) adjustment. Thus in Fig. 307, which is the plan view of two vertical loops set up in the same plane and distant d apart, extinction must occur at $\theta = \pm 90°$, since then neither loop receives any E.M.F. There is (in general) another pair of extinction directions controlled,

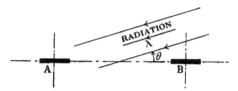

Fig. 307. Two spaced loop antennas.

Fig. 308. Directivity curves of Fig. 307.

7. TWO SPACED ANTENNAS

as in Figs. 305 and 306, by the values of the phase angles $\pm \beta$. Fig. 308 shows a set of polar directivity curves (halves) for the special loop antennas of Fig. 307 of the same nature as the curves in Fig. 306 for the spaced non-directive antennas. The dotted half-circles show the figure-of-eight curve of one loop alone when tuned to the signal.

There is an unlimited complexity of possible antenna combinations from which useful directive results of this nature can theoretically be got. The next step is, obviously, to convert each of the antenna systems A, B into a cardioid, instead of figure-of-eight, system. Moreover the directivity curves of the separate systems A, B may themselves be rotated by the Bellini-Tosi principle, or adjusted in the manner exhibited in Figs. 294 (p. 450), 295 and 306. Further, there is no need to limit the antenna systems A, B ... to two; theoretically any number may be used, with the object of multiplying extinction directions or of further concentrating the sensitivity around one or more favoured directions. The long-wave transatlantic telephony reception is carried out with the aid of a set of 4 spaced Beverage antennas in the United States (Houlton)*, and a set of 6 spaced cardioid combinations (loop and open antenna) in Great Britain (Cupar)†. Fig. 309 is the directivity curve (amplitude ration) of the latter; it shows almost complete concentration within $\pm 30°$ of the most favoured direction. The practical requirement that some or all of the adjustments of these distributed antenna systems shall be effected from a single spot is met by the use of high-frequency transmission lines connecting each antenna system with the operating room.

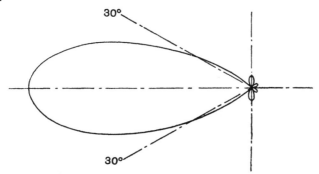

Fig. 309. Directivity curve of antenna combination at Cupar.

* See Bailey, Dean and Wintringham, *loc. cit.* p. 400.
† See A. G. Lee, *loc. cit.* p. 309. Fig. 309 is prepared from Fig. 18 of that paper.

8. Antenna Arrays

8 (a). *Nature of an array*. In the last Section we have looked at the directive action of combinations of two or a small number of antenna units spaced apart by distances which are a considerable fraction of the wavelength. With increasing number of units more and more concentration in a favoured direction can be obtained; but unless some simple relationship exists between the units as regards the amplitude and phase of their contributions to the combining circuit, both theoretical analysis and empirical manipulation become too troublesome. A numerous group of like spaced antenna units, forming a series with simple geometrical and electrical relations between successive units, is called an *antenna array*. Arrays of this kind, commonly built up of units which individually are non-directive, have come into widespread use during the last few years. Historically, antenna arrays and short-wave working have been closely connected; but the relation is primarily an economic one. Short waves have been, and are, widely used with great effect without either directive transmission or directive reception; and arrays of antennas for long waves are absent from engineering practice only because the cost of high masts and large area of ground on which to space them is prohibitive.

8 (b). *Closely packed array*. Consider an array of closely and evenly spaced vertical wires, standing in one vertical plane on flat horizontal ground, excited from a generator so that every wire carries current of the same amplitude and time-phase—a vertical sheet of uniform current density. Fig. 310 shows a plan view of the array AB, of length l. P is a point R, θ at a horizontal distance large compared with l.

If the current per unit length of the array is $I \sin pt$, the field at P due to the element dx of the array is

$$dF = kI\,dx \cos p\left(t - \frac{r}{c}\right),$$

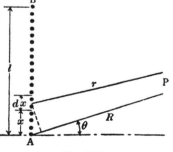

Fig. 310.

8. ANTENNA ARRAYS

where c is the velocity of light, and k is a constant of the given array and frequency $\frac{p}{2\pi}$. Now

$$r = R - x \sin \theta;$$

$$\therefore dF = kI\, dx \cos\left[p\left(t - \frac{R}{c}\right) + \frac{x \sin \theta}{\lambda} 2\pi\right],$$

$$= kI \cos\left(\tau + \frac{2\pi \sin \theta}{\lambda} x\right),$$

where $\tau \equiv p\left(t - \frac{R}{c}\right).$

Hence the field at P from the whole array is

$$F = kI \int_0^l \cos\left(\tau + \frac{2\pi \sin \theta}{\lambda} x\right) dx,$$

$$= \frac{kI\lambda}{2\pi \sin \theta}\left[\sin\left(\tau + \frac{2\pi l \sin \theta}{\lambda}\right) - \sin \tau\right],$$

$$= \frac{kI\lambda}{\pi \sin \theta} \cdot \sin \frac{\pi l \sin \theta}{\lambda} \cdot \cos\left(\tau + \frac{\pi l \sin \theta}{\lambda}\right).$$

As we are interested only in the amplitude of the field F, we will henceforth drop the factor $\cos\left(\tau + \frac{\pi l \sin \theta}{\lambda}\right)$.

If the total current Il had been concentrated in one wire, the field at P would have been

$$F_1 = kIl.$$

Hence the field at P from the antenna array is

$$F = F_1 \frac{\sin \dfrac{\pi l \sin \theta}{\lambda}}{\dfrac{\pi l \sin \theta}{\lambda}}.$$

As the direction θ varies, F vanishes when

$$\frac{l \sin \theta}{\lambda} = \pm 1 \text{ or } \pm 2 \text{ or } \pm 3 \dots,$$

i.e. when $\sin \theta = \pm \dfrac{\lambda}{l}$ or $\pm \dfrac{2\lambda}{l}$ or $\pm \dfrac{3\lambda}{l} \dots.$

If the length of the array exceeds one wavelength, i.e. if $l > \lambda$, there are one or more pairs of equal and opposite values of θ which are extinction directions ($F = 0$).

470 XIV. ANTENNAS AND ANTENNA COMBINATIONS

Fig. 311 is the polar curve showing how F varies with θ with an array two wavelengths wide ($l = 2\lambda$). The dotted circle is the curve which would have been given by the whole current concentrated in one wire.

8 (c). *Any spacing.* Results not very different from this limiting case of wires indefinitely close together are, of course, obtained by the use of a finite number of vertical wires, say over 3 or 4 per wavelength. But there is no difficulty in calculating the field with any spacing between the wires.

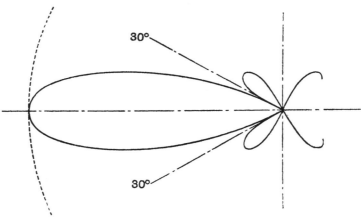

Fig. 311. Directivity curve of Fig. 310 when $l = 2\lambda$.

Let there be N wires $A_1, A_2 \ldots A_N$ (Fig. 312) in the array, spaced a apart, so that the length l of the array is $(N-1)a$. At a distance large compared with l, any antenna A produces a field sensibly the same in amplitude as the next antenna in the series, A', but lagging in phase by an angle

$$\alpha = 2\pi \frac{a \sin \theta}{\lambda}.$$

The total field is the vector sum

$$F = F_1 + F_2 + \ldots + F_N,$$

where F_1 is the side PQ,
  F_2 is the side QR,
    etc.,

and F is the closing side PZ, of the polygon PQRSZ (Fig. 312).

8. ANTENNA ARRAYS

Now
$$PZ = 2 \cdot OP \cdot \sin \frac{N\alpha}{2},$$

and
$$PQ = 2 \cdot OP \cdot \sin \frac{\alpha}{2}.$$

Hence, in magnitude,
$$F = F_1 \frac{\sin \frac{N\alpha}{2}}{\sin \frac{\alpha}{2}} *.$$

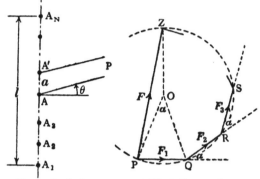

Fig. 312. Antenna array with finite spacing.

Now as θ rises from 0 to 90°;

$\sin \theta$,, ,, 0 to 1;

α ,, ,, 0 to $\frac{2\pi a}{\lambda}$;

$\frac{\alpha}{2}$,, ,, 0 to $\pi \frac{a}{\lambda}$;

$\frac{N\alpha}{2}$,, ,, 0 to $\pi \frac{Na}{\lambda}$.

As θ is changed, $\sin \frac{N\alpha}{2}$ passes through zero several times while $\sin \frac{\alpha}{2}$ is rising from 0 to 1 and falling to 0 again. Each of these values of θ is an extinction direction†. Intermediate maxima of F are NF_1 at $\theta = 0$, and lower values elsewhere.

* Cf. the calculation of E.M.F. in a slot-wound A.C. dynamo.
† Examination shows that, with any given number N of wires, the greatest number of extinction directions is got if the spacing is made $\frac{a}{\lambda} = \frac{N-1}{N}$.

472   XIV. ANTENNAS AND ANTENNA COMBINATIONS

As an example, we take an array of 4 wires spaced $\frac{3}{4}\lambda$ apart; i.e. $N=4$, $\frac{a}{\lambda}=\frac{3}{4}$, $l=2\frac{1}{4}\lambda$. Fig. 313 is the directivity curve plotted from the above formula. Its general resemblance to Fig. 311, where the

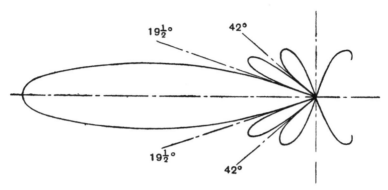

Fig. 313. Directivity curve for an array with finite spacing.

number of wires was indefinitely great, is obvious. The shape of the beam in the favoured direction depends mainly on the length l of

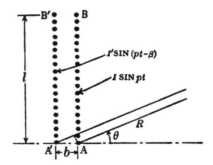

Fig. 314. Reflecting array.

the array, and little on the spacing of the wires in it if there are more than very few.

8(d). *Reflecting array.* In order to suppress the radiation (say) from left to right, i.e. to restrict the beam to only one sense along the normal to the array, a second array may be erected behind the first array, to serve as a reflector.

8. ANTENNA ARRAYS

In Fig. 314, AB is the disposition of Fig. 310; and A'B' is a like parallel array distant b behind, but carrying current

$$I' \sin(pt - \beta).$$

The field at the remote point is the sum of the fields from AB and from A'B'; viz.

$$F + F' = \frac{kI\lambda}{\pi \sin \theta} \cdot \sin \frac{\pi l \sin \theta}{\lambda} \cdot \cos\left(pt - \frac{pR}{c} + \frac{\pi l \sin \theta}{\lambda}\right) *$$

$$+ \frac{kI'\lambda}{\pi \sin \theta} \cdot \sin \frac{\pi l \sin \theta}{\lambda} \cdot \cos\left(pt - \beta - \frac{p.\overline{R + b\cos\theta}}{c} + \frac{\pi l \sin \theta}{\lambda}\right),$$

$$= \frac{k\lambda}{\pi} \cdot \frac{\sin \frac{\pi l \sin \theta}{\lambda}}{\sin \theta} \left[I \cos(pt + \gamma) + I' \cos\left(pt + \gamma - \beta + \overline{\frac{pb \cos \theta}{c}}\right) \right],$$

where $\gamma \equiv -\dfrac{pR}{c} + \dfrac{\pi l \sin \theta}{\lambda}$.

This cannot vanish for any value of θ (except such as makes F itself vanish) unless $I = I'$; and then it vanishes for the direction in which

$$\beta + \frac{pb \cos \theta}{c} = \pi, 3\pi, 5\pi \ldots,$$

i.e. $\cos \theta = \dfrac{c}{pb}(-\beta + \overline{\pi \text{ or } 3\pi}\ldots) = \dfrac{\lambda}{b} \cdot \dfrac{-\beta + \overline{\pi \text{ or } 3\pi}\ldots}{2\pi}$.

To give extinction of left-to-right rays ($\theta = 0$), we therefore make $\dfrac{\lambda}{b} \cdot \dfrac{-\beta + \pi}{2\pi} = 1$; e.g. if $b = \tfrac{1}{4}\lambda$, $\beta = \dfrac{\pi}{2}$; i.e. if the reflector array is $\tfrac{1}{4}\lambda$ behind the first array, its current should lag by 90°. When the reflector is thus adjusted to give extinction for $\theta = 0$, the field in the opposite direction ($\theta = 180°$) is increased $\sqrt{2}$ times.

The reflector can, of course, be separately excited to the exact values of I' and β desired; but it is usual for the sake of simplicity to be content with the current induced in it from the excited array, the natural frequency of the reflector wires being adjusted to suit the spacing b. This is often about $\tfrac{1}{4}\lambda$.

* As found on p. 469.

474   XIV. ANTENNAS AND ANTENNA COMBINATIONS

8 (e). *Height exceeding the half-wavelength.* Advantage is taken of the shortness of wavelength to use arrays of antennas distributed not only horizontally but also vertically. As we noticed in Subsection 1 (c) of this Chapter, if a vertical wire is more than half a wavelength high, parts of it must carry currents in opposition as regards the production of horizontal radiation. In high antenna arrays, the portions of wire in which the current is in the wrong direction are prevented from contributing to the radiation in either

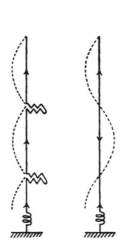

Fig. 315. Franklin's half-wavelength suppression.

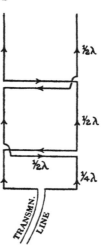

Fig. 316. Half-wavelength suppression by meshwork.

of two ways. First, the negative half-wavelengths of wire are coiled or bunched into a small space, as indicated in Fig. 315*, thus in effect allowing any number of half-wavelength vertical antennas to be piled one on the other. Second, a square meshwork of $\frac{1}{2}\lambda$ side may be used, so arranged that a side occupied by a wire from which radiation is to be suppressed is also occupied by another wire carrying current in anti-phase. The method can be carried out in a variety of patterns of mesh. The principle is illustrated in Fig. 316, where the arrowhead on each wire is placed at the antinode of current, a node (with reversal of direction of current) occurring at each right-angle bend. Right-angle bends may be taken in planes

* This simple and ingenious device is due to C. S. Franklin of the Marconi Co., a pioneer designer of short-wave beam installations.

perpendicular to the paper*; as well as in the plane of the paper, as shown in the Figure.

It is obvious that arrays in 2 or in 3 dimensions can be contrived in great variety. Any superiority of one pattern of array over another seems to reside more in the convenience of its mechanical suspension than in its electrical properties.

8 (f). *Example of antenna array.* Plate XXX shows part of three antenna arrays at the Grimsby short-wave transmitting station†. These arrays are of the Franklin type, explained with reference to Figs. 310, 314 and 315. Two tiers of $\frac{1}{4}\lambda$-suppression coils (Fig. 315) are visible on the right. The three masts at the back of the photograph are 260 feet high with cross-arms 90 feet long. The two bays so formed, each 650 feet wide, together contain the 32 vertical wires, rather over 2λ high and spaced about $\frac{1}{2}\lambda$ apart, which constitute a plane antenna array for transmission to Australia on a wavelength of 26 metres. There are two such arrays; they are suspended on triatics which for one array are attached to the left ends of the cross-arms, and for the other to the right ends. A curtain of insulated reflector wires ($\frac{3}{8}\lambda$ long, $\frac{1}{4}\lambda$ from each antenna) is suspended midway between them. The two arrays are used alternatively, projecting the beam towards Australia either eastwards or westwards; for which is the better route depends upon the distribution of daylight and darkness round the globe.

All the members of the array are excited from a 25 kW triode oscillator through a concentric-tube copper transmission line emerging from the generator and bifurcating repeatedly. Each member is fed through the same total length of line, so that synphase currents may flow. Plate XXXI shows the termination of one of these transmission lines at a junction box containing a high-frequency transformer, by which the impedance of the load is fitted to the characteristic impedance of the transmission line. The leads to the antennas are seen at the top. There is an earth connection for each antenna at its junction box.

* As in the "T.W." arrays used, amongst other patterns, by the British Post Office. See T. Walmsley, "Beam arrays and transmission lines," *Journ. I.E.E.* Vol. 69 (1931), p. 299.

† Designed and erected by the Marconi Co., and now controlled by Imperia and International Communications Ltd.

For the Australian transmitter at Grimsby, some calculations and measurements of the distribution of radiation, in a vertical plane perpendicular to the array, are given by R. M. Wilmotte, "The radiation distribution of antenna in vertical planes," *Journ. I.E.E.* Vol. 68 (1930), p. 1191.

Australia, because of its proximity to the antipodes, is exceptional in being served by a single wavelength. Thus at the Grimsby station there are two other similar arrays (carried on other masts, of which one is seen in Plate XXX), both of them for transmission to India in the shorter great-circle direction. The wavelengths are about 16 m and 33 m, and the two arrays are used alternatively according to the time of day.

Antenna arrays have here been described as from the transmission aspect. But the arrays show the same directional curves in reception, and they are equally used for this purpose. As explained in XIV–1(b), high potentials are not reached in short-wave antennas; as a consequence it may be impossible to tell, from a cursory inspection of the antenna and feeder systems, whether the array is designed for transmission or reception.

CHAPTER XV

DISTRIBUTION OF HIGH-FREQUENCY CURRENT IN CONDUCTORS

1. Slabs

It may be deduced* from the Maxwell equations for the propagation of a wave in a homogeneous isotropic medium that if P (Fig. 317) is a point distant s from a surface AB at which the field is the steady alternating quantity
$$F_0 \sin 2\pi nt,$$
the field at P is
$$F = F_0 \epsilon^{-\frac{2\pi nas}{c}} \sin 2\pi n \left(t - \frac{\beta s}{c}\right) \quad \ldots\ldots\ldots(1),$$
where
$$\alpha^2 = \frac{\mu}{2}\left(\sqrt{\kappa^2 + \frac{4}{n^2\rho^2}} - \kappa\right) \quad \ldots\ldots\ldots\ldots(2),$$
$$\beta^2 = \frac{\mu}{2}\left(\sqrt{\kappa^2 + \frac{4}{n^2\rho^2}} + \kappa\right) \quad \ldots\ldots\ldots\ldots(3),$$
c = velocity of light (*in vacuo*),

and μ, κ, ρ are the permeability (assumed constant), dielectric constant and resistivity of the medium†.

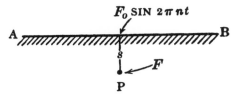

Fig. 317. Penetration into conducting slab.

The amplitude of the disturbance in the medium thus decreases with increasing depth s below the surface according to the factor
$$\epsilon^{-\frac{2\pi nas}{c}} = \epsilon^{-\frac{2\pi as}{\lambda}};$$
and there is an increasing lagging phase-angle
$$\frac{2\pi n\beta s}{c} = \frac{2\pi \beta s}{\lambda};$$

---

* See, e.g., Pierce, *loc. cit.* p. 11, Book II, Chapter VII.
† ρ in C.G.S. E.S. units; $\mu=1$, $\kappa=1$ *in vacuo*.

where λ is the wavelength in empty space. The amplitude is reduced to $\frac{1}{\epsilon} = 0\cdot37$ of its value at the surface at a depth $\frac{\lambda}{2\pi a}$; and the phase is reversed (lag π radians) at a depth $\frac{\lambda}{2\beta}$.

It is unnecessary to specify what aspect of the disturbance at P is under consideration. It might be the component, in any direction, of magnetic field, potential gradient or current density.

AB in Fig. 317 might be the surface of conducting ground, or of a slab of metal, in which currents are set up by the existence of radiation propagated through the atmosphere above, as in Figs. 14 and 15 (pp. 29 and 30). If AB is the surface of the ground, in which $\mu = 1$, $\kappa = 10$, $\rho = 10\,\mathrm{k}\Omega$ per cm cube $= 1\cdot1 \times 10^{-8}$ C.G.S. E.S. units*, for the frequency $n = 10^6$ c/s ($\lambda = 300$ m), formulae (2) and (3) give

$$\alpha = 9\cdot3 \text{ and } \beta = 9\cdot8.$$

The $\frac{1}{\epsilon}$-attenuation depth is therefore $5\cdot1$ m; and the phase reversal depth is $15\cdot3$ m, where the amplitude is already reduced to about $\frac{1}{20}$ of the surface value. In sea water ($\kappa = 80$, $\rho = \frac{1}{10}\,\mathrm{k}\Omega$ per cm cube), the penetration is much less. A practical illustration of this fact is the rapid decrease in strength of the wireless signals receivable by a submarine as the depth of submergence is increased.

If AB is the surface of a copper slab, in which $\mu = 1$, $\rho = 1\cdot78\,\mu\Omega$ per cm cube $= 2 \times 10^{-18}$ C.G.S. E.S. units†, for the same frequency

$$\alpha = \beta = 7 \times 10^5;$$

and the $\frac{1}{\epsilon}$-attenuation depth is $0\cdot068$ mm. If the slab is of iron in which $\mu = 400$‡, $\rho = 16 \times 10^{-18}$ C.G.S. E.S. units,

$$\alpha = \beta = 5 \times 10^6;$$

and the $\frac{1}{\epsilon}$-attenuation depth is $0\cdot0095$ mm.

These calculations show that at 10^6 c/s the depth of sensible penetration into a conducting slab is several metres in moist soil (and is deeper in dry soil); but that in metals only an extremely thin layer next the surface carries any appreciable current: say $0\cdot3$ mm in copper and $0\cdot04$ mm in iron. Fig. 318 shows the $\frac{1}{\epsilon}$-attenuation

---

* These are representative figures for moist soil; see table given by J. A. Fleming, "The principles of electric wave telegraphy and telephony" (1916), p. 800.

† As is customary, we take κ to be negligibly small in metallic conductors.

‡ As found with very weak fields; see p. 235.

2. SCREENING ENVELOPES

depths for these three conducting substances at frequencies between 10 kc/s and 10,000 kc/s.

Consideration of the penetration of high-frequency disturbances into conducting slabs touches wireless technique at two distinct points. Firstly, there is the question of the loss accompanying the currents near the surface of the ground. For a given intensity of radiation above ground, the total current below the surface is fixed.

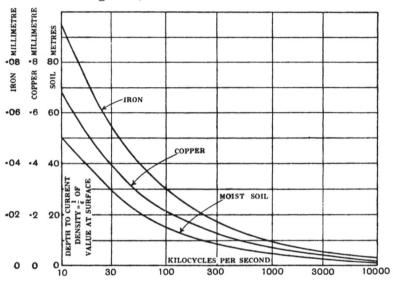

Fig. 318. Penetration in various materials.

The drain of power from the radiant field in the atmosphere above is greater as the crust in which this current flows is thinner; but also it is proportional to the resistivity, and the thickness of the crust increases with the resistivity. The power is fed into the ground by virtue of the forward tilt θ of the wave-front, shown in Fig. 289 (p. 445) with reference to the Beverage antenna. Of the radiant energy in a specified volume near the ground, the proportion $\sin^2\theta$ may be regarded as entering the soil and the proportion $\cos^2\theta$ as being propagated parallel to the ground.

2. SCREENING ENVELOPES

Secondly, there is the question of the screening of one portion of transmitting or receiving apparatus from another, in order that currents flowing in the one shall produce no E.M.F. in the other.

480   XV. DISTRIBUTION OF CURRENT IN CONDUCTORS

Examples of screening between a conducting circuit and the radiant field carrying the signal have been seen in the D.F. antenna system of p. 457 and Plate XXIX, and in the feeder lines of antenna arrays, p. 475 and Plate XXXI.

Wherever a transmitting or receiving disposition embodies a chain of triodes, it is a practical necessity to pay great attention to the screening between one circuit and another. The short-wave receiver, Fig. 281 (p. 427) and Plate XXVIII, provides an example. The screening is effected by segregating the several circuits in separate copper boxes.

We have seen that the metal, whether copper or iron, may be very thin; but the joints of lids and doors must be adequately electrically closed—in extreme cases by the use of mercury seals. Although a closed envelope of iron is more effective than an equal thickness of copper, it is common to use copper because with the latter at high frequencies less power is absorbed from the enclosed apparatus, e.g. an inductance coil located near part of the screening envelope. A quite closed envelope of material (however thin) of infinite conductivity would constitute a perfect screen without introducing any additional resistance into the screened circuits.

3. Skin effect in a wire

Non-uniformity of current density in a conductor forming part of a circuit carrying a high-frequency current must make the dissipation of power larger than when an equal steady current is carried; the alternating-current resistance R_n exceeds the direct-current resistance R_0. Qualitatively, the process of penetration of a steady alternating disturbance from the surface of a wire inwards is of the same nature as the penetration into a slab already discussed. The quantitative analysis is, of course, different. An exact solution for a single straight conductor of circular cross-section was obtained by Kelvin in terms of his ber and bei functions (see p. 484), values of which have been tabulated. We will limit our analysis to an approximate treatment of such a conductor—a treatment which, on account of its mathematical simplicity, is perhaps physically the more illuminating. The disturbance of current distribution we seek to ascertain is caused by the magnetic field of the current; our approximation consists in ignoring the change in that magnetic field which occurs when the current distribution changes*.

* As done by E. B. Moullin, *loc. cit.* p. 73, p. 93. In like manner we are

3. SKIN EFFECT IN A WIRE

Consider a long straight wire of radius a, uniform permeability μ, and resistivity ρ, carrying a current $i = I \cos 2\pi nt$. If the current density were uniform (as with steady direct current) it would be

$$\frac{i}{\pi a^2} = \sigma_0 \text{ (say)};$$

and the magnetic field at a point within the wire distant x from the axis (Fig. 319) would be h, where

$$h \cdot 2\pi x = 4\pi \cdot \sigma_0 \pi x^2;$$

i.e. $\qquad h = 2\pi x \sigma_0 \quad \ldots\ldots\ldots(4).$

The flux threading unit length of a rectangle ABCD, with one side AB on the axis of the wire and of width r, would be

$$\int_0^r \mu h \, dr = \pi r^2 \mu \sigma_0;$$

and there would be an E.M.F. round it

$$e = -\frac{d}{dt}(\pi r^2 \mu \sigma_0),$$
$$= -\pi r^2 \mu \cdot \frac{d\sigma_0}{dt},$$
$$= \frac{2\pi n r^2 \mu}{a^2} I \sin 2\pi nt \ldots(5).$$

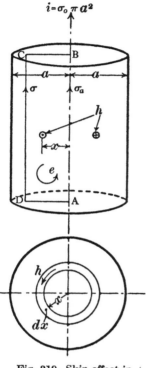

Fig. 319. Skin effect in a wire.

Along AB, therefore, there would be an E.M.F. $\dfrac{\pi n r^2 \mu}{a^2} I \sin 2\pi nt$ per unit length upwards, and along CD an equal E.M.F. downwards.

Consequently the current density can not be uniform across the section. Let it be σ_a at the axis and σ at a distance r therefrom. We continue, however, to take the magnetic field as given by (4), viz. that due to the uniform distribution σ_0 with which we started.

Equating to zero the total change of potential around ABCD, we have

$$-\pi r^2 \mu \frac{d\sigma_0}{dt} + \rho \sigma - \rho \sigma_a = 0,$$

i.e. $\qquad \sigma = \sigma_a + \dfrac{\pi r^2 \mu}{\rho} \dfrac{d\sigma_0}{dt} \quad \ldots\ldots\ldots\ldots\ldots\ldots\ldots\ldots(6).$

accustomed, in calculating the eddy loss in a laminated iron core, to neglect the magnetic field of the eddy currents.

482 XV. DISTRIBUTION OF CURRENT IN CONDUCTORS

But since the total current is $\pi a^2 \sigma_0$,

$$\pi a^2 \sigma_0 = \int_0^a \sigma \cdot 2\pi r\, dr,$$

$$= 2\pi \int_0^a \left(\sigma_a r + \frac{\pi \mu}{\rho}\frac{d\sigma_0}{dt} r^3\right) dr;$$

i.e.
$$\sigma_a = \sigma_0 - \frac{\pi a^2 \mu}{2\rho}\frac{d\sigma_0}{dt} \quad\ldots\ldots\ldots\ldots\ldots\ldots\ldots(7).$$

From (6) and (7),

$$\sigma = \sigma_0 - \frac{\pi a^2 \mu}{2\rho}\left(1 - \frac{2r^2}{a^2}\right)\frac{d\sigma_0}{dt} \quad\ldots\ldots\ldots\ldots(8).$$

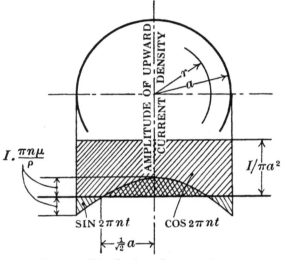

Fig. 320. Distribution of current density.

The actual distribution of current density σ is thus seen to be the uniform distribution

$$\sigma_0 = \frac{1}{\pi a^2} I \cos 2\pi n t \quad\ldots\ldots\ldots\ldots\ldots\ldots(9)$$

which would obtain if n were indefinitely small, with an added distribution

$$\frac{\pi n \mu}{\rho}\left(1 - \frac{2r^2}{a^2}\right) I \sin 2\pi n t \ldots\ldots\ldots\ldots\ldots(10).$$

The second term (10) is an eddy current density which varies across the section of the wire from $\dfrac{I}{\pi a^2} \cdot \dfrac{\pi n \mu}{\rho} \sin 2\pi n t$ at the centre,

3. SKIN EFFECT IN A WIRE

through 0 at $x = \frac{1}{\sqrt{2}} a$, to $-\frac{I}{\pi a^2} \cdot \frac{\pi n \mu}{\rho} \sin 2\pi n t$ at the periphery (Fig. 320).

Since (10) is in time-quadrature with (9), the heat losses of the two terms of current density are independent*. Hence the eddy heat power per unit length of wire at any instant is

$$\int_0^a \left[\frac{\pi n \mu}{\rho} \left(1 - \frac{2r^2}{a^2}\right) I \sin pt \right]^2 \rho \cdot 2\pi r \, dr,$$

$$= \frac{\pi^3 n^2 a^2 \mu^2}{3\rho} I^2 \sin^2 2\pi n t.$$

The mean eddy power per unit length is therefore

$$\frac{\pi^3 n^2 a^2 \mu^2}{6\rho} I^2 \quad \ldots\ldots\ldots\ldots\ldots(11).$$

The mean heat power from the uniform distribution is

$$\frac{\rho}{2\pi a^2} I^2 \quad \ldots\ldots\ldots\ldots\ldots(12).$$

Hence if R_n is the resistance of the wire at frequency n, and R_0 is the D.C. resistance,

$$\left. \begin{array}{l} \dfrac{R_n}{R_0} = \dfrac{(11)+(12)}{(11)}, \\[1em] = 1 + \dfrac{\pi^4 n^2 a^4 \mu^2}{3\rho^2}, \\[1em] = 1 + \dfrac{z^4}{192}, \end{array} \right\} \quad \ldots\ldots\ldots\ldots(13)$$

where $z \equiv 2\pi a \sqrt{\dfrac{2n\mu}{\rho}}.$

The formula (13) is applicable when the eddy distribution (10) is weak compared with the uniform distribution (9), since it is derived by assuming a magnetic field due to (9) alone instead of (9) and (10) together. It will serve to ascertain the relations between size of wire a, permeability μ and resistivity ρ of the metal, and the frequency n, which must subsist if R_n is only slightly to exceed R_0. It shows that, for any specified ratio $\dfrac{R_n}{R_0}$ only slightly exceeding unity, the thickness of the wire is proportional to the square

* $\int_0^{\frac{2\pi}{p}} (A \sin pt + B \cos pt)^2 \, dt \equiv \int_0^{\frac{2\pi}{p}} [(A \sin pt)^2 + (B \cos pt)^2] \, dt.$

root of the wavelength, and to the square root of the resistivity. A constantan or eureka wire would accordingly be about 5 times as thick as a copper wire at the same wavelength.

If the calculated eddy distribution (10) is not weak in comparison with the uniform distribution σ_0, a second approximation might be made by recalculating along the same lines the effect of the neglected magnetic field belonging to the eddy distribution (9). Since the magnetic field of the uniform upward current density σ_0, proportional to $\cos 2\pi nt$ (Fig. 320), was found to produce near the middle of the wire an upward eddy-current density proportional to $\sin 2\pi nt$, the effect of the neglected magnetic field of this eddy-current density must be to introduce near the middle of the wire an upward density proportional to $(-\cos 2\pi nt)$. In this way it may be seen that with rising frequency—more generally, with rising $z \equiv 2\pi a \sqrt{\dfrac{2n\mu}{\rho}}$ —the inner region of the wire becomes more and more devoid of current density; the larger is z, the more completely is the current in the wire confined to a skin of small depth. This condition is known as the *skin effect* in a wire; and the excess of the resistance R_n over the resistance R_0 is called the skin-effect resistance. Writing it R_s, we have

$$R_n = R_0 + R_s.$$

The exact analysis of the flow of alternating current along a straight wire of circular cross-section led, in the hands of Kelvin, to the result that $\dfrac{R_n}{R_0}$ is a function of z only (where z has the meaning we have already attached to it), and is

$$\frac{R_n}{R_0} = \frac{z}{2} \frac{\operatorname{ber} z \cdot \dfrac{d}{dz} \operatorname{bei} z - \operatorname{bei} z \cdot \dfrac{d}{dz} \operatorname{ber} z}{\left(\dfrac{d}{dz} \operatorname{ber} z\right)^2 + \left(\dfrac{d}{dz} \operatorname{bei} z\right)^2} \quad \ldots\ldots(14),$$

where
$$\operatorname{ber} z \equiv 1 - \frac{z^4}{2^2 \cdot 4^2} + \frac{z^8}{2^2 \cdot 4^2 \cdot 6^2 \cdot 8^2} - \cdots,$$

$$\operatorname{bei} z \equiv \frac{z^2}{2^2} - \frac{z^6}{2^2 \cdot 4^2 \cdot 6^2} + \cdots *.$$

* (14) is a troublesome expression for numerical computations; these are much facilitated by simple approximations for certain ranges of z, and by tabulated values of certain functions of ber and bei.

3. SKIN EFFECT IN A WIRE

From the exact result (14) it may be calculated* that the value of $\dfrac{R_n}{R_0}$ given by our approximate formula (13) is too large by 0.4% when $z = 2$, and by 2.2% when $z = 2.5$. But an approximate formula developed by S. Butterworth, viz.

$$\frac{R_n}{R_0} = \frac{\sqrt{2}\,z + 1}{4} \quad\dotfill(15),$$

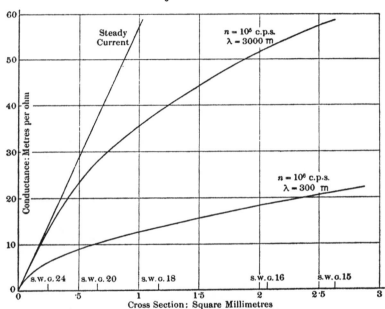

Fig. 321. Conductance of straight copper wire at high frequencies.

gives values less than 4.3% too low when z exceeds 2.5, and less than 1% too low when z exceeds 3.

For carrying large high-frequency current, thick solid conductors would be wasteful, since only a thin outer layer would be of appreciable service. Copper tubing or strip is therefore commonly employed, of which an example is seen in Plate IV. Even in thin wires the effective sectional area may be only a small fraction of the whole area. This is well shown in Fig. 321, where the conductance of straight copper wires expressed as metres per ohm is plotted

* See E. B. Moullin, *loc. cit.* p. 73, p. 96. Moullin gives a table which is very convenient for practical calculations.

against the cross-section in square millimetres, for steady current and for the two frequencies 100 kc/s and 1000 kc/s ($\lambda = 3000$ m and 300 m).

4. Proximity effect in wires

In the single straight wire, the increase of resistance at high frequency is due to the non-uniform distribution of current-density along a radius of the cross-section. But the distribution along every radius is the same. When two straight wires lie alongside each other, or when the wire is formed into an inductance coil, this symmetry no onger obtains. This is known as *proximity effect*, and the consequent

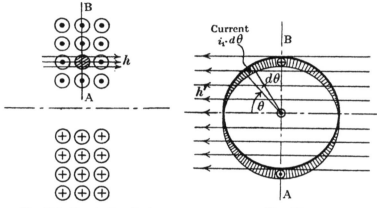

Fig. 322. Proximity effect.   Fig. 323.

further increase in the resistance is called the proximity resistance R_p. $R_n = R_0 + R_s + R_p$. We have already seen that R_s may be several times R_0, and we shall see below that R_p may be several times R_s. Calculations of proximity effects present much difficulty; but they have been carried out fairly completely by various workers, notably G. W. O. Howe, S. Butterworth and C. L. Fortescue.

While skin effect is caused by the magnetic field in the wire associated with the current in the wire itself, proximity effect is caused by the magnetic field in one piece of the wire associated with the current in other pieces of the wire. Thus in a coil of several layers one convolution, such as that with a section shaded in Fig. 322, finds itself in a nearly uniform magnetic field h perpendicular to the wire and proportional to the current in the coil

4. PROXIMITY EFFECT IN WIRES

and (for a given shape of coil) to the number of turns in the coil. Now a uniform magnetic field h' within the cross-section of wire in Fig. 323 would be produced by some such current distribution around the periphery of the cross-section as that indicated by the shaded areas*. At sufficiently high frequencies, therefore, when the penetration must be very slight and the total magnetic field within the cross-section must be sensibly zero, the proximity of the other turns of the coil causes an increase of current density near A and a reduction near B.

The skin resistance R_s and the proximity resistance R_p would both be zero if the current were constrained to flow so that the density were uniform over every cross-section. Theoretically this would be effected on replacing the solid conductor by a cable of very fine insulated strands of the same total copper cross-sectional area, provided that these strands were interchanged in their relative positions in the cross-sections of the cable in such manner that electrically each strand were like every other. Conductors approaching this ideal are constructed by twisting together 3 fine insulated wires (e.g. S.W.G. 38 enamelled and/or S.S.C.) with (say) right-handed lay, so forming group A; twisting together 3 groups A with left-handed lay, so forming group B; twisting together 3 groups B; and so on n times. A cable of 3^n strands is thus formed, in which every strand wanders equally between the centre and the periphery of the whole cross-section, and therefore carries an equal share of the total current. An extreme example of such a cable is that at Rugby, which contains $3^8 = 6561$ separately insulated strands of S.W.G. 36. Typical stranded cables used in British Navy designs† are $3^6 = 729$ strands of S.W.G. 38 enamelled for transmitting coils, and 9 strands of S.W.G. 36 enamelled and S.C.C. for receiving coils.

The principles governing the design of inductance coils of low power factor are now well understood, largely owing to the work of S. Butterworth. It is impossible to summarise briefly the results of analyses containing so many independently variable factors—size and shape of coil; number of layers; spacing between turns and

---

* C. L. Fortescue shows that if i_1 is the current per unit of angle θ the requisite distribution is $i_1 = i_{1\max}.\sin\theta$, giving $h' = 2\pi i_{1\max}$. See "The design of inductances for high-frequency circuits," *Journ. I.E.E.* Vol. 61 (1923), p. 938.

† G. Shearing and J. W. S. Dorling, "Naval wireless communications," *Journ. I.E.E.* Vol. 68 (1930), p. 237.

488 XV. DISTRIBUTION OF CURRENT IN CONDUCTORS

layers; solid wire or stranded; number, size and spacing of strands*—but the examples in the next Section will exhibit the kind of effects obtained and the necessity for instructed design.

The project of a good coil involves so fine a balance between the several components of copper loss—which, as Butterworth shows, is the greater part of the loss in a good coil—that attempts at 'designing by eye' are destined to meet with little success. Designs must be based on numerical computations from formulas derived by intricate mathematical analysis.

If infinitely fine stranding were practicable, and without excessive occupation of the cross-section by the insulation, any coil would be improved at any frequency by the substitution of stranded for solid wire. Under practical conditions the gain to be had from the use of stranded wire proves to be more marked at long than at short wavelengths. Further, to obtain the benefit of stranding, high quality of material and workmanship is requisite. Broken strands, imperfect soldering at the ends, and leakage between the strands may easily cause a well designed costly stranded coil to have a resistance higher than that of a much cheaper coil of the same size wound with solid wire.

5. Examples of inductance coils

(i) A coil of 2000 μH is required for use at $\lambda = 1600$ m. Convenient size and efficient shape are chosen, as in Fig. 324. The Table shows the results of winding the coil with wire of various gauges.

Wire S.W.G.	Ohms				$\dfrac{R_n}{R_0}$
	R_0	R_s	R_p	R_n	
36	21·7	0·0	1·2	22·9	1·05
32	10·8	0·1	4·7	15·6	1·45
28	5·7	0·3	13·7	19·7	3·45
24	2·6	0·5	37·6	40·7	16

There is room in the coil for any thickness of wire up to S.W.G. 18; but already at S.W.G. 24 the total resistance R_n is nearly 3 times the minimum obtainable by the choice of thinner wire.

* Butterworth has summarised the analysis, and presented the results (with experimental confirmations) in a form as convenient for the designer as the complicated nature of the eddy-current phenomena will admit, in his series of articles, "Effective resistance of inductance coils at radio frequency," *Experimental Wireless*, Vol. 3 (1926), pp. 203, 309, 417, 483. Examples (i), (ii) and (iii) in the next Section are taken therefrom, pp. 485, 490 and 491 of those articles.

5. EXAMPLES OF INDUCTANCE COILS

(ii) A certain coil of 202 μH, containing 36 turns of 81/S.W.G. 40 S.S.C., at 600 kc/s shows a calculated resistance of 2·9 Ω. Substitution of wire with 27 strands of the same gauge (in place of 81 strands) reduces the resistance to 1·8 Ω.

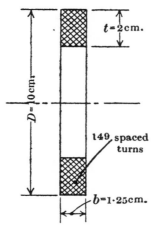

Fig. 324. Multi-layer coil of 2000 μH.

(iii) A series of coils of the same external dimensions—viz. $D = 8·3$ cm, $b = 1·5$ cm, $t = 3·0$ cm—to be used with a condenser of 500 $\mu\mu$F, are wound with the best solid and the best 9-strand wire for each inductance. The Table shows the coils obtained.

Inductance, μH	73	243	2170
Wavelength, m	362	786	1970
Power factor $\dfrac{R}{2nL}$:			
Solid wire	0·75 %	0·86 %	0·93 %
Stranded wire	0·52 %	0·49 %	0·45 %

(iv) The four curves A, B, C, D in Fig. 325 show the calculated resistances R_n* at frequencies between 100 and 1000 kc/s ($\lambda = 3000$–300 m) of 20 metres of solid copper wire of diameter 0·1 cm. (nearly S.W.G. 19). The D.C. resistance R_0 is 0·43 Ω.

* Excluding dielectric and leakage losses; and neglecting the effect of self-capacitance, which becomes important when the frequency approaches the natural frequency of the coil (see Fig. 63, p. 104).

Curve A is for the wire when straight (or, say, bent into a large circle of 20 m perimeter). The excess over 0·43 Ω is the skin resistance R_s.

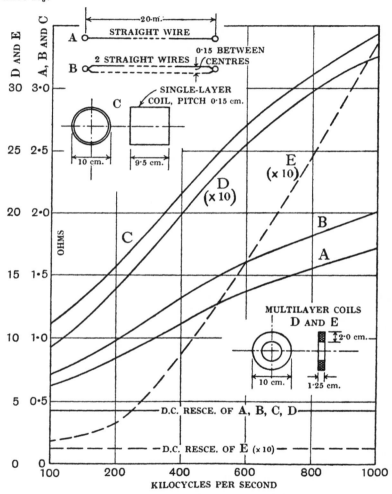

Fig. 325. Resistance of a piece of wire in various configurations.

Curve B is for the wire when connected in parallel with a second like wire, the distance between their centres being 0·15 cm. The excess of an ordinate of curve B over the corresponding ordinate of curve A is the proximity resistance R_p, due to the eddy current in

5. EXAMPLES OF INDUCTANCE COILS

one wire produced by the magnetic field of the current in the other wire.

Curve C is for the wire coiled into a single-layer solenoid of diameter 10 cm and length 9·5 cm, with the same spacing 0·15 cm. The proximity resistance of C exceeds that of B because the eddy current in any one portion of the wire is now produced by the magnetic field of the current in all the rest of the coil. The inductance is about 320 μH, and the power factor $\frac{R}{2nL}$ at 100 kc/s is about 0·55 °/$_o$.

Curve D is for the wire when wound as a multilayer coil of the same external diameter, the winding space being 1·25 cm wide and 2·0 cm deep. The coil is now very compact; consequently the proximity resistance is much increased. The inductance is about 580 μH, and the power factor at 100 kc/s is about 1·87 °/$_o$.

Curve E is for the same coil as D except that the solid wire is now replaced by an equal length of stranded wire 9/S.W.G. 36. The D.C. resistance is increased to 1·28 Ω. The improvement over coil D is very marked at the lower frequencies. At 100 kc/s the power factor is about 0·49 °/$_c$.

CHAPTER XVI

FILTERS

1. General

THE frequency-selective properties of an oscillatory circuit of low damping, and of several such circuits in cascade without retroaction between them, have been examined in XII–1 and in XIII–5 (b). The peaked frequency-response curves obtained can theoretically be given any desired steepness of sides; but only with a corresponding narrowness of the peak (e.g., see Fig. 243, p. 357). By these dispositions sustained signals within a narrow frequency band can be segregated from all outside that band for special treatment—for acceptance or rejection according to the circuit arrangement. In modern high-frequency practice, however, it is becoming more and more necessary to accord privileged treatment, not to a single frequency (or extremely narrow band of frequencies), but to a band of frequencies of finite extent. An instance of such practice is found in the single-side-band transmission described in XIII–4 (g); and many recent developments depend upon means of discriminating sharply between bands of frequencies rather than single frequencies.

The devices by which such band discrimination is achieved are circuital configurations which, in performance, in underlying conception, and in mathematical analysis, contrast somewhat sharply with other branches of wireless technique and theory. They consist of networks of recurrent sections known as *filters**. They might be regarded as chains of closely coupled oscillatory circuits, but they are analysed more conveniently in the terms used in the theory of telephonic propagation along cables. Frequency discontinuities are absent in a uniform line; their suppression is sought in Pupin-loaded lines; they are cultivated in filter networks.

The subject of filters is a large one†. Networks can be designed,

---

* Electric filters of this character are the invention of G. A. Campbell (U.S. patent 1,227,113 of 1917), and appear to have resulted from the study of loaded telephone lines. They have been developed mainly by the American Telephone and Telegraph Co.; chiefly, perhaps, because of the very important part they play in carrier-wave wire telephony.

† Numerous papers on the subject will be found in the *Bell System Technical Journal* since 1922. An authoritative treatise deriving from the Bell Labora-

1. GENERAL

and are indeed used, in great variety; and the analysis is necessarily lengthy. No general treatment of filter networks is here attempted; but the basic ideas are exposed by an examination of certain simple forms, generality being sacrificed in favour of analytical simplicity and brevity.

Consider a chain of n repeated impedances x, y* (Fig. 326), fed at the input end with current I_0 of frequency $\frac{p}{2\pi}$, and delivering current I_n to the terminal apparatus of impedance z_t. The impedances are of any form, but the chain is *passive*, i.e. contains no source of power apart from the generator connected to one end.

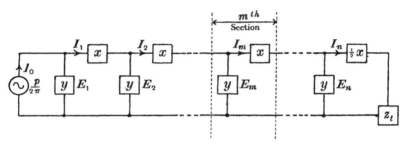

Fig. 326. A general filter chain.

The currents I_0, I_1, I_2, ... are all steady sinoidal currents of frequency $\frac{p}{2\pi}$.

We will suppose first that the terminal impedance z_t is such as would be offered by a continuation of the chain of impedances x, y to infinity. This is called the *image impedance* z_i of the filter. The relation of each section of the filter to the next is then the same throughout; i.e.

$$\frac{I_2}{I_1} = \frac{I_3}{I_2} = \ldots = \frac{I_m}{I_{m-1}} = \ldots = \frac{I_n}{I_{n-1}} = \epsilon^{-k} \text{ (say)};$$

and likewise

$$\frac{E_m}{E_{m-1}} = \epsilon^{-k}.$$

......(1)

---

tories is that of T. E. Shea, "Transmission networks and wave filters" (1929). See also M. Reed, "Electrical wave filters," *Experimental Wireless*, Vol. VII (1930), pp. 122, 190, 256, 315, 382 and 440.

* The reason for choosing the impedance $\tfrac{1}{2}x$ at the end will be seen in Section 5 of this Chapter.

For any two successive sections (Fig. 327) we can write

$$E - E' = xI \quad \ldots\ldots\ldots(2),$$

and $\quad E' = y(I - I') \ldots(3).$

Dividing (2) by (3),

$$\frac{E}{E'} - 1 = \frac{x}{y\left(1 - \dfrac{I'}{I}\right)};$$

i.e. $\quad \epsilon^k - 1 = \dfrac{x}{y(1 - \epsilon^{-k})};$

Fig. 327. Two sections of Fig. 326.

i.e. $\quad \epsilon^k + \epsilon^{-k} = 2 + \dfrac{x}{y};$

i.e. $\quad k = \cosh^{-1} g, \text{ and } \epsilon^{-k} = g - \sqrt{g^2 - 1},\quad \ldots\ldots(4)$

where g is the positive value of $\pm \left(1 + \dfrac{x}{2y}\right).$

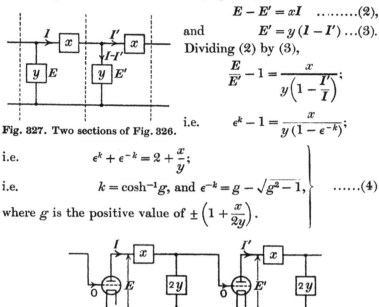

Fig. 328. Chain of non-reacting elements.

The propagation ratio for each section of the filter is the complex number ϵ^{-k}, where k is an ascertained function, given by (4), of the constituent impedances x and y of which the filter is built. The selective property of each filter section depends upon the influence of frequency on ϵ^{-k}; and whatever is accomplished by one section is raised to the power of n by the complete filter containing n sections.

It is instructive at this stage to compare the performance of the n reactively related sections of the filter with that of n oscillatory circuits coupled as a chain without appreciable reaction on each circuit from the next following. (See p. 357 and XIII-5 (b).) Fig. 328 shows diagrammatically two sections of such a chain arranged for treatment analogous to that accorded the filter. The triodes are supposed to prevent reaction between one x, y section and the next preceding, serving merely to pass the signal on without amplification*.

* x may be taken as including the anode resistance ρ of the triode; or ρ may be taken as negligibly small, with $\mu = 1$.

2. PURE-REACTANCE ELEMENTS

Here $E - E' = xI$;
and $E' = 2yI$;

$$\therefore \frac{E'}{E} = \frac{2y}{x+2y} = \frac{1}{g} \quad \ldots\ldots\ldots\ldots\ldots\ldots(5).$$

The chain of n such sections gives the propagation ratio

$$\left(\frac{1}{g}\right)^n \quad \ldots\ldots\ldots\ldots\ldots\ldots\ldots(6);$$

whereas the filter gives

$$\epsilon^{-nk} = (g - \sqrt{g^2-1})^n \ldots\ldots\ldots\ldots\ldots(7).$$

The contrast between the frequency-selective properties of the two chains is the contrast between the manners in which the respective functions of g in (6) and (7) depend upon frequency.

In general k is a complex number, say

$$k \equiv \alpha + j\beta,$$

where α and β are real numbers and $j \equiv \sqrt{-1}$. Then from (4)

$$g = \cosh(\alpha + j\beta),$$
$$= \cosh\alpha \cdot \cos j\beta + \sinh\alpha \cdot \sin\beta \quad \ldots\ldots\ldots(8).$$

For any specified impedances x, y, the real and imaginary parts of g can be written down and equated respectively to the first and second terms of the R.H.S. of (8); whence α and β can be evaluated. Since

$$\epsilon^{-k} I \sin pt \equiv \epsilon^{-\alpha} \cdot \epsilon^{-j\beta} I \sin pt,$$

and $\epsilon^{-j\beta}$ is the operator $(\cos\beta - j\sin\beta)$, the amplitude is attenuated by the factor $\epsilon^{-\alpha}$ in each section, and the phase changed by the lag β per section.

2. PURE-REACTANCE ELEMENTS

When x and y are mixed impedances, the numerical evaluation of α as a function of frequency is in general laborious. We confine our analysis henceforth to filters in which x and y are both either pure resistances[*] or pure reactances, so that g is real.

---

[*] Dispositions with x and y pure resistances exhibit, of course, no filtering action; but it is often useful to make x and y pure resistances in checking formulae. It may thus be verified that in (4) the positive sign of the square root must be taken. Practical filters are ordinarily made as devoid of resistance as can be contrived.

XVI. FILTERS

Then from (8),

$$\left.\begin{array}{r}\sinh \alpha = 0 \\ \cos \beta = g\end{array}\right\} \quad \ldots\ldots\ldots\ldots\ldots\ldots(9);$$

or

$$\left.\begin{array}{r}\sin \beta = 0 \\ \cosh \alpha = g\end{array}\right\} \quad \ldots\ldots\ldots\ldots\ldots\ldots(9').$$

If $\dfrac{x}{2y}$ lies within the range -2 to 0, $(9')$ is impossible and (9) gives

$$\alpha = 0 \ldots\ldots\ldots\ldots\ldots\ldots\ldots\ldots(10).$$

The frequencies corresponding with this no-attenuation range are called *pass frequencies*.

If $\dfrac{x}{2y}$ lies without the range -2 to 0, (9) is impossible and $(9')$ gives

$$\alpha = \cosh^{-1} g,$$
$$\epsilon^{-\alpha} = g - \sqrt{g^2 - 1} \ \ldots\ldots\ldots\ldots\ldots(10').$$

In (10) the peculiar property of the filter is disclosed, viz. that over the finite range of frequencies making $\dfrac{x}{2y}$ lie between -2 and 0, there is no attenuation*. Outside that range there is an attenuation ratio $\epsilon^{-\alpha}$ for each section, where $\epsilon^{-\alpha}$ is the function of x and y given in $(10')$.

The performance of real filters, in which some resistance impurity must be present, can be calculated from the general formula (8); but the pure-reactance formulas (10) and $(10')$ provide a fair approximation. In practical filters resistance has little effect outside the pass frequencies; it has most effect near the limits of the range of pass frequencies; it rounds off the discontinuities given by the theoretical pure-reactance elements.

---

* It is convenient to use the phrase *no attenuation* when, more exactly, unity attenuation ratio is meant. $\alpha = 0$; $\epsilon^{-\alpha} = 1$.

3. Low-pass filter

Let the impedance x be inductive, viz. jpL, and let the impedance y be capacitative, viz. $\dfrac{1}{jpC}$ (Fig. 329). Then

$$\frac{x}{2y} = -\frac{p^2 LC}{2} = -\frac{1}{2}\left(\frac{p}{p_r}\right)^2,$$

where $\dfrac{p_r}{2\pi}$ is the resonance frequency of L and C closed on each other. It follows from (10) and (10′) that there is no attenuation with any values of p lying between

$$0 \text{ and } 2p_r \quad \ldots\ldots\ldots\ldots\ldots\ldots(11);$$

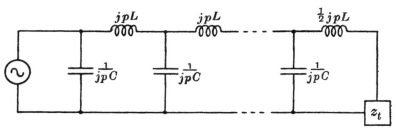

Fig. 329. A low-pass filter.

and at frequencies higher than $2\dfrac{p_r}{2\pi}$, the attenuation ratio for each section is

$$g - \sqrt{g^2 - 1},$$

where g is defined by (4) as the positive value of

$$\pm \left[1 - \frac{1}{2}\left(\frac{p}{p_r}\right)^2\right] \quad \ldots\ldots\ldots\ldots(11').$$

The behaviour, shown by (11) and (11′), of the filter of Fig. 329 should be compared with that of the corresponding non-reactive chain of Fig. 328 with the same elements $x = jpL$ and $y = \dfrac{1}{jpC}$. From (5) the attenuation ratio for each section is

$$\pm \frac{1}{1 - \frac{1}{2}\left(\dfrac{p}{p_r}\right)^2} \quad \ldots\ldots\ldots\ldots\ldots(12).$$

498 XVI. FILTERS

In Fig. 330 the full-line curve shows the attenuation ratio $\frac{E'}{E}$ for each section of the filter (Fig. 329), and the dotted-line curve the ratio $\frac{E'}{E}$ for each section of the corresponding non-reactive chain (disposed as in Fig. 328), plotted on a frequency base. Both the con-

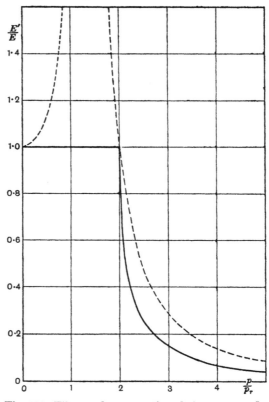

Fig. 330. Filter and non-reactive chains compared.

figurations were supposed to be sensibly devoid of resistance. Accordingly, the dotted-line curve rises to infinity at $p = \sqrt{2}p_r$. But the full-line curve remains at unity until $p = 2p_r$, and then drops down continuously towards zero. The presence of a small amount of resistance would limit the resonance peak of the dotted-line curve, and would round off the discontinuous corner of the full-line curve at $p = 2p_r$.

Because the filter has the property of passing all frequencies below the critical frequency $2\frac{p_r}{2\pi}$ without attenuation, it is called a *low-pass filter*. The critical frequency at which attenuation begins is called the *cut-off frequency*.

4. HIGH-PASS FILTER

On interchanging the elements of the filter in Fig. 329, making $x = \frac{1}{jpC}$ and $y = jpL$, we get the filter in Fig. 331. Here

$$\frac{x}{2y} = -\frac{1}{2}\left(\frac{p_r}{p}\right)^2;$$

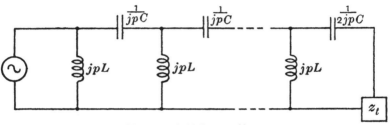

Fig. 331. A high-pass filter.

i.e. p and p_r are interchanged. Equation (10) shows that there is now no attenuation if p lies between

$$\tfrac{1}{2}p_r \text{ and infinity} \ldots\ldots\ldots\ldots\ldots\ldots(13);$$

and equation (10′) shows that at frequencies below $\frac{1}{2}\frac{p_r}{2\pi}$ the attenuation for each section is

$$g - \sqrt{g^2 - 1},$$

where g is defined by (4) as the positive value of

$$\pm \left[1 - \frac{1}{2}\left(\frac{p_r}{p}\right)^2\right] \ldots\ldots\ldots\ldots(13').$$

The attenuation for the corresponding non-reactive chain (cf. (12)) is

$$\pm \frac{1}{1 - \frac{1}{2}\left(\frac{p_r}{p}\right)^2} \ldots\ldots\ldots\ldots(14).$$

Fig. 332 shows the curves of (13) and (13') for the filter, and of (14) for the non-reactive chain, corresponding with the curves of (11) and (11') and (12) in Fig. 330.

Because all frequencies above the critical frequency $\frac{1}{2}\frac{p_r}{2\pi}$ are passed without attenuation, a filter of this character is called a *high-pass filter*.

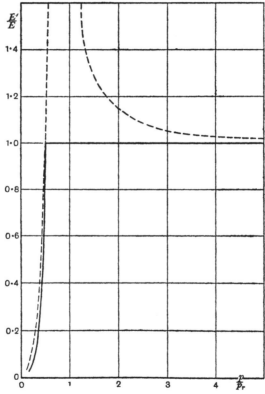

Fig. 332. Filter and non-reactive chains compared.

5. THE TERMINAL IMPEDANCE

In the foregoing analysis it was assumed that the terminal impedance z_t attached to the output terminals of the nth section was the image impedance z_i, viz. the same as the input impedance of an extension of the filter continued indefinitely. When z_t has not this

5. THE TERMINAL IMPEDANCE

image value z_i, the action is much more complex. It may be shown*
that in the filter of Fig. 326 (p. 493) the current at the mth section is

$$I_m = \frac{E_0}{y(\epsilon^k - 1)} [\epsilon^{-km}$$
$$+ B\epsilon^{-k(2n-m)} \quad + AB\epsilon^{-k(2n+m)}$$
$$+ AB^2\epsilon^{-k(4n-m)} \quad + A^2B^2\epsilon^{-k(4n+m)}$$
$$+ A^2B^3\epsilon^{-k(6n-m)} \quad + A^3B^3\epsilon^{-k(6n+m)}$$
$$+ \ldots\ldots\ldots\ldots\ldots\ldots] \quad \ldots(15),$$

where A is a function of x and y only, viz. $-\dfrac{\epsilon^{-k}-1}{\epsilon^k-1}$, and B is a function of x, y and z_t, viz.

$$B = \frac{\dfrac{z_t}{y} - (g - \epsilon^{-k})}{\dfrac{z_t}{y} - (g - \epsilon^k)} \quad \ldots\ldots\ldots\ldots\ldots(16).$$

A physical interpretation of (15) is that the current is multiplied by the propagation factor, the complex number ϵ^{-k}, at each section, and that an infinite series of reflections occurs at the two ends of the filter, thus:

1st term = direct transmission from source (with propagation over m sections);

2nd term = reflection from right end (with propagation over $\overline{2n-m}$ sections; reflection coefficient B at right end);

3rd term = reflection from left end (with propagation over $\overline{2n+m}$ sections; reflection coefficient A at left end);

4th term = reflection from right end (with propagation over $\overline{4n-m}$ sections);

etc.

By a suitable choice of z_t, making $B = 0$, all the reflection terms vanish. From (16) and (4) this occurs when $z_t = z_i$, where

$$\frac{z_i}{y} = g - \epsilon^{-k},$$
$$= \sqrt{g^2 - 1};$$
i.e. $z_i = \tfrac{1}{2}\sqrt{x^2 + 4xy}$ † $\quad\ldots\ldots\ldots\ldots(17).$

---
* See G. W. Pierce, *loc. cit.* p. 11, p. 291.
† This result can be obtained directly from (2) and (3) by writing
$$z_i - \tfrac{1}{2}x = \frac{E'}{I} \text{ (Fig. 327).}$$

For the low-pass filter of Fig. 329 (p. 497), (17) gives

$$z_i = \frac{1}{2}\sqrt{-p^2 L^2 + 4\frac{L}{C}} \quad \ldots\ldots\ldots\ldots\ldots\ldots(18);$$

and for the high-pass filter of Fig. 331 (p. 499),

$$z_i = \frac{1}{2}\sqrt{-\frac{1}{p^2 C^2} + 4\frac{L}{C}} \quad \ldots\ldots\ldots\ldots\ldots\ldots(19).$$

In (18) z_i is a pure resistance if $p^2 LC < 4$, i.e. if $p < 2p_r$ (where $p_r^2 LC = 1$, as hitherto); and in (19) z_i is a pure resistance if $p > \frac{1}{2}p_r$. In both these filters, therefore, a pure resistance as the terminal impedance z_t gives the non-reflecting condition sought; that is, it meets the conditions required for the no-attenuation portions of the curves in Figs. 330 and 332. Moreover, by making L large (and C correspondingly small) in the low-pass filter, and C large (and L correspondingly small) in the high-pass filter, the terminal impedances (18) and (19) required to avoid reflection can be made nearly constant pure resistances over any desired working range of the pass frequencies. With fixed terminal impedance, however, the ideal curves of Figs. 330 and 332 cannot be followed near the cut-off frequencies.

Apart from the great simplification of the analysis ensuing from non-reflective termination of the filter, this condition is favourable to good performance in practice. In any real filter, dissipation of power must occur in the unavoidable resistances of the coils. The use of a non-reflecting terminal impedance reduces this loss to a minimum, and therefore gives the nearest approach in the real filter to the ideal zero attenuation within the pass range of frequencies.

If an impedance $(\frac{1}{2}x + z_i)$, where z_i is given by (17), is inserted also in series with the source of E.M.F. in Fig. 326 (p. 493), the filter is symmetrical with respect to the ends; the same performance is then given when filtering from right to left as from left to right.

6. Band-pass filter

The combination of a high-pass filter with cut-off frequency $\frac{p_1}{2\pi}$ and a low-pass filter with cut-off frequency $\frac{p_2}{2\pi}$ results, if $p_2 > p_1$, in a range of pass frequencies $\frac{p_1}{2\pi}$ to $\frac{p_2}{2\pi}$. By the use of mixed re-

6. BAND-PASS FILTER

actances for the impedances x and y, single filters giving this performance can be contrived. They are known as *band-pass filters*.

Consider the filter shown in Fig. 333. Here

$$x = \frac{1 - p^2 LC}{jpC} \quad\quad\quad\quad\quad (20),$$

$$y = \frac{jpS}{1 - p^2 SK} \quad\quad\quad\quad\quad (21);$$

$$\therefore \frac{x}{2y} = -\frac{(1 - p^2 LC)(1 - p^2 SK)}{2p^2 CS} \quad\quad (22).$$

(10) shows that there is no attenuation when (22) lies between -2 and 0. There are in general two ranges of p which meet this condition; i.e. there are two finite bands of frequencies for which

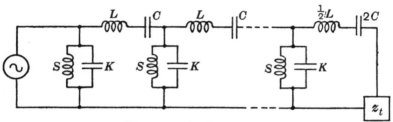

Fig. 333. A band-pass filter.

there is no attenuation. These two pass bands reduce to a single pass band

if $LC = SK$;

or if $C = \infty$ or $K = 0$ or $L = 0$ or $S = \infty$, i.e. if x or y is purely inductive or purely capacitative.

Fig. 334 shows two sections of the single-band-pass filter obtained on making $S = \infty$ in Fig. 333.

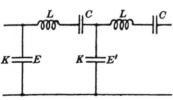

Fig. 334. Two sections of a band-pass filter.

From (22),

$$\frac{x}{2y} = \frac{K}{2C}(1 - p^2 LC) \quad (23).$$

From (10), there is no attenuation when $\frac{K}{2C}(1 - p^2 LC)$ lies between -2 and 0, i.e. when p lies between

$$p_r \text{ and } p_r \sqrt{1 + \frac{4C}{K}} \quad\quad\quad (24),$$

where $\frac{p_r}{2\pi}$ is the resonance frequency of L and C closed on each other; and from (10′), outside this range the attenuation ratio for each section is

$$g - \sqrt{g^2 - 1}, \qquad \ldots\ldots(24').$$

where g is the positive value of $\pm \left[1 + \frac{K}{2C}(1 - p^2 LC)\right]$

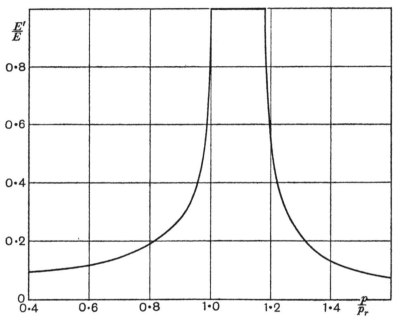

Fig. 335. Frequency characteristic of Fig. 334.

Fig. 335 shows the attenuation ratio $\frac{E'}{E}$ of the filter in Fig. 334 for the case $K = 10C$, calculated from (24) and (24′).

The image impedance of the filter of Fig. 334 is given by (17), (20) and (21) as

$$z_i = \frac{1}{2} \sqrt{\left(\frac{1 - p^2 LC}{jpC}\right)^2 + 4 \cdot \frac{1 - p^2 LC}{jpC} \cdot \frac{j}{-pK}},$$

$$= \sqrt{\frac{L}{C}} \sqrt{\left[1 - \left(\frac{p_r}{p}\right)^2\right] \left[\frac{C}{K} - \frac{(p/p_r)^2 - 1}{4}\right]} \quad \ldots\ldots(25).$$

6. BAND-PASS FILTER

This is a pure resistance for frequencies within the pass band, but falls rather steeply to zero at each of the cut-off frequencies. For

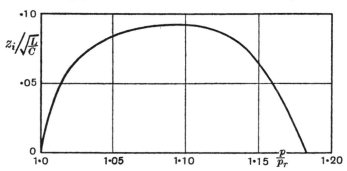

Fig. 336. Image impedance of Fig. 334.

the case $K = 10C$ (as in Fig. 335), its value over the pass band is plotted in Fig. 336.

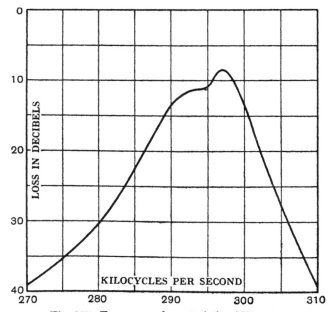

Fig. 337. Frequency characteristic of Fig. 338.

The measured frequency characteristic of a complete real band-pass filter (F in Fig. 281, p. 427) is given in Fig. 337*. The construction of the filter is shown in Fig. 338.

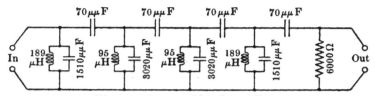

Fig. 338. An actual band-pass filter.

7. Mixed filters

Provided only that the out-end image impedance of one filter section is the same as the in-end image impedance of another filter section, the latter may be connected in succession to the former without causing reflection at the junction. Filters may therefore be constructed of a succession of unlike sections, with unequal attenuating properties, if these sections are individually so constructed as to have correct impedances for avoiding reflection at the junctions.

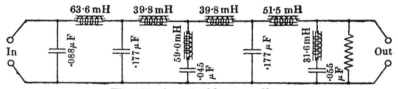

Fig. 339. An actual low-pass filter.

The variety of possible forms of filter is thus unlimited. The practical art of designing them depends on a systematic treatment of a limited number of standard useful forms of filter section. Methods are developed for facilitating the calculation of the critical frequencies and the more significant image impedances, so that trial designs of a composite filter for any specified frequency characteristics and terminal impedances can be prepared without too much labour.

Fig. 339 shows the construction of the low-pass filter K in Fig. 281 (p. 427). The function of this filter is to pass all acoustic frequencies up to 3000 c/s and to suppress all higher frequencies.

* Got as the difference of the ordinates of the two curves in Fig. 282 (p. 428). As explained in XIII–1, an ordinate of n decibels signifies a ratio of $10^{\frac{n}{10}}$ between output and input amplitudes.

CHAPTER XVII

ATMOSPHERICS

1. Nature of atmospherics

If interference were wholly absent, the almost unlimited sensitiveness of modern triode receivers would render a signal, however weak the incoming radiation, competent to actuate the indicating instrument. The presence of interfering disturbance of any sort places a limit on the degree of amplification which can be profitably provided. The chief sources of such disturbance are the waves emitted from wireless stations, and the waves of natural origin called *atmospherics**. The former are systematic, law-abiding, in the double sense that their generation is subject to human regulation, and that their quality approximates to a simple periodic function of time. The latter are unsystematic, random and uncontrollable in origin, quality and quantity. The occurrence of atmospherics is a governing factor in every project for establishing continuous communication on long or medium wavelengths; especially in tropical climates, where their violence is relatively very great.

Until a few years ago not only was the origin of atmospherics unknown, but their nature on arrival at the receiver could only be inferred, with much uncertainty, from their effects on the final indicating instrument, the relay or telephone sounder. By the aid of the cathode-ray oscillograph, however, atmospherics have now been directly measured, in shape of oscillogram, in strength†, and—more recently—in direction of incidence.

The cause of atmospherics is a question belonging more to meteorology than to engineering. Probably they are mainly of terrestrial origin, the result of lightning flashes between cloud and cloud or cloud and earth, perhaps a thousand miles away. The location of the origin, and the form of the wave when it arrives at the wireless receiver, are questions of direct concern to the wireless engineer; for on them must depend whatever defence against atmospherics it may be possible to contrive.

---

* By Americans, often *static*: a curious term, for there seems to be no reason to deny to atmospherics the benefit of Maxwell's equations.

† See Appleton, Watt and Herd, "On the nature of atmospherics," *Proc. Roy. Soc.* A, Vol. 111 (1926), p. 615.

The forms of atmospherics are divers; but they possess the common feature, at least in their gross structure, of non-periodicity. They do not resemble the undamped or moderately damped waves employed for signalling. The majority are quite aperiodic in the sense of showing no reversal of field. A typical form is a rise of field strength from zero to a maximum—often of the order of $\frac{1}{10}$ volt per metre—and a fall to zero again, in a time of order of $\frac{1}{500}$ second. Atmospherics which do show reversals show only one or two; and it is only in rare instances that the positive and negative parts are not very unequal. Even in long-wave receivers the rise and fall of the atmospheric covers very many periods of the free oscillation of the receiving circuit, so that resonance phenomena are not approached. It is for this reason that reception can be carried on through atmospherics with fields thousands of times stronger than that of the working signal.

Closer examination of atmospheric wave forms, however, discloses in some instances a fine-structure ripple with frequency of the order of 10 kc/s and maximum amplitude of the order of $\frac{1}{10}$ of the gross-structure maximum field. It may be that this fine structure provides a large part of the atmospheric interference experienced. But since even such ripples are very far from synchronous with the signalling wave, are irregular and of brief duration, and occur at random instants with random frequencies, interference by atmospherics cannot be analysed in the steady-state terms appropriate to the calculation of interference by wireless transmitters. This conclusion is supported by the experimental observation that atmospherics are not found to afflict particularly severely receivers operating on particular wavelengths; but that, roughly speaking, the violence of atmospheric interference increases more or less continuously and smoothly with increase of wavelength throughout the whole of the range of wavelengths occupied by wireless telegraphy.

2. The Atmospheric as a Pulse

Mathematical examination of the effect on the receiving circuit of pulses of various analytic forms, simulating the oscillographic observations, shows that the effect depends mainly on the initial rate of change of the field strength and on the duration of the main portion of the pulse. In selecting a form for analytical treatment,

2. THE ATMOSPHERIC AS A PULSE

a highly damped sine wave has much to recommend it. If the field is taken as the pulse

$$F \epsilon^{-at} \sin qt,$$

by suitable selection of values for F, a and q, the pulse can be made to resemble fairly closely various typical observed forms of atmospheric. It is a mathematically tractable form, and it will serve at least to give some indication of the way in which receiving circuits are disturbed by atmospherics. For the damping exponent a we adopt the value $\frac{1}{2}q$*. To accord with the duration of $\frac{1}{500}$ second mentioned above, q must be thought of as about 500π sec^{-1}.

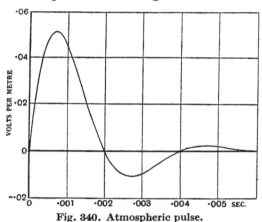

Fig. 340. Atmospheric pulse.

F may, of course, be anything from zero up to enormous potential gradients from a near lightning flash; it may be thought of as 0·1 V/m in typical observations. Fig. 340 shows the atmospheric pulse when $a = \frac{1}{2}q$, scaled for $q = 500\pi$ sec^{-1} and $F = 0·1$ V/m.

Consider, then, an E.M.F.

$$E_a \epsilon^{-\frac{1}{2}qt} \sin qt \quad \dots\dots\dots\dots\dots\dots\dots(1)$$

impressed by an atmospheric in a receiving circuit LRC of damping exponent $b = \dfrac{R}{2L}$, resonance frequency $\dfrac{p}{2\pi}$ where $p^2 LC = 1$, and decrement $\delta = \dfrac{2\pi}{p} b$. $\dfrac{b}{p}$ is not assumed to be small, but $\dfrac{q}{p} \ll 1$. The

---

* As suggested by E. B. Moullin, "Atmospherics and their effect on wireless receivers," *Journ. I.E.E.* Vol. 62 (1924), p. 353. In this Section we begin by following the analysis there given; but, for a reason which will appear, we do not assume (as in that paper) that the receiving circuit is very lightly damped.

current i set up in the circuit is determined by the equation

$$L\frac{d^2i}{dt^2} + R\frac{di}{dt} + \frac{1}{C}i = E_a \epsilon^{-\frac{1}{2}qt}(-\tfrac{1}{2}q \sin qt + q \cos qt) \ldots(2).$$

The particular integral of this equation is approximately*

$$E_a Cq\, \epsilon^{-\frac{1}{2}qt}(-\tfrac{1}{2}\sin qt + \cos qt) \quad \ldots\ldots\ldots\ldots(3);$$

and the complementary function, for $i = 0$ and $\dfrac{di}{dt} = 0$ at $t = 0$, is found to be

$$E_a Cq\, \epsilon^{-bt}\left(\frac{q-b}{p} \sin pt - \cos pt\right)\ldots\ldots\ldots\ldots(4).$$

The P.D. developed across the inductance, say v_a, is

$$v_a = L\frac{di}{dt},$$

where i is the sum of (3) and (4). This gives

$$v_a = -\frac{5}{4} E_a \frac{q^2}{p^2} \epsilon^{-\frac{1}{2}qt} \sin(qt + \psi)$$

$$+ E_a \frac{q}{p^3} \sqrt{(p^2+b^2)(p^2+\overline{b-q}^2)}\, \epsilon^{-bt} \sin(pt+\phi) \ldots\ldots(5),$$

where ψ and ϕ are phase angles (whose values do not interest us). Since $p \gg q$ and $b \not> p$, the radical expression in the second term in (5) simplifies to approximately $\sqrt{(p^2+b^2)^2}$, and (5) becomes

$$v_a \doteqdot E_a \frac{q}{p}\left[-\frac{5}{4}\frac{q}{p}\epsilon^{-\frac{1}{2}qt}\sin(qt+\psi) + \left(1+\frac{b^2}{p^2}\right)\epsilon^{-bt}\sin(pt+\phi)\right](6).$$

Whether $\dfrac{b}{p}$ is small or not, the initial amplitude of the low-frequency (the first) term in (6) is of the order $\dfrac{q}{p}$ times that of the high-frequency (the second) term; its frequency is widely different from that of any subsequent selective circuits in the receiver, and it is moreover heavily damped. The low-frequency term may therefore be neglected. We are left with the high-frequency P.D. of amplitude

$$E_a \frac{q}{p}\left(1+\frac{b^2}{p^2}\right)\epsilon^{-bt} \quad\ldots\ldots\ldots\ldots\ldots\ldots(7).$$

---

* See Moullin, *loc. cit.* p. 509, p. 356.

2. THE ATMOSPHERIC AS A PULSE

If there are N random atmospherics per second, and the effect of each has died away in the LRC circuit before the next begins, the mean value of (7) is

$$\frac{N}{b} E_a \frac{q}{p} \left(1 + \frac{b^2}{p^2}\right),$$

$$= \frac{NE_a q}{p^2} \left(\frac{p}{b} + \frac{b}{p}\right) \quad \ldots\ldots\ldots\ldots\ldots\ldots(8).$$

This mean value (8) measures the rectified current produced by the atmospherics if the P.D. (6) is applied to a linear-law detector (which represents the usual working condition fairly well).

To obtain a measure of the interference caused by the atmospherics, (8) must be compared with the mean P.D. produced by the desired signal. This we will represent by the side-wave

$$E_s \sin(p + \omega) t \quad \ldots\ldots\ldots\ldots\ldots\ldots(9),$$

corresponding with a telephonic (or Morse keying) modulation at frequency $\frac{\omega}{2\pi}$ of a carrier-wave of the frequency $\frac{p}{2\pi}$ to which the circuit is tuned. Assuming that the steady state is sensibly reached, the amplitude of current produced is $\frac{E_s}{R} \cos\theta$ (see p. 71), where

$$\tan\theta = \frac{\pi}{\delta} \left(\frac{p+\omega}{p} - \frac{p}{p+\omega}\right),$$

$$\doteqdot \frac{\pi}{\delta} \frac{2\omega}{p} = \frac{\omega}{b};$$

$$\therefore \cos\theta = \frac{b}{\sqrt{b^2 + \omega^2}}.$$

The P.D. across the inductance is therefore of amplitude

$$pL \frac{E_s}{R} \cos\theta = E_s \frac{p}{2\sqrt{b^2 + \omega^2}} \quad \ldots\ldots\ldots\ldots(10).$$

We may assess the interference as the ratio of the rectified currents from atmospherics and from the signal. Accordingly the $\frac{\text{atmospheric}}{\text{signal}}$ ratio, say A, is

$$A \equiv \frac{(8)}{(10)},$$

$$= \frac{2NE_a q}{E_s p^3} \left(\frac{p}{b} + \frac{b}{p}\right) (b^2 + \omega^2)^{\frac{1}{2}} \quad \ldots\ldots\ldots(11).$$

The chief interest of the expression (11) found for A is that it indicates how atmospheric interference is likely to depend on the damping of the receiving circuit. If the signal is in tune with the receiving circuit (as it might be with slow-speed Morse signalling), $\omega = 0$, and (11) becomes

$$A_{\omega=0} = \frac{2NE_a q}{E_s p^2}\left(1 + \frac{\delta^2}{4\pi^2}\right) \quad \ldots\ldots\ldots\ldots(12).$$

This decreases indefinitely with decrease of the decrement δ, but not towards zero; no appreciable advantage is to be got by decreasing δ below (say) 1*. That is, the much lower decrements ordinarily provided for other reasons are of no assistance in avoiding atmospheric interference.

When ω is not zero, reduction of the decrement beyond a critical value δ_c actually increases the atmospheric interference. For choice of b, minimum interference is given when $\dfrac{dA}{db} = 0$. From (11) this is found (for $p \gg \omega$) to be when b has the critical value, say b_c, given by

$$b_c^2 \doteqdot \frac{p\omega}{\sqrt{2}};$$

$$\text{i.e. } \delta_c \doteqdot 5\cdot 3\sqrt{\frac{\omega}{p}} \quad \ldots\ldots\ldots\ldots\ldots(13).$$

In practice the damping is often less than this optimum for avoiding atmospheric interference†. Thus if $\dfrac{p}{2\pi} = 100 \text{ kc/s} (\lambda = 3000 \text{ m})$ and $\dfrac{\omega}{2\pi} = 1 \text{ kc/s}$, $\delta_c = 0\cdot 53$; whereas in practice the decrement would be kept much lower in order to avoid interference from other wireless transmitters.

---

* In E. B. Moullin's paper, *loc. cit.* p. 509, the contrary conclusion is reached that the interference is proportional to the decrement δ. This is because the significance of the factor ϵ^{-bt} in (7) is there ignored, the *initial* amplitude due to the atmospheric being compared with the amplitude due to the signal.

† This theoretical result is in accord with operating experience. In a receiver where the decrement is controllable by triode retroaction, the operator is prone to reduce the retroaction (increase the decrement) when atmospherics are troublesome, although thereby the desired signals are weakened as well as the atmospherics.

3. THE ATMOSPHERIC AS A CONTINUOUS SPECTRUM

Putting $b^2 = \dfrac{p\omega}{\sqrt{2}}$ in (11), we find (for $p \gg \omega$)

$$A_{\min} \fallingdotseq \frac{2NE_a q}{E_s p^2}\left(1 + \frac{1}{\sqrt{2}}\frac{\omega}{p}\right)^2,$$

$$\fallingdotseq 2\frac{E_a}{E_s}\frac{Nq}{p^2} \quad\quad\quad\quad\quad\quad\dots\dots\dots(14).$$

The formulas (11), (12) and (14), expressing the amount of interference, A, to be expected from atmospheric conditions specified by E_a, q and N, are in accord with the well-established fact that atmospherics are more troublesome the longer the wavelength of the receiver. Thus (12) and (14) both show interference which increases as the square of the wavelength. Moreover, it is only with short waves that the decrements imposed by considerations of selectivity and of circuit design do not lie far below the optimum decrements (13) for avoiding atmospheric interference. One of the strongest recommendations of short-wave working has been the almost complete escape from atmospheric interruption. It is roughly true that long-wave working is subject to interruption from atmospherics but not from fading; and that short-wave working is subject to interruption from fading but not from atmospherics.

3. THE ATMOSPHERIC AS A CONTINUOUS SPECTRUM

Another approach to an understanding of the problems involved in attempting to accept the signal and reject the atmospheric is from the standpoint of the assertion that an atmospheric contains components of all frequencies. This loose statement conveys a useful physical idea, which is not easily expressed in precise terms.

A periodic disturbance of any form, recurring at equal intervals of time $\dfrac{1}{n}$, is equivalent to a Fourier series consisting of an infinite number of terms of finite amplitudes and discrete frequencies n, $2n$, $3n$...; and the effect on oscillatory circuits of an E.M.F. of this form is often conveniently calculated as the sum of the effects of the component sustained sinoidal E.M.Fs. But a non-recurrent disturbance can be expressed in sine or cosine functions of time only as a continuous spectrum, a Fourier integral, the sum of a series of terms such as

$$A_0 + \dots + A_p \cos pt + (A_p + dA_p)\cos(p + dp)t + \dots$$
$$+ B_p \sin pt + (B_p + dB_p)\sin(p + dp)t + \dots \text{ ad inf.}$$

In this series there are an infinite number of terms between any two frequencies $\frac{p_1}{2\pi}$ and $\frac{p_2}{2\pi}$, and the amplitude $\sqrt{A_p^2 + B_p^2}$ of the component of any frequency $\frac{p}{2\pi}$ is infinitesimal. Every frequency is represented by a term of infinitesimal amplitude; but these amplitudes have finite ratios, so that if we take out the first term A_0 as a common factor the pulse is represented by

$$A_0[1 + \ldots + a_p \cos pt + (a_p + da_p)\cos(p + dp)t + \ldots$$
$$+ b_p \sin pt + (b_p + db_p)\sin(p + dp)t + \ldots],$$

where a_p, b_p, etc. are finite numbers.

The energy is distributed throughout the spectrum according to the values of a and b, and these depend on the shape of the pulse which the Fourier integral represents. The energy within the range of frequency $\frac{p}{2\pi}$ to $\frac{p+dp}{2\pi}$ is measured by $(a_p^2 + b_p^2)\,dp$. The plot of $(a_p^2 + b_p^2)$ or $\sqrt{a_p^2 + b_p^2}$ against p or $\frac{p}{2\pi}$ is called the *periodogram* of the pulse. Fig. 341 is the periodogram (plotted logarithmically) of the pulse $E_a \epsilon^{-\frac{1}{2}qt} \sin qt$*, which was taken in the last Section to represent the E.M.F. of an atmospheric.

In Fig. 342 a portion of the periodogram in Fig. 341 is replotted, but now with a linear scale of ordinates proportional to the square of the amplitude, and with the abscissae scaled in wavelengths, e.g. 3 km to 10 km†. This curve shows how the energy of the atmospheric pulse is distributed through the spectrum; the energy lying between any two wavelengths is the area below the corresponding portion of the curve. Thus the energy within a 5 kc/s band is represented by the smaller or the larger of the two shaded areas in

---

* Calculated from the formula given by C. R. Burch and J. Bloemsma, "Application of periodogram to wireless telegraphy," *Phil. Mag.* Vol. XLI (1925), p. 486. This important paper deals in a very general manner with the disturbance in selective circuits produced by recurrent and non-recurrent pulses.

The choice of any other value, less than q, in place of $\frac{1}{2}q$, in the damping factor $\epsilon^{-\frac{1}{2}qt}$ would affect the periodogram inappreciably at frequencies for which $\frac{p}{q} > 10$.

† Any multiple of these wavelengths may be substituted; for at any practical wavelength (sensibly straight part of graph in Fig. 341) the ordinates in Fig. 342 are sensibly proportional to λ^4.

3. THE ATMOSPHERIC AS A CONTINUOUS SPECTRUM

Fig. 342 according as the bottom wavelength of the band is 4 km or 8 km.

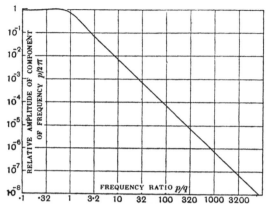

Fig. 341. Periodogram of atmospheric pulse.

The area beneath the curve between any two wavelengths in Fig. 342 is the energy which would be received from our assumed

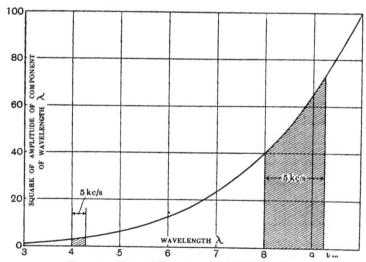

Fig. 342. Part of Fig. 341 plotted differently.

atmospheric pulse of E.M.F. by an imaginary circuit possessing uniform impedance (i.e. accepting uniformly) over the band between those wavelengths, and infinite impedance (i.e. rejecting wholly) outside that band. But circuits, even filters sensibly devoid of resistance, cannot be constructed to reject wholly all but a selected band in a continuous spectrum; and it is not easy to see how far it is permissible, with finite sharpness of cut-off, to disregard the continuous spectrum of impressed E.M.F. lying to left and right of the pass band.

This presentation of an atmospheric pulse as a continuous spectrum does, however, very strongly suggest that it is impossible to cut out the atmospheric from any receiving disposition whatsoever, unless the pass band of the receiver is made indefinitely narrow. Since it is the very modulation of the carrier-wave which constitutes the signal, a receiver with an indefinitely narrow band is ruled out on theoretical grounds: we could not use it even if we could construct it.

The optical parallel is very close. Light from an incandescent solid body is the aggregate of random pulses. It is representable statistically by a Fourier integral, and is commonly and rightly thought of as possessing a continuous spectrum in which every colour is present. Such light is like our atmospherics. Colour filters may be constructed which are very transparent to a narrow band of the spectrum, and very opaque outside that band. Such filters are like the selective circuits in a wireless receiver. The light filter passes freely a beam of nearly monochromatic light, if of the right colour; and the wireless receiver accepts freely the signals of the transmitter to which it is tuned. But although the filter may pass a red beam and reject a green, it passes *some* of the light from any incandescent solid; and the light that comes through is red. In like manner, the wireless receiver must find in the atmospheric pulse periodic components of just those frequencies to which it is sensitive.

There have been many inventions for combatting atmospherics, for sifting them out in divers ways from the desired signals; and some have seemed—not only to the inventor—full of promise. But the history of anti-atmospheric devices has been for the most part a story of disappointment, often of puzzled disappointment. The penetrating analyses of Burch and Bloemsma[*], and of J. R. Carson[†], have shown that much of the effort has been directed to achieving

[*] *Loc. cit.*, p. 514.
[†] "Selective circuits and static interference," *Bell System Technical Journal*, Vol. IV (1925), p. 265.

3. THE ATMOSPHERIC AS A CONTINUOUS SPECTRUM

the impossible. The position is summed up in the following extract from the latter paper: "The analysis of a number of representative schemes, such as the introduction of resistance to damp out disturbances, balancing schemes designed to neutralize static without affecting the signal, detuning to change the natural oscillation frequency of the circuit, demodulation through several frequency stages, etc., has shown that they are one and all without value in increasing the ratio of mean square signal to interference current. In the light of the general theory the reason for this is clear, and the limitation imposed on the solution of the static problem by means of selective circuits is seen to be inherent in the nature of the interference itself."

In the face of this situation it seems that the reduction of atmospheric interference must be sought mainly in one or more of the following directions:

(i) The provision of very high signalling field-strength at the receiver. (E.g., the British Broadcasting Corporation's aim of 5 or 10 millivolts per metre; and the use of antenna arrays in short-wave transmitters.)

(ii) The placing of the signalling wavelength in a region of the spectrum where atmospherics are intrinsically weak. (E.g., the use of short waves.)

(iii) The restriction of the frequency range over which the receiver is sensitive to as narrow and as uniform a pass band as the nature of the signals will permit. (E.g., the use of filters.)

(iv) The restriction of the direction of incident ray over which the receiver is sensitive, as narrowly as can be managed. (E.g., the use of antenna arrays in reception.)

In the problem of atmospherics, as in so many engineering problems, perfection is not to be attained; to strive for it is to waste effort. Atmospherics are a natural phenomenon presumably for ever outside our control, and against which, it seems, no complete defence can be contrived. They are an enemy to be avoided rather than opposed. The nearly uninterrupted telephony service between this country and America, with directive antennas and alternative short and long waves, is evidence of no small measure of success.

INDEX

Titles of Sections and Subsections are not, as such, entered in the Index. See the Table of Contents.

Authorship of publications referred to is indicated by the use of italic type.

Abraham, M., 33
Acoustic conditions in rectifier, 346
Acoustic measurements, amplifier for, 251
Acoustic vibrations, theory of, 416
Action at a distance, 12, 14
Adcock, F., 460
Adherent air, 420
Aerial, 432
Aether, 1, 6
Air condensers, variable, 64, 65
Air friction, heating by, 137
Air gap in steel cores, 236
Air reaction, 416, 420
Alexanderson, E. F. W., 135
Alexanderson, E. F. W., 137, 378
Alice and the looking glass, 47
Alternating current, meters for, 144
Alternators in cascade, 138
American Telephone and Telegraph Co., 405, 492
Ammeter for A.C., types of, 144
Ampère's law, 14, 19
Amplification factor, 196, 202, 203, 208
Amplification factor, inverted, 340
Amplifier instability, 257
Amplifier triode, 225
Amplitude: effect of—on damping, 356, 361
Anode, 172, 185
Anode-anode conductance, 202
Anode battery: economy in—by use of reactance, 233
Anode current, discontinuous, 281
Anode dissipation, 211, 274, 277, 281, 286
Anode-grid capacitance, 219, 220, 255
Anode-grid conductance, 202
Anode rectification in grid rectifier, 328, 341, 350
Anode rectifier, 321
Anode resistance, 203
Anode tap, 278
Antenna, 432

Antenna array, 468
Antenna, effect of atmospherics on, 78
Antenna, effective height of, 33, 433
Antenna, electrical dimensions of, 78
Antenna, patterns of, 79
Antenna resistance, components of, 152, 440
Antennas, capacitance values of, 80, 82
Antennas: loop and open—compared, 441
Antennas, resistance values of, 82, 83
Anti-atmospheric devices, 516
Antipodes, 39, 476
Aperiodic retroaction, 260
Appleton and Barnett, 59
Appleton, E. V., 47
Appleton, E. V., 48, 59, 60
Appleton Watt and Herd, 507
Arc characteristics, 122, 124, 126
Arcing in spark-gap, 110, 112, 116
Arc lengths, ignition and extinction, 130
Arc, nature of, 120
Arc transmitters, examples of, 131, 132
Armstrong, E. H., 365
Armstrong, E. H., 365, 367
Array of antennas, 468
Ashbridge—see *Eckersley*
Assel, transmitter at, 141
Atmospheric interference, measure of, 511
Atmospherics, 507
Atmospherics and short waves, 513
Atmospherics: effect of—on antenna, 78
Atmospherics, optical parallel to, 516
Attenuation by dispersion, 36
Attenuation by dissipation, 41
Attenuation constant, 66
Attenuation, no, 496
Auditive reception, optimum damping for, 356
Austin-Cohen formula, 41

INDEX

Austin, L. W., 41
Australia, dark and light routes to, 475
Autoheterodyne, 360
Autoheterodyne, limitation of amplitude in, 361

Backlash, 187
Bailey Dean and Wintringham, 373, 400, 446, 467
Balanced demodulator, 414
Balanced modulator, 395
Balancing no-signal current, 226
Baldock, beam receiver at, 425
Ballantine, S., 82, 438
Ballast resistance of arc lamp, 121
Band-pass filter, 503
Barfield—see *Smith-Rose*
Barnett—see Appleton
Beacons, wireless, 455
Beam, oscillating or rotating, 462
Beam stations, Marconi, 211
Beam transmitter of very short wavelength, 443
Beat reception, 167
Beats in coupled circuits, 91
Bei, 484
Bel, 372
Bell, A. G., 372
Bellini-Tosi system, 456
Bell Telephone Laboratories, 383, 416, 492
Ber, 484
B.E.S.A. nomenclature, 203
Bessel function, 335
Best—see *Turner*
Beverage Rice and Kellogg, 446
Bijl, H. J. van der, 173, 175, 275
Bligh, N. R., 220
Bloemsma—see *Burch*
Bordeaux, arc generator at, 132
Bordeaux, signals from, 43
Boucherot, P., 138
Branly, 16
Breit and Tuve, 59
Breit Tuve and Dahl, 60
British Broadcasting Corporation, vii, 278, 300, 423, 425, 517
Brookman's Park station, 292, 423
Burch, C. R., and J. Bloemsma, 514, 516
Butterworth, S., 485, 486, 487
Butterworth, S., 488
Buzzer, 117

Cady, W. G., 301, 306
Caernarvon, antenna at, 442
Caernarvon, signals from, 40, 43
Cambridge Instrument Co., 66
Campbell, G. A., 492
Campbell, G. A., 492
Capacitance adjustment, 64
Capacitance coupling, 94
Capacitance of vertical wire, 436
Capacitance retroaction, 345
Capacitance, stray, 8, 76, 233, 241
Carborundum rectifier, 154
Cardioid directivity, 450
Cardioid systems, spaced, 467
Carrier-wave, 390
Carrier-wave multiplex telegraphy, 405
Carrier-wave unnecessary, 393
Carson, J. R., 516
Cathode, 172, 185
Cathode, manners of heating, 186
Centre Radioélectrique de Paris—see Sainte Assise
Centripetal force in alternators, 133, 136
Chain of selective circuits, 357, 408
Chalcopyrite-zincite rectifier, 154
Characteristic impedance of line, 446, 475
Charges, electric and magnetic, 17
Choke control, 385
Choke, high-frequency, 234
Choke, steel-cored, 235
Chronometer accuracy, 264
Circular polarisation, 55
Clapping, sound of, 391
Classical mechanics, 1
Clock as oscillator, 264
Cobbold, G. W. N., and A. E. Underdown, 301
Coils, losses in, 488
Coils, variables in design of, 487
Colebrook, F. M., 311, 334, 411
Collision between electrons and massive ions, 52
Colour filters, 516
Colpitts circuit, 318
Common battery, 238, 430
Compagnie Générale de T. S. F., 137
Complementary function, 66
Complex triode circuit, 310
Condenser connection between triodes, 239
Condenser retroaction, 345
Conductors, nature of, 35, 174
Cone of loud-speaker, 417
Constantinesco water-wave transmission, 2

INDEX

Contact E.M.F., 198, 214
Contact resistance, 153
Continuous spectrum, 516
Continuous wave, 119
Control electrode, 192, 197
Conversation, range of acoustic power in, 374
Cooled-anode triodes, 210
Copper versus iron for screening, 480
Copper wires, graphs of conductance of, 479
Corona loss, 433
Corpuscles, 172
Cosens, C. R. G., 420, 421
Coupled circuit in arc generator, 131
Coupled circuits, 85, 94
Coupling, coefficient of, 91
Coupling, internal and external, 96
Coupling, methods of, 94
Crandall, I. B., 384, 416
Critical velocities for emission, 176
Croix d'Hins—see Bordeaux
Crossed-antenna D.F., 461
Crystal rectifiers, 154
Cumulative grid rectifier, 351
Curvature of ray, 51
Cut-off frequency, 499
Cut-off of high tones, 392, 407
C.W., 119
C.W. and spark reception compared, 170
C.W. generator, types of, 120
Cylindrical electrodes, 192, 196

Dahl—see Breit
Damped oscillation, 84
Damping caused by rectifier, 151
Damping for auditive reception, optimum, 356
Damping, microphonic control of, 380
Damping: rectifier—and strength of signal, 342
Damping, very low, 356
Dash, Morse, 374
Daventry (5XX), frequency characteristic of, 392
Dean—see *Bailey*
Decibel, 372
Decrement, 66
Decrement from resonance curve, 71
Degrees Kelvin, 177
Depth of modulation, 389
Detector as rectifier-amplifier, triode, 323, 326
Detector sluggishness, 162, 351
D.F., 453

D.F. aberrations, 45, 458, 461
D.F. at night, 460
Dielectric constant, 17
Dielectric constant of upper atmosphere, 47
Dielectric current, 15
Differential coefficients in grid characteristic, 328, 332, 334
Differential conductance, 202
Differential resistance, 121
Diffraction formula, 40
Diode, 172, 186
Direction: effect of—on propagation, 45
Discontinuity in anode current, 281
Discontinuity in arc current, 126
Dispersive medium, 56
Distributed capacitance and inductance, 77
Dorling—see *Shearing*
Dot, Morse, 374
Double-ended triode, 212
Double modulation, 403
Duddell, W., 120
Dull filaments, emission voltages of, 334
Duplex working, 430, 442, 443
Dye, D. W., 300
Dyke, K. S. van, 303
Dynamic theory of gases, 175, 176
Dynamo, high-tension, 292
Dynamometer ammeter, 144
Dynatron, 222, 261

Ear, change of intensity perceived by, 373
Earth connection, 432
Earth wires, 34
Eccles, W. H., vii, 48, 197
Eccles, W. H., 10, 16, 173, 183, 188, 196
Eccles, W. H., and F. W. Jordan, 300
Echoes, 58
Echoes of long delay, 60
Eckersley, P. P., 392
Eckersley, P. P., and N. Ashbridge, 423
Eckersley—see *Round*
Eckersley, T. L., 42
Eckersley, T. L., 37, 38, 439, 460
Eddy-current heating of anode, 209
Eddy loss in iron cores, 63, 133, 143, 248, 481
Edison effect, 172, 174
Edison, T. A., 174

INDEX

Eiffel Tower, spark transmitter at, 112
Eilvese, alternator at, 139
Electrodes, cylindrical form of, 192
Electrodes, plane form of, 188
Electrodynamics, first law of, 19
Electrodynamics, second law of, 20
Electron, 17, 172, 174
Electron affinity, 175
Electron evaporation constant, 175
Electron theory of matter, 174
Elliptic path of electron, 54
Elmen, G. W., 235
Elwell, C. F., 210
Elwell-Poulsen arcs, 131
Emde—see *Jahnke*
Emission and temperature, graphs of, 180, 182
Emission, heating power for, 181
Emission, maximum, 180
Emission velocity, effect of, 195, 198
Encyclopaedia Britannica, 3
Energy of field, 19
Engineering, electrical and mechanical, vi
Epstein-Joly frequency doubler, 139
Equivalent concentrated circuit, 78
Equivalent volts, 176
Escapement, 264
Extinction directions, maximum number of, 471
Extinction from two spaced antennas, 463
Extinction in arc oscillation, 126, 131
Extinction of arc, 126

Facsimile reproduction, 374
Fading, 44, 426
Fahie, J. J., 3, 16
Faraday, 1
Faraday's law, 20
Faraday tube, 17
Farnborough, rotating beacon at, 454
Faulkner—see *Hansford*
Ferranti resistor, 263
Ferranti transformers, 246, 249
Fessenden, R. A., 167
Field strengths, received, 38, 43, 44, 426, 434, 435
Filament, types of, 182
Filaments in grid rectifier: dull and bright—compared, 336
Filter, 408, 492
First wireless telegraphy, 107
Flat ground, 36
Flat top of antenna, 28, 433, 445
Fleming, J. A., 172, 319

Fleming, J. A., 319, 478
Fleming valve, 193, 319
Fletcher, H., 416
Flexural waves on rod, 56
Flux-density, electric and magnetic, 18
Flux-density in air-gap of moving-coil speaker, 420
Flux, electric and magnetic, 17
Forced oscillation, 65
Forced synchronisation, 362
Forest, L. de, 172, 193
Fortescue, C. L., 486
Fortescue, C. L., 487
Fourier integral, 513
Fourier series, 99, 142, 287, 513
Four-wire working, 430
Franklin, C. S., 474
Franklin suppression coils, 474, 475
Freak ranges, 39
Free electricity, 174, 179
Free oscillation, 66
Frequencies of coupled circuits, 90
Frequency and wavelength, 75
Frequency, effect of grid current on, 269
Frequency, effect of resistance on, 67
Frequency meter, 75
Frequency reduction by heterodyne, 362
Frequency spacing between stations, 309, 357
Frequency, unit of, 76
Frequency wandering, effects of, 306
Frequency wandering, observations of, 308
Funkentelegraphie, 109

Gain, amplification expressed as, 372
Gain control, automatic, 429
Gecosite rectifier, 158
General Electric Co., 158
Generalised triode circuits, 310
German Post Office, vii, 143
Gettering, 209
Gill—see *Lee*
Gill, T. H., and N. F. S. Hecht, 454
Goldschmidt, R., alternator of, 138
Goldsmith, A. N., 375
Goniometer, 456
Gramophone record, reproduction of, 374
Gravity, pendulum for measuring, 66
Great-circle track of ray, 52, 459, 476
Grid, 192
Grid bias, automatic, 286

Grid capacitance, 212
Grid current: dependence of—on anode potential, 340
Grid current: effect of—on frequency, 269
Grid leak, 239, 329
Grid leak in rectifier, value of, 328
Grid leak line, 329
Grid mesh, 196, 205-7
Grid, physical action of, 193
Grid power, 270
Grid rectifier, 321
Grid rectifier, coefficients of performance of, 328
Grid rectifier, no-signal state-point in, 329
Grimsby short-wave station, 475
Groenveld van der Pol and Posthumus, 335
Ground: effect of—on non-earthed antenna, 7, 439
Ground loss, 34, 445, 479
Group velocity, 57
Growth, condition for, 269
Growth of oscillation, rate of, 365
Gutton, C., 275

Half-wave suppression, 474
Hals—see Störmer
Hanover, signals from, 43
Hansford, R. V., and H. Faulkner, 33, 34, 64, 83, 293
Harley—see *Hodgson*
Harmonics, cultivation of, 289, 299
Harmonics, elimination of, 288
Harmonics in oscillator, 269, 287
Hartley circuit, 279
Hartley, R. V. L., 395, 405
Heaviside layer, 36
Heaviside layer imagined non-existent, 40
Heaviside layers, two, 60
Hecht—see *Gill*
Height of antenna, effective, 33, 433
Heising, R. A., 403
Heising Schelleng and Southworth, 44
Herd—see *Appleton*
Hertz as unit of frequency, 76
Hertz, H. R., 16, 432, 439
Heterodyne, 167
High-pass filter, 500
High receiving antennas unnecessary, 440
Hodgson Harley and Pratt, 181
Holweck demountable valve, 210
Homopolar alternator, 134

Hot-wire ammeter, 144
Howe, G. W. O., 486
Howe, G. W. O., 23, 58
Hughes, D. E., 16, 107
Hughes—see *Morris-Airey*
Hull, A. W., 221, 222, 261
Hydrogen in arc, 125
Hyper-acoustic frequency, 362
Hyper-acoustic quenching, 367
Hypo-acoustic frequency, 362

Image impedance, 493
Imperial and International Communications Ltd., 475
Incandescent solid, 516
Inchkeith, rotating beacon at, 443, 455
Inductance adjustment, 64
Inductance coupling, 94
Inductor alternator, 112, 133
Inertias in moving-coil speaker, 419
Infinity, engineer's grasp on, 356
Input impedance of amplifier, 258
Insulation of masts, 83
Interference a controlling factor, 440, 507
Interference from echoes, 59
Interference, reduction of, 443, 446, 450
Intermediate frequency in superheterodyne, 364
Intruder frequencies from rectifier, 410
Inverted amplification factor, 340
Ion, 174
Ionisation gradient in atmosphere, 48
Ionisation in arc, 121
Ions in atmosphere, recombination of, 48
Iron cores, eddy loss in, 63, 133, 143, 248, 481
Iron cores for high-frequencies, 63
Iron tape in saturated chokes, 143
Iron versus copper for screening, 480

$j \equiv \sqrt{-1}$, steady-state analysis by, 99
Jahnke und Emde, 335
Jeans, J. H., 11, 15
Jordan—see Eccles

Kallirotron, 260
Kay see ess, 76
Kellogg—see *Beverage*
Kelvin, 480
Keying arc oscillator, 131
Kilocycle, 76

INDEX

Kirke, H. L., 351
Königs Wusterhausen, transmitter at, 143, 379
Kühn, L., 378

Landale, S. E. A., 338, 339, 340, 411
Langmuir, I., 172, 179
Langmuir, I., 179, 188, 192
L antenna, inverted, 29, 442
Larmor, J., 48
Lateral deviation of ray, 459
Latour, M., 137
Latour, M., 134, 138
Leafield, arc generators at, 131
Lee, A. G., 300, 309, 373, 429, 447, 467
Lee, A. G., and A. J. Gill, 131
Lepel, E. von, 114
Light, visible, 7
Lighthouses, wireless, 454
Lightning, 507
Line, infinite, 2
Line, short, 2
Linear performance of detectors, 160
Lodge, O. J., 432
Logarithmic decrement, 66
Loop abandoned for D. F., 460
Loop acting as open antenna also, 452, 457
Loop antennas, spaced, 464
Lorenz Aktiengesellschaft, 143, 379
Loss, attenuation expressed as, 372
Lower side-band, 391
Lower side-wave, 390
Low-pass filter, 499
Lucas, H. J., 301
Lumped characteristic, 273, 285
Lumped potential, 197
Lunnon—see Round

Magnetic alloys, 235
Magnetic amplifier, 137, 378
Magnetic blast in arc, 125
Magnetic field of filament current, 199
Magnetic saturation, use of, 136, 139, 141
Magnetic shell, 14
Magnetomotive force, 14
Maintenance, condition for, 269
Malabar, transmitter at, 141
Mallett, E., 275, 423, 430
Mallock, R. R. M., 397
Marconi Co., vii, 42, 112, 118, 212, 371, 381, 442, 456, 475
Marconi, G., 432

Marconi International Marine Commn. Co., 294, 370
Margin of stability in triode circuits, 356
Marine service, receiver for, 370
Mark, Morse, 374
Marking and spacing waves, 131
Martin, W. H., 372
Master oscillator, 292
Masts, 33, 433
Masts, effect of insulating, 83
Mathematician, description of, vii
Maxwell, Clerk, 1, 14
Maxwell equations, 15, 477, 507
Meacham, L. A., 263
Meacham—see Turner
Mechanical distortions in telephony, 415
Mechanical model of oscillatory circuit, 105
Mechanical oscillators, 66
Mercury seal, 480
Meshwork antenna arrays, 474
Metal-glass envelope, 210
Mica condenser, 64, 65
Microhenry, 8
Micromicrofarad, 8
Microphone, carbon, 376
Microphone, condenser, 383
Microphone, electromagnetic, 383, 417
Microphone, inventor of, 16
Modulation, double, 403
Modulation ratio, 389, 410
Mögel—see Quäck
Morris-Airey Shearing and Hughes, 210
Morse code, 374
Morse signals, frequency of, 356, 374
Morse signals, frequency wandering in, 307
Morse switch, position of, 112, 131, 136, 293
Most probable speed, 176
Motional impedance, 422
Moullin, E. B., vii
Moullin, E. B., v, 11, 13, 73, 83, 145, 159, 441, 480, 485, 509, 510, 512
Moullin voltmeters, 145
M.O. Valve Co., 211, 220, 278
Moving-iron ammeter, 144
Mullard Wireless Service Co., 218, 223, 263, 297
Mumetal, 235
Munich, broadcasting station at, 143
Mush in arc oscillation, 130
Music, frequencies in, 374

INDEX

Music, range of acoustic power in, 374
Music, reproduction of, 244
Musical spark transmitter, 111
Mutual conductance, 203
Mutual coupling, 94
Mutual formula, 94
Mutual inductance, variation of, 93

National Physical Laboratory wavemeter, 300
Natural frequency, 66
Nauen, alternator transmitters at, 161
Nauen, signals from, 43
Navy, stranded wire used by, 487
Negative conductance, 221, 261, 268, 295
Negative dielectric constant, 47
Negative resistance of arc, 121
Neper, 373
Neutrodyne, 259, 345
New Brunswick, alternators at, 136
Nichols, H. W., and J. C. Schelleng, 55
Night effect, 458
Night errors, 458
No attenuation, 496
Nodes of current along antenna, 80, 436, 474
Non-dispersive medium, 56
Northolt station, 300
No-signal state-point of grid rectifier, 329

Ohm's law not followed, 120, 145
Optical parallel to atmospherics, 516
Optical refraction, 48
Orontes, S. S., wireless cabin of, 371
Oscillation: α and β—, of arc, 125
Oscillator for very short waves, 279
Oscillatory valve relay, 361
Oswald, A. A., and J. C Schelleng, 402
Our reference triode, 225
Output triode, 225
Oxide-coated filaments, 181

Pacific Ocean, attenuation over, 42
Palmer, L. S., 97, 99
Parabolic rectifier characteristic, 148
Parabolic reflector, 443
Particular integral, 65, 99
Pass frequencies, 496
Passive chain, 493
Peak voltmeter, Moullin's, 159
Peakiness of resonance curve, 69
Pedersen, P. O., 167

Pedersen, P. O., 57, 58, 59, 61, 120
Pendulum for gravity measurement, 66
Pendulum model, 105
Pendulums, damping of, 66
Penetration into conductors, graphs for, 479
Pentode, 223
Perfect rectifier, 146, 347
Periodic representation of speech and music, 391
Periodogram, 514
Permalloy, 235
Permeability, 16, 17
Phantom telephone circuit, 393
Phase displacement by distuning, 72, 463
Phase velocity, 57, 446
Photo-electric emission, 176
Picken, W. J., 210, 211, 278
Pidduck, F. B., 15
Pierce, G. W., 11, 15, 29, 33, 87, 92, 477, 501
Piezo-electric property, 301
Pitch: changing—, effect of, 400
Plane polarisation, 45
Pliodynatron, 261
Pol, van der, 40, 362
Pol, van der—see *Groenveld*
Polarisation changes in propagation, 45, 55, 459
Polarisation, circular, 55
Polarisation of aeroplane signals, 460
Polarisation, plane, 45
Portable receiver, antenna for, 441
Posthumus—see *Groenveld*
Post Office, viii, 131, 309, 426, 455, 475
Potential, electric and magnetic, 18
Potential gradient along filament, 198
Poulsen, V., 120, 125
Power consumed in ammeters, 144
Power factors of coils, 64
Power factors of condensers, 65
Power output from anode, 227
Power transmission systems, 1
Power valve, 225
Pratt—see *Hodgson*
Preece, W. H., 2
Primary anode rectifier, 350
Primary rectifier, 319
Profane point of circuit, 8
Propagation distortions in telephony, 395
Proximity effect, 486
Puddle, drying of, 180

Pupin-loaded line, 492
Push-pull disposition, 249, 347, 395, 413, 426

Quäck and Wagner, 58
Quäck, E., and H. Mögel, 59
Quartz crystal, cut of, 301
Quartz slab, natural frequencies of, 302
Quartz slab, shapes of, 303
Quenched spark gap, 114
Quenching an oscillation, 114, 366
Quenching, methods of, 367

Radiated field, 26
Radiated field, alternative ways of regarding, 31
Radiation in vertical plane, distribution of, 437
Radiation neglected at low frequency, 11
Radiation resistance, 32
Radiation resistance of loop, 13, 441, 454
Radiation resistance of open antenna, 13
Radio Corporation of America, 444
Rayleigh, Lord, 420
Receiver, overall efficiency of, 10
Rectangular frequency characteristic, 407
Rectified current, 149
Rectifier detector, 145
Rectilinearity of triode characteristic, 199, 202
Reed, M., 493
Reference triode, our, 225
Reflection at end of line, 446
Reflections at ends of filter, 501
Refraction, 48
Rejector circuit, 101
Rejector circuit, graphs for resistance and reactance of, 104
Residual gas, 185, 319
Resistance: effect of—on frequency, 67
Resistance, location of, 89
Resistance, meaning of term, 120
Resistance of retroacted circuit, 353, 358
Resistor for high frequency, 233
Resistor, reactance impurity in, 263
Resistor, self-capacitance in, 233
Resonance curve, 69, 72
Resonance curves of coupled circuits, 98
Resonance effect of Earth's magnetic field, 54

Resonance frequency, 66
Resonance in low-frequency circuit, 112, 116
Retroaction, avoidance of, 259
Retroaction, capacitance, 345
Retroaction in quartz oscillator, 305
Rice—see *Beverage*
Richardson, O. W., 172, 175
Richardson, O. W., 175, 179, 183
Ringing of signals, 364
Röntgen rays, 23
Rotary spark gap, 111
Rotating beacon, 454, 455
Round Eckersley Tremellen and Lunnon, 42
Rugby station, 33, 34, 64, 65, 82, 286, 292, 293, 300, 309, 434, 487
Rukop—see *Zenneck*
Rundfunk, 109
Russell, Bertrand, vii
R valve, 173

Sacred point of circuit, 8
Sainte Assise, alternators at, 137
Saturation current, 180
Schelleng—see *Heising*
Schelleng—see *Nichols*
Schelleng—see *Oswald*
Schottky, W., 219
Schreihage—see *Schuchmann*
Schuchmann and Schreihage, 308
Screened Bellini-Tosi antennas, 457
Screening box, joints in, 480
Screening D.F. antennas, 457, 460
Secondary emission, 176, 221, 223, 261, 275
Secondary rectifier, 320
Selective circuits, chain of, 357
Selectivity, acoustic, 253
Selectivity, adjustment of, 253
Self-inductance, 11
Self-supporting field, 21
Sense discrimination, 449, 458, 472
Series condenser in antenna, 82
Severn, early wireless across, 3
Sharpness of tuning, 69, 357
Shaughnessy, E. H., 82, 300
Shea, T. E., 493
Shearing, G., and J. W. S. Dorling, 487
Shearing—see *Morris-Airey*
Ships, D.F. apparatus on, 455, 457
Shortest wave for terrestrial signalling, 46, 51
Short-wave beam transmitter, very, 443

INDEX

Short-wave propagation, 44
Short waves, amplifier for, 253
Side-bands, 391
Side-waves, 390
Side-waves unequally treated, 396
Signal current, 149
Silicia envelope, 210
Singing arc, 120, 122, 124
Skin effect in wire, 484
Skip, 44
Slide-back in detector, 158, 347
Slide-back of grid potential, 331
Slope conductance, 202
Slope permeability, 235
Slope resistance, 121
Slot pitches in inductor alternator, 134
Slot-wound A.C. dynamo, 471
Smith, C. G., 247
Smith-Rose, R. L., and R. H. Barfield, 460
Soft triodes, 185
Sound, analysis of, 374
Sounder, telephone, 375, 422
Southworth— see *Heising*
Space, Morse, 374
Space-charge, 188, 193
Spacing wave in arc generator, 131
Spark and C.W. reception compared, 170
Spark gap, current in, 113
Spark, nature of, 107
Spark signals, 119
Spark transmitter, obsolescence of, 107
Sparking P.D., 108
Speech, intelligible, 374
Speed of electrons in conductors, 26
Speeds of electrons, distribution of, 177, 183
Stalloy core, 235
Static, 507
Steadiness of arc oscillation, 130
Steady state, algebraic treatment of, 99
Steel masts, effect of, 33
Steel-quartz oscillator, 306
Stokes, G., 16
Störmer, G., and J. Hals, 60
Stranded wire, 64, 454, 487, 489, 491
Stray capacitance, 8, 76, 233, 241
Stress, electric and magnetic, 17
Striking an arc, 131
Strong signals for linear rectification, 347
Submarine, reception by, 478

Submarine telegraph, 2, 9
Superheterodyne, 362, 426
Superimposed telephone circuit, 393
Super-regenerative receiver, 367
Symonds, A. A., 236
Synchronisation, forced, 362
Synthetic crystal rectifier, 157

T antenna, 442
Tariffs and carrier-wave suppression, 393
Telefunken Co., vii, 114, 141
Telegraphy, 373
Tele-kinematography, v
Telephone line, 9
Telephone sounder, 375
Telephony, 374
Television, v, 374
Temperature gradient along filament, 197
Temperatures of filaments, 182
Tetrode, 214
Thackeray, E. St J., viii
Then, meaning of, 374
Thermionic emission, 176
Thermionic tube, history of, 172
Thermionic voltmeter, 145
Thermo-couple ammeter, 144
Thomson, J. J., 23, 172
Three-dimensional array, 474
Three-halves-power law, 192, 195, 199
Tikker, 167
Tilt of wave-front, 445, 479
Time-constant in rectifier, 346, 351, 412
Time lag in amplifier, 228
Towers, 433
Transatlantic telephony, 403, 426, 467, 517
Transformer for high frequencies, 241
Transformer, resistance of, 248
Transients in sound, 391
Transmission line, 425, 467, 475, 480
Transmission unit, 372
Tremellen—see *Round*
Triode, 172
Triode and transformer compared, 197
Triode, combined rectifying-amplifying function of, 323, 326
Triode oscillator, alternating values in linear, 272
Triode oscillator: linear and high-efficiency cycles in—compared, 283
Triode oscillator, total values in linear, 273
Triode rectifier dispositions, 320

INDEX

Trumpet of loud speaker, 417
Tube noise, 373
Tuckerton, alternator at, 139
Tungsten, 181
Tuning curves, general, 72
Tuning-fork, constancy of, 300
Tuning-fork restraint on triode oscillator, 299
Tuning, manners of, 88
Turner, L. B., 37, 117, 236, 260, 275, 361, 364, 405
Turner, L. B., and F. P. Best, 356
Turner, L. B., and L. A. Meacham, 277
Turner, P. K., 421
Tuve—*see* Breit

Ultra-violet light, 48
Underdown—see *Cobbold*
Unidirectional property, 449, 472
Unipivot instrument, 144, 226
Units, electrostatic and electromagnetic, 22
Unsteadiness in triode, 356, 368
Upper side-band, 391
Upper side-wave, 390

Vacuum in thermionic tube, 172
Vacuum, production of, 209
Vacuum-tight seal, 210
Valve, thermionic, 187
Velocity acquired by electrons, 176, 185, 216
Velocity exceeding velocity of light, 48
Velocity of light, 21, 75
Velocity of wave in wire, 446
Verne, Jules, 61

Vertical-plane directivity, 437
Vigoureux, J. E. P., 301
Voice-operated switching, 430
Voltmeter, Moullin, 145

Wagner—see Quäck
Walmsley, T., 475
Washington Conference of 1927, 107
Watson, G. N., 40, 42
Watson-Eckersley formula, 42, 434
Watt—see *Appleton*
Wave antenna, 444
Wave-band theory of modulation, 390
Wavelength and frequency, 75
Wavelength, choice of, 13, 434
Wavemeters, 75, 76, 117
Wave motion, 3
Wave terms, definitions of, 5
Wells, H. G., 61
Wells, N., 443
Wente, E. C., microphones of, 383
Western Electric Co., 293
Wien, M., 114
Will and control electrode, 197
Wind, drying action of, 180
Wind, on antenna, effect of, 292
Wintringham—see *Bailey*
Wire resistor, 234
Wired wireless, 405, 492
Wires, resistance at high frequency, 480
Work function, 175

Young's modulus, 19

Zenneck, J., and H. Rukop, 275
Zincite-chalcopyrite rectifier, 154

www.ingramcontent.com/pod-product-compliance
Ingram Content Group UK Ltd.
Pitfield, Milton Keynes, MK11 3LW, UK
UKHW040700180125
453697UK00010B/311